D0640735

Gene Structure and Regulation in Development

The Forty-First Symposium of
The Society for Developmental Biology
Cambridge, Massachusetts, June 17–19, 1982

Gene Structure and Regulation in Development

Stephen Subtelny
Volume Editor
Department of Biology
Rice University
Houston, Texas

Fotis C. Kafatos
Coeditor and Symposium Organizer
Department of Cellular and Developmental Biology
Harvard University
Cambridge, Massachusetts

Alan R. Liss, Inc. • New York

Address all Inquiries to the Publisher
Alan R. Liss, Inc., 150 Fifth Avenue, New York, NY 10011

Printed in the United States of America.

Library of Congress Cataloging in Publication Data

Main entry under title:

Gene structure and regulation in development.

(Symposium of the Society for Developmental Biology ; 41st)
"The forty-first symposium of the Society for Developmental Biology, Cambridge, Massachusetts, June 17–19, 1982"—P.
Includes bibliographies and index.
1. Developmental genetics—Congresses. 2. Genetic regulation—Congresses. I. Subtelny, Stephen Stanley, 1925– II. Kafatos, Fotis C. III. Society for Developmental Biology. IV. Series. [DNLM: 1. Genetics—Congresses. 2. Gene expression regulation—Congresses. W3 SO59H 41st QH 453 G326 1982]

QH511.S6 41st [QH453] 574.3s [574.3] 83–11285
ISBN 0–8451–1502–2

Contents

III. OOGENESIS AND EARLY DEVELOPMENT

IV. GENOMIC INSTABILITIES

V. THE CONTROL OF GENE ACTIVITY

Contributors

F.M. Ausubel, Department of Molecular Biology, Massachusetts General Hospital, Boston, MA 02114 **[3]**

Lawrence Bogorad, Department of Cellular and Developmental Biology, Harvard University, Cambridge, MA 02138 **[13]**

Ralph L. Brinster, Laboratory of Reproductive Physiology, School of Veterinary Medicine, University of Pennsylvania, Philadelphia, PA 19104 **[235]**

Roy J. Britten, Division of Biology, California Institute of Technology, Pasadena, CA 91125 **[147]**

Peter C. Brown, Department of Biological Sciences, Stanford University, Stanford, CA 94305 **[197]**

S.E. Brown, Department of Molecular Biology, Massachusetts General Hospital, Boston, MA 02114 **[3]**

W.J. Buikema, Department of Molecular Biology, Massachusetts General Hospital, Boston, MA 02114 **[3]**

Benjamin Burr, Biology Department, Brookhaven National Laboratory, Upton, NY 11973 **[185]**

Frances A. Burr, Biology Department, Brookhaven National Laboratory, Upton, NY 11973 **[185]**

Lucy Cherbas, Department of Cellular and Developmental Biology, Harvard University, Cambridge, MA 02138 **[95]**

Peter Cherbas, Department of Cellular and Developmental Biology, Harvard University, Cambridge, MA 02138 **[95]**

Mary Collins, Department of Embryology, Carnegie Institution of Washington, Baltimore, MD 21210 **[213]**

Eric H. Davidson, Division of Biology, California Institute of Technology, Pasadena, CA 91125 **[147]**

Igor B. Dawid, Laboratory of Molecular Genetics, National Institute of Child Health and Human Development, National Institutes of Health, Bethesda, MD 20205 **[171]**

Manuel O. Diaz, Department of Biology, Yale University, New Haven, CT 06511 **[137]**

C.D. Earl, Department of Molecular Biology, Massachusetts General Hospital, Boston, MA 02114 **[3]**

Eleanor Erikson, Department of Pathology, University of Colorado School of Medicine, Denver, CO 80262 **[77]**

R.L. Erikson, Department of Pathology, University of Colorado School of Medicine, Denver, CO 80262 **[77]**

J.G. Foulkes, Massachusetts Institute of Technology, Center for Cancer Research, Cambridge, MA 02139 **[77]**

The number following each author's affiliation is the opening page number of that author's article.

Joseph G. Gall, Department of Biology, Yale University, New Haven, CT 06511 **[137]**

Tona M. Gilmer, Department of Pathology, University of Colorado School of Medicine, Denver, CO 80262 **[77]**

Earl J. Gubbins, Amoco Research Center, P. O. Box 400, Mail Station B1, Naperville, IL 60566 **[13]**

Ira Herskowitz, Department of Biochemistry and Biophysics, University of California, San Francisco, CA 94143 **[65]**

Barbara R. Hough-Evans, Division of Biology, California Institute of Technology, Pasadena, CA 91125 **[147]**

Randal N. Johnston, Department of Biological Sciences, Stanford University, Stanford, CA 94305 **[197]**

Setsuko O. Jolly, Universal Foods Technical Center, 6143 North 60th Street, Milwaukee, WI 53218 **[13]**

Fotis C. Kafatos, Department of Cellular and Developmental Biology, Harvard University, Cambridge, MA 02138 **[xi,33]**

Roger E. Karess, Department of Embryology, Carnegie Institution of Washington, Baltimore, MD 21210 **[213]**

Brian K. Kay, Laboratory of Molecular Genetics, National Institute of Child Health and Human Development, National Institutes of Health, Bethesda, MD 20205 **[171]**

M. Macy D. Koehler, Department of Cellular and Developmental Biology, Harvard University, Cambridge, MA 02138 **[95]**

Enno T. Krebbers, Department of Cellular and Developmental Biology, Harvard University, Cambridge, MA 02138 **[13]**

Ignacio M. Larrinua, Department of Cellular and Developmental Biology, Harvard University, Cambridge, MA 02138 **[13]**

Alf Larsen, Hutchinson Cancer Research Center, 1124 Columbia Street, Seattle, WA 98104 **[241]**

Robert Levis, Department of Embryology, Carnegie Institution of Washington, Baltimore, MD 21210 **[213]**

S.R. Long, Department of Biological Sciences, Stanford University, Stanford, CA 94305 **[3]**

Kathleen A. Mahon, Department of Biology, Yale University, New Haven, CT 06511 **[137]**

Beatrice Mintz, Institute for Cancer Research, 7701 Burholme Avenue, Fox Chase, Philadelphia, PA 19111 **[113]**

Christine Murphy, Department of Embryology, Carnegie Institution of Washington, Baltimore, MD 21210 **[213]**

Karen M.T. Muskavitch, Department of Cellular and Developmental Biology, Harvard University, Cambridge, MA 02138 **[13]**

Kevin O'Hare, Department of Embryology, Carnegie Institution of Washington, Baltimore, MD 21210 **[213]**

Richard D. Palmiter, Howard Hughes Medical Institute Laboratory, Department of Biochemistry, University of Washington, Seattle, WA 98195 **[235]**

Janice Pero, Department of Cellular and Developmental Biology, Harvard University, Cambridge, MA 02138 **[227]**

Steven R. Rodermel, Department of Cellular and Developmental Biology, Harvard University, Cambridge, MA 02138 **[13]**

Gerald M. Rubin, Department of Embryology, Carnegie Institution of Washington, Baltimore, MD 21210 **[213]**

G.B. Ruvkun, Department of Biology, Massachusetts Institute of Technology, Cambridge, MA 02139 **[3]**

Thomas D. Sargent, Laboratory of Molecular Genetics, National Institute of Child Health and Human Development, National Institutes of Health, Bethesda, MD 20205 **[171]**

Charalambos Savakis, Department of Cellular and Developmental Biology, Harvard University, Cambridge, MA 02138 **[95]**

Robert T. Schimke, Department of Biological Sciences, Stanford University, Stanford, CA 94305 **[197]**

Andre Steinmetz, IBMC, 15, rueRene Descartes, F-67084, Strasbourg, CEDEX, France **[13]**

Edwin C. Stephenson, Department of Biology, Yale University, New Haven, CT 06511 **[137]**

Alap Subramanian, Department of Cellular and Developmental Biology, Harvard University, Cambridge, MA 02138 **[13]**

Harold Weintraub, Hutchinson Cancer Research Center, 1124 Columbia Street, Seattle, WA 98104 **[241]**

Preface

Ever since it was founded in 1939 by well-known scientists working on a wide variety of problems in "Growth and Development," the Society now known as Society for Developmental Biology has remained an open, diverse, and productive meeting ground for developmentalists. It has always attempted to maintain broad coverage, and to be receptive to new ideas. Perusal of the annual Symposium volumes dating back to 1939 will show preoccupation with plants, animals and microbes, and with levels of biological organization ranging from the molecule to the organism. The Society was mindful of the importance of genetics when much of the community of embryologists was not. More generally, it identified major advances and new fields early, sometimes before their importance to development was generally recognized.

The 41st Symposium was conceived and organized within that tradition. It is quite obvious that the recent revolutionary advances in molecular biology are of central importance for progress in developmental biology. The invited talks of the Symposium gave an overview of how the tools of molecular biology currently are helping illuminate one of the fundamental aspects of development: the regulated expression of specific genes. The meeting attracted a large group of working scientists and students concerned with the molecular biology of gene expression and related developmental problems–a group which we hope will find a permanent home in the Society and in its journal, Developmental Biology. At the same time, the diversity of the Society was manifested in the composition of the group attending the meeting, as well as the breadth of the material discussed in poster presentations and in three symposia.* We hope that, within the Society, recent converts to developmental biology will profit from interactions with scientists from diverse backgrounds, who are intimately familiar with the wealth of developmental phenomena and who are studying these phenomena with different methodologies and at different levels of biological organization.

The Symposium sessions, and the volume sections, roughly correspond to the main topics considered, as follows.

*The Symposia were: Dynamics and change in multigene families (organized by G.A. Dover); Cell-cell recognition and cohesion during development (organized by E.A. Berger); and Replication-transcription coupling (organized by H.W. Sauer).

Developmental regulation affects sets of genes, often of considerable complexity. In Sessions I and III, we considered the structural organization and possible modes of regulation for selected examples of such gene sets, representing both classical embryological subjects and favorable model systems of differentiation: the nitrogen fixation and symbiotic-state genes of *Rhizobium,* the genes of the chloroplast genome, the chorion genes of silkmoths and *Drosophila,* the mammalian globin genes (T. Maniatis) and the genes active in oogenesis and early development of sea urchins and amphibia.

Differentiation entails commitment to and subsequent manifestation of alternative cell states. The mechanisms of developmental commitment (determination) remain a mystery, but we considered favorable systems from which new insights are being derived, through the combination of genetic and molecular analyses: the switching between cell types in yeast, and the bithorax gene complex which specifies segmental type in *Drosophila* (M. Goldschmidt-Clermont and D.S. Hogness). Radical changes in cellular activity are known to be activated by simple switches, and we examined two such examples which are favorable for mechanistic analysis: the transformation-inducing products of avian sarcoma viruses and their cellular analogues, and the steroid-controlled genes in *Drosophila.*

In the last few years the eucaryotic genome has been shown to be in considerable flux, and this fact has profound implications for developmental regulation, as B. McClintock first indicated. We considered selected cases that operate in nature: transposable elements that control gene activity in maize, selective gene amplification in cultured mammalian cells, transposable elements in *Drosophila*. Furthermore, we discussed our current ability to transfer genes into tissue culture cells or into the germ lines of mice and *Drosophila,* and thus dissect their regulatory regions; in addition to the papers printed here, this Session included a presentation by R. Mulligan.

Finally, we considered temporal transcriptional regulation. The most complete understanding to date is available in the phage SP01 system. For higher organisms, we were introduced to two approaches: studies of chromatin conformational changes associated with specific gene expression, and analysis of factors necessary for specific transcription in reconstituted *in vitro* systems (R. Roeder).

The Society is grateful to the NSF for financial support of the Symposium, and to the new Department of Cellular and Developmental Biology at Harvard for its co-sponsorship and multiple support of the meeting. The members of the local committee (Tom Eickbush, Jim Ciotti, Lucy Cherbas and Bill Gelbart) were very helpful, as was the staff of the Science Center and the Law School Housing Office. The success of the meeting depended on the organizational genius of Rebecca Sarkisian and Holly Schauer, and the volume on the patient and dedicated editing of Steve Subtelny and the excellent work of the staff at Alan R. Liss, Inc.

Fotis C. Kafatos

Young Investigator Awards — 1982

First Place Award

Jo Ann Render
Graduate Student
Department of Zoology
University of Texas, Austin
Gary Freeman, Sponsor

Evidence for an Apical Tuft Inhibitor in the Second Polar Lobe of the Sabellaria cementarium Embryo

Second Place Awards

Paul W. Sternberg
Graduate Student
Department of Biology
Massachusetts Institute of Technology, Cambridge
H. Robert Horvitz, Sponsor

Evolutionary Divergence and Genetic Programming of Nematode Cell Lineage

William J. Wolfgang
Graduate Student
Department of Zoology
University of Washington, Seattle
Lynn M. Riddiford, Sponsor

Control of Cuticular Morphogenesis in the Caterpillar Manduca sexta

Abstract of the First Place Young Investigator Award

Evidence for an Apical Tuft Inhibitor in the Second Polar Lobe of the Sabellaria cementarium Embryo. Jo Ann Render, Department of Zoology, University of Texas, Austin, TX 78712.

The embryo of Sabellaria cementarium forms a polar lobe at each of the first two cleavage divisions which becomes absorbed into one of the blastomeres at the end of the division. The polar lobe preceding first cleavage contains determinants for the apical tuft and the post-trochal region of the trochophore larva. The polar lobe preceding second cleavage is smaller than the first polar lobe and contains determinants only for the post-trochal region of the larva. Post-trochal region development is indicated by the formation of post-trochal chaetae. In blastomere isolation experiments, isolates containing the C but not the D blastomere form apical tufts. Isolates containing the D but not the C blastomere do not form apical tufts. When the polar lobe preceding second cleavage is removed and the C and D blastomeres separated and raised in isolation, each can form an apical tuft. When second cleavage is equalized such that both C and D blastomeres receive second cleavage polar lobe material, no apical tuft is formed. These results suggest that the apical tuft determinants are distributed to both the C and D blastomeres at second cleavage but that the second polar lobe contains an inhibitor for apical tuft formation which is shunted to the D blastomere after the completion of second cleavage.

This work was supported in part by a Sigma Xi Grant-in-Aid of Research.

Photograph by Rick Stafford

Participants at the Forty-First Symposium of the Society for Developmental Biology

I. Structure of Developmentally Regulated Gene Sets

Gene Structure and Regulation in Development, pages 3–12

Genetic Analysis of Symbiotic Nitrogen Fixation Genes

G.B. Ruvkun, F.M. Ausubel, W.J. Buikema, S.E. Brown, C.D. Earl, and S.R. Long

Department of Cellular and Developmental Biology, Harvard University, Cambridge, Massachusetts 02138

Introduction

Bacteria in the genus Rhizobium fix nitrogen in specialized nodules that form on the roots of legumes upon the symbiotic interaction of bacterium and plant. The formation of an effective nitrogen-fixing nodule involves differentiation of both plant and bacterial cells and is characterized by a series of morphologically distinct developmental stages: (1) The Rhizobium species and host plant specifically recognize each other; each species of legume is nodulated by a particular Rhizobium species. (2) The Rhizobia penetrate the root-hair cell wall and gain entry to the interior of the root via an infection thread that progressively penetrates into the root cortex as the bacteria multiply. (3) Root cortical cells proliferate and differentiate to form the root nodule. (4) Rhizobia multiply and enlarge to become the differentiated "bacteroid" form and begin to express the nitrogen fixation genes.

Within the nodule, enzymes and substrates necessary for nitrogen fixation are supplied by both the bacteria and plant; the nitrogen-fixing enzyme, nitrogenase, is located within bacteroids, and the energy requirement for nitrogen fixation is supplied by the plant as photosynthate transported to the roots. It is known that bacterial genes control at least some of the steps in the nodule developmental pathway, because in several species of Rhizobium, bacterial mutants have been isolated that cause a spectrum of mutant root and nodule phenotypes, ranging from no nodules at all (Nod$^-$ phenotype) to nodules that appear to be morphologically normal but fail to fix nitrogen

Current addresses: Department of Biology, Massachusetts Institute of Technology, Cambridge, Massachusetts 02139 (G.B.R.), Department of Molecular Biology, Massachusetts General Hospital, Boston, Massachusetts 02114 (F.M.A., W.J.B., S.E.B., C.D.E.), and Department of Biological Sciences, Stanford University, Stanford, California 94305 (S.R.L.).

(Fix$^-$ phenotype). Because this developmental process is at least in part dependent on bacterial genes, it is amenable to genetic analysis using the bacterial endosymbiont as a genetic probe. Our focus has been to use transposon Tn*5* mutagenesis to identify Rhizobium genes specifically involved in this symbiotic developmental process. The major questions we are attempting to answer are the following: (1) Which steps in the development of the plant root nodule depend on Rhizobium gene expression? (2) Are bacterial nodulation genes expressed coordinately with nitrogen-fixation genes? (3) How are these bacterial developmental genes regulated?

The identification and manipulation of symbiotic (*sym*) and nitrogen-fixation (*nif*) genes in Rhizobium has been complicated by the fact that these genes are not normally expressed when the bacteria are in the free-living state. In order to overcome this difficulty and to identify Rhizobium *sym* and *nif* genes, we have developed a variety of molecular genetic techniques that greatly simplify the isolation of *sym* mutations and the subsequent cloning of *sym* genes. In this chapter, we describe how these techniques have been used to identify nitrogenase (*nif*) genes in R. meliloti, the endosymbiont of alfalfa. We also summarize experiments which show that the *nif* genes are located on a large indigenous plasmid and are closely linked to other *sym* genes.

Localization of Rhizobium *nif* Genes by Interspecies DNA Homology

Because nitrogenase genes in R. meliloti are not expressed in the free-living state, it is extremely tedious to identify these genes by screening R. meliloti clones for defects in the ability to symbiotically fix nitrogen; individual clones in a mutagenized culture would have to be individually tested on individual alfalfa plants to screen for symbiotic mutants (Sym$^-$ phenotype), and then analyzed biochemically to identify mutations in nitrogenase genes. To circumvent this cumbersome experimental strategy, we sought a direct method to identify the R. meliloti nitrogenase genes based on our previous work on the cloning and characterization of the *nif* genes from Klebsiella pneumoniae. K. pneumoniae is closely related to Escherichia coli and fixes nitrogen in a free-living state utilizing N_2 as its sole nitrogen source. The K. pneumoniae nitrogenase genes were identified using a combination of classical bacterial genetic techniques and molecular cloning techniques. Because nitrogenases from all species examined share many biochemical and structural features, we hypothesized that nitrogenase genes had been conserved in evolution and that we might be able to utilize the cloned K. pneumoniae nitrogenase genes to identify the corresponding R. meliloti nitrogenase genes.

In fact, we found that the DNA sequences of nitrogenase gene(s) are sufficiently conserved between R. meliloti and K. pneumoniae that a 6.3-kilobase (kb) K. pneumoniae EcoRI fragment carrying *nifK*, *nifD*, and *nifH*, specifically hybridizes to a 3.9-kb EcoRI fragment from total R. meliloti

DNA [Ruvkun and Ausubel, 1980]. This interspecies homology was used in a colony hybridization procedure [Grunstein and Hogness, 1975; Hanahan and Meselson, 1980] to clone the R. meliloti 3.9-kb EcoRI fragment. A restriction map of a portion of the R. meliloti genome containing the 3.9-kb EcoRI fragment is shown in Figure 1.

To further establish that the cloned 3.9-kb EcoRI fragment contained R. meliloti nitrogenase genes, as well as to locate the *nif* genes on the fragment, we used the method of Maxam and Gilbert [1980] to determine the DNA sequence of both strands of the 3.9-kb EcoRI fragment between Bgl II site (b), and Xho I site (a) (see Fig. 1), which, based on the interspecies DNA hybridization data, was expected to contain the N terminal coding region of R. meliloti *nifH*. The DNA sequencing showed that the presumptive R. meliloti *nifH* gene contains 61% DNA homology and 72% amino acid homology to the K. pneumoniae *nifH* gene over the 360-bp region analyzed [Ruvkun et al, 1982a; Sundaresan and Ausubel, 1981; Scott et al, 1981]. These data indicated that we had indeed cloned the R. meliloti *nifH* gene and that transcription of this gene proceeds right-to-left as drawn in Figure 1. We also showed, by DNA sequence analysis, that the R. meliloti nifD gene, as in K. pneumoniae, is located directly distal to the *nifH* gene and is transcribed in the same direction as *nifH*. Results similar to ours have also been obtained by Torok and Kondorosi [1981], who determined the entire DNA sequence of the *nifH* gene from a different strain of R. meliloti. Their sequence exactly matches our partial sequence in the protein coding region.

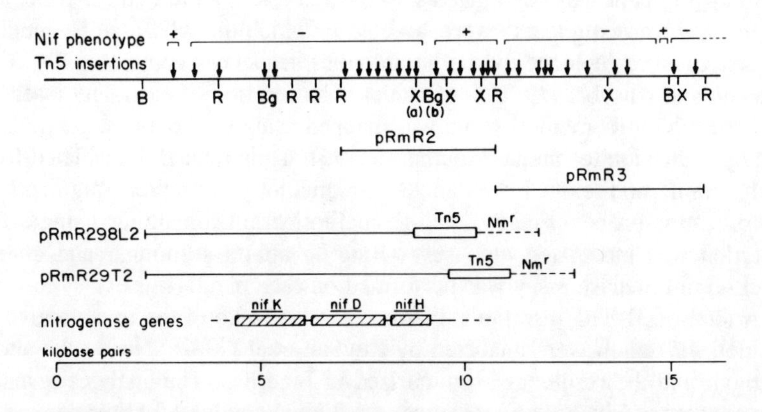

Fig. 1. The physical genetic map of the *R. meliloti nif* region. (Top): Arrows denote Tn5 insertions, and the symbiotic phenotypes appear above the arrows. Plasmids pRmR298L2 and pRmR29T2 were used in the complementation analysis (see text). (Bottom): Orientation of *nif*KDH based on homology to *K. pneumoniae* sequences and complementation analysis. B, *Bam*HI; R, *Eco*RI; Bg, *Bgl* II; X, *Xho* I.

The analysis of the presumptive R. meliloti *nif* genes described above was based on DNA sequence analysis and comparison of R. meliloti DNA sequences with the known DNA and amino acid sequences of K. pneumoniae nitrogenase genes and proteins. This analysis clearly showed that R. meliloti contains genes that are very similar to the K. pneumoniae *nif* genes. However, this type of physical analysis did not prove that the presumptive R. meliloti *nif* genes are used in, and are essential for, symbiotic nitrogen fixation. Therefore, in order to provide genetic evidence that these presumptive R. meliloti *nif* genes are essential symbiotic genes, we devised a directed method to introduce mutations into the *nif* region of the Rhizobium genome [Ruvkun and Ausubel, 1981]. The technique we developed involves the replacement of wild-type R. meliloti genomic sequences with the corresponding cloned sequences that had been mutagenized with transposon Tn*5* in E. coli.

Specifically, a R. meliloti restriction fragment cloned on a plasmid vector such as pBR322 was mutagenized with Tn*5* in order to generate a collection of plasmids with Tn*5* inserted at different locations. The location of each Tn*5* insertion ($\pm$ 100 bp) was determined for each Tn*5*-containing plasmid by performing a standard restriction enzyme analysis. Because pBR322 does not replicate in R. meliloti, R. meliloti fragments containing Tn*5* were recloned into the low-copy number, broad host-range vector pRK290 [Ditta et al, 1980], which confers tetracycline resistance and contains a single EcoRI site for cloning.

Each pRK290 recombinant plasmid containing a Tn*5*-mutagenized R. meliloti fragment was conjugated into the Fix$^+$ R. meliloti strain 1021, and R. meliloti transconjugants were selected in which the wild-type R. meliloti sequences were replaced by the homologous sequences containing Tn*5* (see Ruvkun and Ausubel [1981] for details). The location of each Tn*5* insertion in the resulting R. meliloti strain was mapped using the Southern gel transfer and hybridization technique [Southern, 1975], using total DNA isolated from each strain, and cloned ^{32}P-labeled R. meliloti restriction fragments as hybridization probes. Finally, each R. meliloti strain containing a single Tn*5* insertion was inoculated onto several sterile alfalfa seedlings, and after 4 weeks a nitrogenase assay was performed on each plant.

A total of 31 Tn*5* insertions distributed over 14 kb of the presumptive R. meliloti *nif* region were analyzed by Ruvkun et al [1982a], using the above method, and the results are summarized in Figure 1. The analysis revealed two clusters of Fix$^-$::Tn*5* insertions of 6.3 kb and at least 5 kb, separated by a 1.6 kb cluster of Fix$^+$ insertions. The 6.3-kb cluster of Fix$^-$ insertions corresponds exactly to the location of R. meliloti sequences which are homologous to the K. pneumoniae nitrogenase genes (see Fig. 1), verifying that the presumptive R. meliloti *nif* genes are indeed required for symbiotic nitrogen fixation.

The R. meliloti *nifH*, *nifD*, and *nifK* Genes Are in an Operon

By analogy with K. pneumoniae, we expected the R. meliloti *nifD* and *nifK* genes to be situated within a single operon. To determine whether this is the case, we performed complementation tests between selected *nif*::Tn*5* insertions and appropriate plasmids, as described below. Because Tn*5* normally causes polar mutations in an operon, insertion of Tn*5* into any gene in an operon leads to inactivation of all distal genes in that operon. Therefore, complementation analysis using transposon mutations will establish the boundaries of transcription units but not the boundaries of genes within an operon.

The strategy we adopted to perform the complementation analysis was to construct two conjugative plasmids carrying the *nifHDK* genes. One of these plasmids contained the presumptive *nifHDK* promoter and the other plasmid lacked this promoter due to the insertion of Tn*5* near the NH_2 terminal end of the *nifH* gene. These plasmids were then conjugated into a variety of *nif*::Tn*5* R. meliloti strains, and the resulting merodiploid R. meliloti strains were assayed for the ability to fix nitrogen after nodule formation on alfalfa. This analysis showed that the plasmid that contained the presumptive promoter region was able to complement all of the genomic *nifHDK*::Tn*5* mutations, whereas the plasmid that lacked the promoter region failed to complement any of these mutations. These data are consistent with the interpretation that the *nifH*, *nifD*, and *nifK* genes in R. meliloti are present in an operon that is transcribed in the direction *nifH* to *nifK*.

Nitrogenase Genes Are Located on a Large Plasmid in R. meliloti Strain 1021

Because large plasmids in several Rhizobium species have been shown to carry genes essential for symbiotic nitrogen fixation [Johnston et al, 1978; Zurkowski and Lorkiewicz, 1979; Beynon et al, 1980; Banfalvi et al, 1981; Rosenberg et al, 1981], including nitrogenase genes, we sought to determine whether nitrogenase genes were located on plasmids in R. meliloti strain 1021. Using the Eckhardt procedure of "in gel" lysis of the bacteria [Eckhardt, 1978], a single extremely large (approximately 500 kb) plasmid (megaplasmid) was identified in strain 1021. In order to determine whether the plasmid band carries nitrogenase genes, we devised the following internally controlled experimental strategy. The overall idea was to directly compare strains with Tn*5* insertions in nitrogenase genes (presumptive plasmid location) with strains containing Tn*5* insertions in genes necessary for amino acid biosynthesis (chromosomal location) (see Buikema et al [1982]).

Tn*5* insertions inside the nitrogenase structural genes were constructed as described above using the site-directed mutagenesis procedure [Ruvkun and

Ausubel, 1981]. Tn*5* insertions within amino acid biosynthetic genes were constructed using a suicide plasmid technique for generating random Tn*5* insertions in the R. meliloti genome [Meade et al, 1981]. The Tn*5* insertions were correlated with the auxotrophic phenotype by showing that the neomycin resistant phenotype of Tn*5* in each strain was 100% linked to the auxotrophic phenotype. In addition, the neomycin resistance phenotype was found to be linked to a *pan* marker, which is part of a single circular linkage group that includes at least 20 auxotrophic and drug-resistance mutations and which is presumed to represent the R. meliloti chromosome [Meade and Signer, 1977]. In contrast to the auxotrophic strains, the Tn*5* insertions in the nitrogenase genes showed less than 1% linkage to several selected chromosomal markers widely dispersed around the R. meliloti chromosome.

The wild-type strain, the strains containing Tn*5* in the nitrogenase genes, and the strains containing Tn*5* in auxotrophic genes, all contained a single megaplasmid DNA band after electrophoretic separation, as described by Eckhardt [1978]. When transferred to nitrocellulose and hybridized with ^{32}P-labeled Tn*5* probe, the megaplasmid band from strains containing Tn*5* insertions in the nitrogenase genes hybridized to the top of the gel where chromosomal DNA is trapped, in the case of the strains containing Tn*5* insertions in the auxotrophic genes (data not shown). This experiment demonstrates on the basis of both genetic and physical criteria that the megaplasmid carries the nitrogenase genes. These data confirm results published recently by Rosenberg et al [1981] and by Banfalvi et al [1981].

Identification of *R. meliloti* Symbiotic Genes

In contrast to the *nif* genes, many R. meliloti *sym* genes most likely have no functional analogues in K. pneumoniae and cannot be identified by interspecies homology with cloned K. pneumoniae *nif* genes. We have therefore adopted two general methods to isolate mutations, and thereby identify R. meliloti *sym* genes. First, based on the results obtained recently in several laboratories that a variety of symbiotic genes are located on plasmids [Johnston et al, 1978; Zurkowski and Lorkiewicz, 1979; Beynon et al, 1980; Banfalvi et al, 1981; Rosenberg et al, 1981], we made the assumption that interesting symbiotic genes would be closely linked to the *nifHDK* genes already identified and cloned on the R. meliloti megaplasmid. Working on this assumption, we cloned approximately 90 kb of the R. meliloti megaplasmid using the cosmid cloning vector pHC79. Starting at the point of the *nifHDK* locus, overlapping clones were identified using a standard stepwise hybridization “walking” protocol. Some of the cosmid clones used to generate a restriction map of the 90-kb region are shown in Figure 2. Selected clones were then mutagenized with Tn*5*. Using the gene replacement technique described above for mutagenesis of the R. meliloti *nif* genes, we have

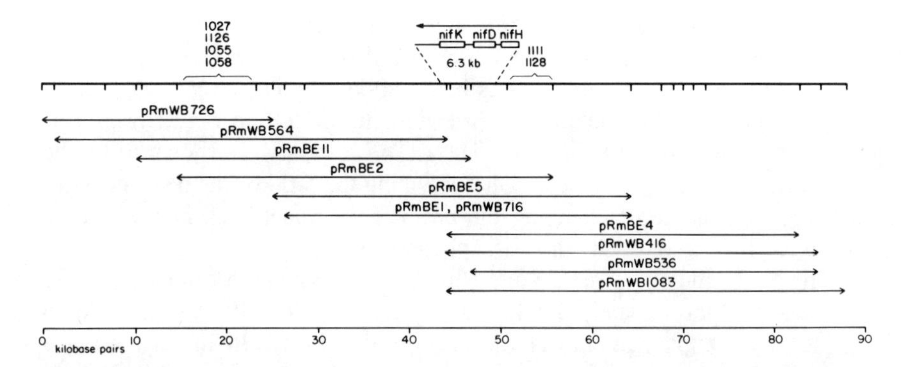

Fig. 2. Cloned region of the *R. meliloti* megaplasmid. The *Eco*RI restriction map of the 90-kb region is shown with the position of the *nif*KDH operon and the overlapping cosmids used to generate the map. The positions of six insertion mutations (see text) appear above the map.

identified several regions in the 90-kb *nif* region that contain gene(s) required to form nodules (*nod* genes), and genes required for effective fixation in formed nodules (*fix* genes).

At the present time, we have examined a total of 90 Tn*5* insertions in selected areas of the 90-kb *nif* region and the results are summarized in Figure 2. The most important results from this series of experiments are that both *nod* and *fix* genes are closely linked on the megaplasmid and that the 90-kb region appears to contain significant regions that do not contain essential *sym* genes. Perhaps the most interesting symbiotic region identified is approximately 20 kb distal to the *nifHDK* operon where gene(s) are located that are required for one of the earliest, if not *the* earliest, steps in plant-bacterial interaction. Bacteria that contain Tn*5* insertion mutations in this region fail to induce root-hair curling, and inoculated plants show no visible sign of having interacted with the bacteria (also see Long et al [1982]).

The second general method we used for identifying *sym* genes involved the use of a random Tn*5* mutagenesis procedure developed by van Vliet et al [1978] and Beringer et al [1978] to generate R. meliloti *sym* mutants presumably caused by Tn*5* insertions with symbiotic defects [Meade et al, 1981].

Based on the results, described above, that symbiotic genes appear to be clustered on the megaplasmid, we screened 14 of 19 presumptive Tn*5*-induced symbiotic mutants for the presence of Tn*5* on the megaplasmid, using the Eckhardt in-gels lysis technique. In addition, working on the assumption that many symbiotic genes might be linked on the R. meliloti megaplasmid, we determined whether the 19 mutants contained an insertion within the 90-kb region surrounding the nitrogenase genes. This was accomplished by probing a Southern blot of EcoRI digested total DNA from the mutants with ^{32}P-labeled pRmBE11 and pRmWB536 DNA (see Fig. 2). Six strains contained identifiable inserts in the 90-kb region (see Fig. 2). Strains Rm1111 and Rm1128 contain Tn*5* insertions in the 5.0 kb *Eco*RI fragment, just to the right of the *nifH* gene. These Tn*5* inserts were shown to be the actual cause of the symbiotic defect by cloning the 5.0-kb fragments containing Tn*5* and marker-exchanging the *nif*::Tn*5* fragment back into wild-type strain Rm1021 to re-create the Nif$^-$ phenotype.

The Nod$^-$ mutants Rm1126 and Rm1027 also contain insertions in the 90 kb region, approximately 20 kb to the left of *nifK*. It was possible to genetically complement the Nod$^-$ phenotype of Rm1126 in vivo using a specially constructed cosmid clone (pRmSL26) containing the wild-type version of the mutated region of Rm1126 [Long et al, 1982]. The other Nod$^-$ mutant, Rm1027, was shown to contain an insertion sequence, IsRm1 [Ruvkun et al, 1982b], in the same EcoRI fragment as Rm1126 (by hybridization with a cloned IsRm1 probe) and could also be complemented genetically by pRmSL26.

Concluding Remarks

The results discussed in this chapter argue strongly for an extrachromosomal location for a number of genes involved in nitrogen fixation in R. meliloti. Utilizing the Eckhardt procedure for plasmid visualization, we have demonstrated that transposon Tn*5* insertions within structural genes for the nitrogenase enzyme and within gene(s) involved in nodulation are all located on a very large megaplasmid in strain Rm1021. Recently, Banfalvi et al [1981] showed that R. meliloti strain 41 also contains a megaplasmid that carries nitrogenase and nodulation genes. Moreover, Rosenberg et al [1981] have demonstrated the presence of similar large megaplasmids that carry nitrogenase genes in a large number of R. meliloti isolates from different geographic regions.

The use of cosmid cloning techniques has allowed us to clone approximately 90 kb of DNA surrounding the *nifHDK* operon. Because of the large size of the inserts, we were able to construct a restriction map of most of the 90-kb region by taking advantage of overlap between different cosmid clones.

The physical mapping of insertion mutations has allowed us to show that nodulation genes are closely linked to fixation genes. By physically mapping the insertions in the Nod$^-$ mutants Rm1027 and Rm1126 to an 8.7-kb EcoRI fragment, we have shown that nodulation gene(s) are approximately 20 kb distal to the *nifKDH* operon. We have also shown that the nodulation deficiency of Rm1126 and Rm 1027 can be complemented by a plasmid containing the 8.7-kb EcoRI fragment [Long et al, 1982].

The phenotypes of the mutations we have isolated allow us to establish which bacterial genes are involved in each nodule developmental stage. All of the *nif*::Tn*5* mutations have no effect on the ability of R. meliloti to form nodules. Therefore, fixed nitrogen itself is not involved in the development. We do find, however, that *nif*::Tn*5* strains form many more nodules and these nodules do not senesce as soon. Scanning and transmission electron microscopic examination of nodules from many of the *nif*::Tn*5* mutants showed some defects in nodule ultrastructure associated with *nif*::Tn*5* mutations in a variety of *nif* operons but no gross change in nodule development [A. Hirsch, unpublished observations].

The isolation of *nod*::Tn*5* mutations shows that the bacteria are not passive in the nodulation process; a functional *nod* gene is necessary for the first stages of nodulation. We have no data on the products of these genes or how their expression is controlled. Clearly the next step is to identify the gene products and their function and localization in the nodule.

The functional significance of the close linkage of *nif* and *nod* genes on a large plasmid R. meliloti is still not understood. It must be assumed that the close linkage is related in some way to the coordinate regulation of *nif* and *nod* genes. However, it is still not known how the expression of these genes is regulated. Furthermore, an outstanding unanswered question is how the Rhizobium senses the plant root environment to trigger its developmental program.

At this point we have only isolated mutations with very gross phenotypes, ie, totally Nod^- or Nif^-. However, the reverse genetic approach of creating the mutant genotype from cloned DNA sequences allows us to survey for the presence of symbiotic genes which, when mutated, might go unnoticed in a general screen because they play subtle roles in the nodulation process. In addition, because transposon insertion creates null mutations, these can be used in the isolation of nonnull mutant genes on chemically mutagenized plasmid clones. These nonnull alleles should greatly aid in the genetic analysis of symbiotic gene function.

REFERENCES

Banfalvi Z, Sakayan V, Koncz C, Kiss A, Dusha I, Kondorosi A (1981): Location of nodulation and nitrogen fixation genes on a high molecular weight plasmid of R. meliloti. Mol Gen Genet 184:318–325.

Beringer JE, Beynon JL, Buchanon-Wollaston AV, Johnston AWB (1978): Transfer of the drug resistance transposon Tn5 to Rhizobium. Nature 276:633–634.

Beynon JL, Beringer JE, Johnston AWB (1980): Plasmids and host-range in Rhizobium leguminosarum and Rhizobium phaseoli. J Gen Microbiol 120:421–429.

Buikema WJ, Long SR, Brown SE, van den Bos RC, Earl C, Ausubel FM (1982): Cosmid cloning of a large region of symbiotic genes from the megaplasmid of Rhizobium meliloti. J Mol Appl Genet (in press).

Ditta G, Stanfield S, Corbin D, Helinski D (1980): Broad host range DNA cloning system for gram negative bacteria: Construction of a gene bank of Rhizobium meliloti. Proc Natl Acad Sci USA 77:7347–7351.

Eckhardt T (1978): A procedure for the isolation of deoxyribonucleic acid in bacteria. Plasmid 1:584–588.

Grunstein M, Hogness DS (1975): Colony hybridization: A method for the isolation of cloned DNAs that contain a specific gene. Proc Natl Acad Sci USA 72:3961.

Hanahan D, Meselson, M (1980): Plasmid screening at high colony density. Gene 10:63–67.

Johnston AWB, Beynon JL, Buchanon-Wallaston AV, Setchell SM, Hirsch PR, Beringer JE (1978): High frequency transfer of nodulating ability between strains and species of Rhizobium. Nature 276:634–636.

Long SR, Buikema WJ, Ausubel FM (1982): Cloning of Rhizobium meliloti nodulation genes by direct complementation of Nod^- mutants. Nature 298:485–488.

Maxam A, Gilbert W (1980): Sequencing end labelled DNA with base specific chemical cleavages. Methods Enzymol 65:499–560.

Meade H, Signer E (1977): Genetic mapping of Rhizobium meliloti. Proc Natl Acad Sci USA 74:2076–2078.

Meade HM, Long SR, Ruvkun GB, Brown SE, Ausubel FM (1981): Physical and genetic characterization of symbiotic and auxotrophic mutants of Rhizobium meliloti induced by transposon Tn5 mutagenesis. J Bacteriol 149:114–122.

Rosenberg C, Boistard P, Denarie J, Casse-Delbart F (1981): Genes controlling early and late functions in symbiosis are located on a megaplasmid in Rhizobium meliloti. Mol Gen Genet 184:326–333.

Ruvkun GB, Ausubel FM (1980): Interspecies homology of nitrogenase genes. Proc Natl Acad Sci USA 77:191–195.

Ruvkun GB, Ausubel FM (1981): A general method for site directed mutagenesis in prokaryotes. Nature 289:85–86.

Ruvkun GB, Sundaresan V, Ausubel FM (1982a): Directed transposon Tn5 mutagenesis and complementation analysis of Rhizobium meliloti symbiotic nitrogen fixation genes. Cell 29:551–559.

Ruvkun GB, Long SR, Meade HM, van den Bos RC, Ausubel FM (1982b): IsRm1: A Rhizobium meliloti insertion sequence which preferentially transposes into nitrogen fixation (nif) genes. J Mol Appl Genet 1:405–418.

Scott KF, Rolfe BG, Shine J (1981): Biological nitrogen fixation: Primary sequence of the Klebsiella pneumoniae nifH and nifD genes. J Mol Appl Genet 1:71–81.

Southern EM (1975): Detection of specific sequences among DNA fragments separated by gel electrophoresis. J Mol Biol 98:503–517.

Sundaresan V, Ausubel FM (1981): Nucleotide sequence of the gene coding for the nitrogenase iron protein from Klebsiella pneumoniae. J Biol Chem 256:2808–2812.

Torok I, Kondorosi A (1981): Sequence of the R. meliloti nitrogenase reductase (nifH) gene. Nucleic Acids Res 9:5711–5723.

van Vliet F, Silva B, van Montagu M, Schell J (1978): Transfer of RP4::Mu plasmids to Agrobacterium tumefaciens. Plasmid 1:446–455.

Zurkowski W, Lorkiewicz Z (1979): Plasmid-mediated control of nodulation in Rhizobium trifolii. Arch Microbiol 123:195–201.

Gene Structure and Regulation in Development, pages 13–32

Maize Plastid Genes: Structure and Expression

Lawrence Bogorad, Earl J. Gubbins, Setsuko O. Jolly, Enno T. Krebbers, Ignacio M. Larrinua, Karen M.T. Muskavitch, Steven R. Rodermel, Alap Subramanian, and Andre Steinmetz

Department of Cellular and Developmental Biology, Harvard University, Cambridge, Massachusetts 02138.

INTRODUCTION

Like other eucaryotic cells, cells of green plants contain distinct sets of separated nuclear and mitochondrial genes, but plant cells also have plastids that carry a third genome. In the electron microscope the DNA molecules of plastids and mitochondria look more like the DNA of procaryotes than the condensed protein-associated chromosomes of the nucleus.

Early in the century it was found that although nuclear genes are involved in the development of chloroplasts, some features are transmitted in a non-Mendelian manner. The "plastome," as a set of genes carried in the cytoplasm, presumably by plastids, is an old idea among plant geneticists. Despite the knowledge that plastid development depends upon both nuclear and plastid genes, the intimacy of the relationship that has been discovered in the past few years was unexpected. Every multimeric structure or enzyme thus far studied in plastids and mitochondria such as ribosomes, multimeric enzymes, and membranes has some components that are products of the nuclear genome—transcribed in the nucleus, translated in the cytoplasm, and transported into the organelle—and others that are products of organelle

Present addresses: Amoco Research Center, P.O. Box 400, Mail Station B 1 Naperville, IL 60566 (E.J.G.); Universal Foods Technical Center, 6143 North 60th Street, Milwaukee, WI 53218 (S.O.J.); IBMC, 15, rue Rene Descartes, F-67084, Strasbourg, CEDEX, France (A. Steinmetz). Permanent address for A. Subramanian: Max-Planck-Institut fur Molekulare Genetik, Berlin-Dahlem.

genes and organelle transcription-translation systems. Understanding the mechanisms for control of gene expression in the nucleus and the organelles and the processes of integration of the expression of these two genomes is a central problem for the student of eucaryotic molecular developmental biology. There are a number of differentiated forms of plastids: green photosynthetic chloroplasts; carotenoid crystal-containing chromoplasts; starch-containing amyloplasts; etioplasts of dark-grown tissues; the differently developed plastids of mesophyll and bundle sheath cells of C4 plants; etc. In some of these cases chloroplast gene expression appears to be regulated at the level of transcription. Do nuclear gene products regulate transcription of chloroplast genes? Is nuclear gene transcription regulated by chloroplast products in some sort of feedback circuit? Are there comparable kinds of translational regulation systems? Or are there independent and separate regulatory systems in each compartment with imbalances corrected by destruction of the polypeptides produced in excess?

The ensuing discussion is a review of some recent work on maize plastids directed at understanding the structure of their genes and genome as well as mechanisms of information processing.

THE MAIZE PLASTID GENOME

The maize plastid chromosome is a circle of approximately 139,000 bp (base pairs) comprising one large and one small single-copy region separated by two inverted repeated sequences each of 22,000 bp [Bedbrook and Bogorad, 1976; Bedbrook et al, 1977]. Each chloroplast may contain DNA equivalent to about 50 such molecules. The locations on the maize chloroplast chromosome of recognition sites for the restriction endonucleases Sal I and Bam HI are shown in Figure 1.

The outermost arcs near the top of the circle in Figure 1 show the positions of two large inverted repeated sequences. The two inner circles in Figure 1 show the positions of genes that have been mapped and sequenced (solid black boxes or lines perpendicular to the circle) or have been mapped to a restriction fragment (open boxes).

Each of the inverted repeated sequences contains genes for a complete set of rRNAs—16S, 23S, 4.5S [Bedbrook and Bogorad, 1976; Bedbrook et al, 1977], and 5S. And, as has been shown by Koch et al [1981], the spacer region between the 16 and 23S rDNAs contains genes for isoleucine and alanine tRNAs. The orientation of the rDNAs and the positions in the spacer of the genes for the two tRNAs resemble the pattern seen in E coli. Genes for other tRNAs and open reading frames that may code for unidentified polypeptides have been located at various positions in the inverted repeated sequences [Schwarz et al, 1981a,b].

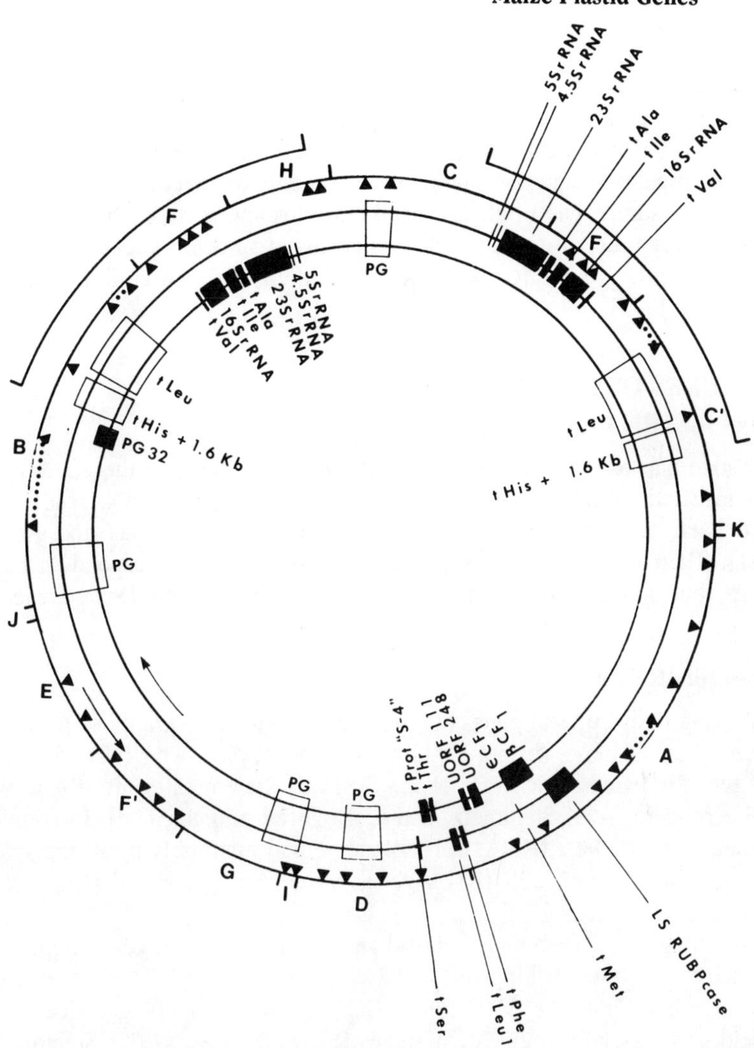

Fig. 1. A map of the circular maize plastid chromosome showing, on the outer circle, recognition sites for restriction enzymes Sal I (fragments generated by this enzyme are designated by capital letters), and by Bam HI (black triangles). It also shows the locations of the two large inverted repeated segments (at the top of this figure), and the positions of genes for rRNAs. The directions of transcription of the two inner circles, representing the two DNA strands, are shown by arrows. Among the other genes shown on this diagram are those for transfer RNAs that have been sequenced, except that the tRNA valine gene that contains a large intron and lies between *cf1BE* and unidentified open reading frame (UORF) 248 is omitted. The location of photogene 32 is shown on the inner circle, and the approximate locations of other light inducible genes that have not yet been assigned to a DNA strand are indicated by boxes labeled PG. rProt "S-4" is a gene with a very strong sequence homology to E coli ribosomal gene S4 [A. Steinmetz, A. Subramanian, and L. Bogorad, unpublished].

The presence of two large inverted repeated sequences, each containing a set of plastid rDNAs, appears to be the most common pattern among plants but there are exceptions. Plastid chromosomes of Vicia faba [Koller and Delius, 1980] and Pisum sativum [Palmer and Thompson, 1981] lack the inverted repeated sequence, and thus there is only one set of genes for rRNAs per chromosome. The Euglena plastid chromosome is strikingly different. It contains three sets of genes for rRNAs arranged tandemly plus an extra copy of the 16S rRNA gene [Gray and Hallick, 1978; Rawson et al, 1978; Jenni and Stutz, 1979].

FEATURES OF PLASTID GENES

Genes for rRNAs

Maize plastid genes for 16S and 23S rRNAs located on the 22,000 base pair inverted repeats [Bedbrook and Bogorad, 1976; Bedbrook et al, 1977] have been sequenced [Schwarz and Kossel, 1980; Edwards and Kossel, 1981]. Their sequences strongly resemble those of the corresponding E coli genes, but some of the flanking sequences are different [Schwarz et al, 1981b].

Genes for tRNAs

Among maize plastid genes that have been sequenced are those for methionine, phenylalanine, serine, histidine, threonine, plus two tRNAs for valine and two for leucine [Schwarz et al, 1981a,b; Steinmetz et al, 1982, 1983; E.T. Krebbers, A.S. Steinmetz, and L. Bogorad, unpublished]. In addition, as mentioned earlier, tRNAs for isoleucine and alanine lying in the spacer between the 16S and 23S rRNA genes have been sequenced [Koch et al, 1981].

Homologies between maize plastid and E coli tRNAs range to about 80%, but there are some striking differences in the genes for tRNAs from these two sources. First, the 3′ terminal CCA sequence is not encoded in the plastid genes as it is in the E coli genes [Schwarz et al, 1981a; Steinmetz et al, 1982, 1983]; in plastids this sequence must be added after transcription as for nuclear-coded cytoplasmic tRNAs. Second, four maize plastid tRNA genes—those for alanine and isoleucine [Koch et al, 1981]; for leucine [Steinmetz et al, 1982], and for valine [E.T. Krebbers, A.S. Steinmetz, and L. Bogorad, unpublished]—have introns ranging from 458 to more than 900 nucleotides inserted in the anticodon loop; the intron in the leucine gene for leucine tRNA splits, whereas the introns of the other plastid tRNA genes are located at different positions in the anticodon loop.

Conserved sequences resembling, in general, "−10" and "−35" promoter regions of procaryotic genes have been found 5′ to the expected 5′

ends of the mature tRNAs [Steinmetz et al, 1983]. In the case of tH-GUG (the histidine tRNA gene with the anticodon GUG), the transcript is estimated to begin 4–8 nucleotides from the conserved "−10-like" sequence [Schwarz et al, 1981a].

Genes for Proteins

rcL, the gene for the large subunit of ribulose bisphosphate carboxylase (LS Rubpcase). The first gene for a plastid protein to be mapped and sequenced was the maize *rcL,* the gene for LS Rubpcase (the large subunit of the carbon dioxide-fixing enzyme ribulosebisphosphate carboxylase) [Coen et al, 1977; Bedbrook et al, 1979; Link and Bogorad, 1980; McIntosh et al, 1980]. A number of features of this gene have been found to be common to the few plastid genes for proteins studied to date: (1) The universal nucleotide code for amino acids is used in plastid genes in contrast to the variable coding systems employed in mitochondria of yeast, *Neurospora,* and mammals [Heckman et al, 1980; Barrell et al, 1980; Bonitz et al, 1980]. (2) Five nucleotides upstream of the codon for translation initiation is a sequence that is transcribed into GGAGG; this is complementary to a sequence near the 3′ terminus of the maize chloroplast 16S rRNA that has been sequenced by Schwarz and Kossel [1980]. This complementary sequence appears to be analogous to "Shine-Dalgarno" sequences believed to serve in binding mRNAs of E coli to 16S rRNA [Shine and Dalgarno, 1974; Steitz and Jakes, 1975]. (3) Conserved sequences resembling "−10" and "−35" sequences of prokaryotic genes are associated with the maize *rcL* but the "−10-like" sequence starts about 25 nucleotides (rather than about 10) 5′ to the transcription initiation site [McIntosh et al, 1980; Jolly et al, 1981] and is separated by 21 rather than the more common 17 to 18 nucleotides [Steinmetz et al, 1983] from the "−35-like" region. A consensus sequence for 13 plastid genes—including seven genes for tRNAs—is shown in Figure 2. (4) Sequences surrounding the 3′ end of the transcript of *rcL* made in vivo (Fig. 3A) can be formed into loop and stem structures that may function in transcription termination or mRNA processing (Fig. 3B).

cf1BE, the gene for subunits beta and epsilon of CF_1. Another DNA sequence that has been analyzed codes for the beta and epsilon subunits of

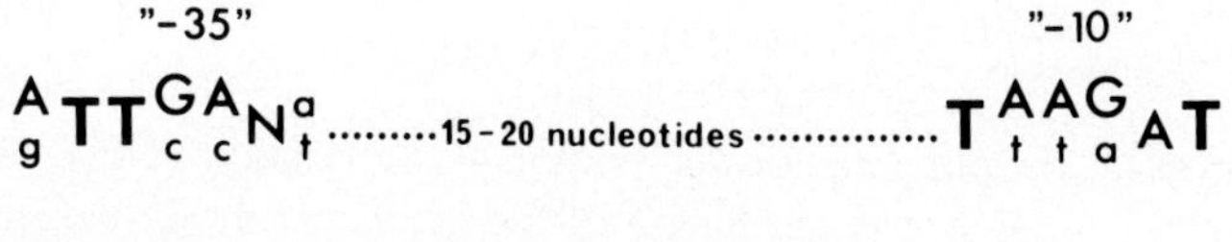

"-35" "-10"

A GA a AAG
TT N ....15-20 nucleotides............... T AT
g c c t t t a

Fig. 2. A consensus sequence for conserved nucleotides in the −10 and −35 regions for 13 plastid genes including several maize plastid tRNA and protein genes.

Ochre
TAA] AATAAAAAAAA AGCAAAAT

ATGAA GTGAA AAAATAAGTT ATGAA ATGAA ATGAA ATGAC

GTA ATTC TTT ATTC CTCTA ATTG ATTG CA ATTC A ATTC

GGCTCATCTTTTCTAAAAAAAAAAAAGACTGAGCCGAAAGA

AAAAGATCT - 3'

A

Ochre
5'-TAA AATAAAAAAAAAGCAAAAT ATGAA GTGAA
AAAATAAGTT ATGAA ATGAA ATGAA ATGAC

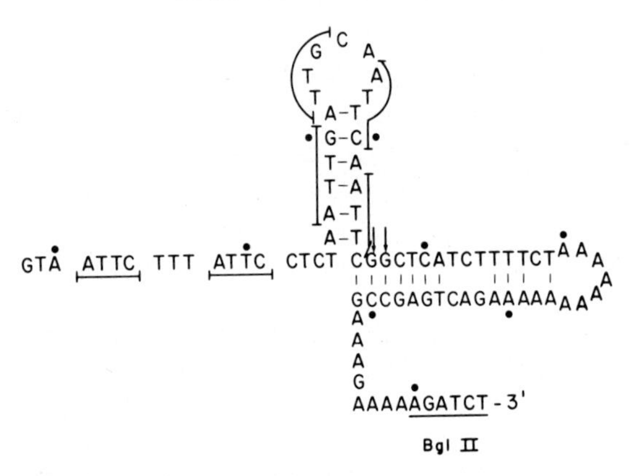

B

Fig. 3. A) The DNA sequence at the 3′ terminus of maize chloroplast gene *rcL* that codes for the large subunit of Rubpcase. Short sequences (ATGAA,GTGAA,ATGAC) that have double underlines bear strong nucleotide homology to one another; those marked by single underlines form another set of homologous sequences (ATTC,ATTG). B) Two possible stem and loop structures are shown; the upright one would probably not be very stable. The transcript formed in vivo terminates at about the position marked with arrows. [C. Poulsen, L. McIntosh, and L. Bogorad, unpublished]. It is not known whether this is the site of termination of transcription or of subsequent RNA processing.

the chloroplast coupling factor for photophosphorylation (CF_1). A notable feature of this sequence is the fusion of the genes for these two subunits of CF_1. As shown in Figure 4, the terminal A of the codon for lysine at the end of the gene for the beta subunit together with the first two nucleotides, U and G, of the translation termination codon in the mRNA constitute the initiating

```
 CF1β
──────►
  lys   Term.
5'---AAA UGA AA UUA---3'
        met  lys  leu
        ──────────────────►
              CF1ε

α2β2γδε2
```

Fig. 4. The region of fusion between the genes for the beta and epsilon subunits of maize CF_1 as they would appear in the RNA transcript. The methionine initiating codon for the epsilon gene, *cf1E*, is formed from the terminal A of the lysine together with the U and G of the termination codon for the beta gene. [Krebbers et al, 1982]. The stoichiometry of the 5 polypeptide subunits of CF_1 [McCarty, 1979] is shown in the lower left-hand portion of the figure.

```
MAIZE CHLOROPLAST GENES.
16S RNA RECOGNITION SEQUENCES(?).
                   ....AAATTATGTGATAATTATG..... 3' β CF1
                            ••••••
16S rRNA 3' End     UUUCCUCCACUAGGUC.............
                        •••••
                ....TTGTAGGGAGGGACTTATG........ 3' LS RuBPcase
```

Fig. 5. Sequences upstream of translation initiation start sites (*ATG*) for *cf1BE* and *rcL* of maize plastids in relation to sequences at the 3′ end of the 16S rRNA. These two sequences are complementary to different nucleotides near the 3′ end of the 16S rRNA.

AUG for the epsilon subunit. A Shine-Dalgarno type sequence is present on *cf1BE* (the fused gene for the beta and epsilon subunits of CF_1) although it is complementary to a slightly different sequence of the 16S rRNA than is *rcL* (Fig. 5). The transcript of *cf1B* begins more than 300 nucleotides upstream of the translation initiation site. A few nucleotides before that is a −10-like sequence. Although there is some shift in the codon usage pattern from *rcL* [Krebbers et al, 1982], the use of the universal code is confirmed.

Maize LS Rubpcase is highly homologous to the spinach [Zurawski et al, 1981] and barley polypeptides but more striking is the 72% amino acid sequence homology between the beta subunits of the photosynthetic CF_1 [Krebbers et al, 1982] and BF_1, part of the bacterial ATP synthase complex, [Saraste et al, 1981]. The amino acid sequence of the nuclear-encoded beta subunit of bovine mitochondrial F_1 has been determined by Walker et al [1982]; the homology to the corresponding polypeptide of E coli BF_1 is 75%. The amino acid homology shared by the beta subunits of maize CF_1, E coli BF_1, and bovine mitochondrial F_1 is a remarkably high 58%. The epsilon subunits of CF_1 and BF_1 are only about 23% homologous with regard to amino acid sequences derived from DNA sequences.

Several other unidentified open reading frames (UORFs) on the maize plastid chromosome have been sequenced by A. A. Steinmetz. These sequences confirm the observations regarding the presence of Shine-Dalgarno sequences and the use of the universal code.

MOLECULAR DEVELOPMENTAL BIOLOGY OF CHLOROPLASTS

Photogenes

Seedlings of maize, like those of other angiosperms, are usually pale yellow when grown in darkness as long as seed reserves last. They contain no chlorophyll, and their plastids, designated etioplasts, lack the characteristic photosynthetic vesicles (thylakoids) and grana of mature functional chloroplasts. Instead they contain one or more paracrystalline prolamellar bodies [Bogorad, 1967, 1981]. Upon illumination, etioplasts mature into chloroplasts by production of chlorophyll, of photosynthetically functional membranes and of thylakoids which, in higher plants, aggregate to form grana. RNA from etioplasts hybridizes to the equivalent of about 70% or 80% of a single strand of the plastid chromosome but in mature photosynthetically competent chloroplasts approximately 90% to 95% of the chromosome is transcribed (again taken as a single strand), if one takes into account the presence of the large inverted repeated sequences [L. Haff and L. Bogorad, unpublished].

Several regions of the maize chloroplast chromosome have now been identified as sites of genes whose transcript level is increased during light-induced development. The locations of these photogenes are shown in Figure 1. Most have been mapped to specific cloned fragments of the maize chloroplast chromosome. They are not all grouped together—we do not know yet whether some photogenes are grouped but they do not occur all together as a single photogene operon. The first discovered and best characterized of these developmentally regulated genes is maize photogene 32 [Bedbrook et al, 1978; Grebanier et al, 1978, 1979; Bogorad et al, 1980]. This gene codes for a 32-kdal polypeptide that is found in the thylakoid membrane. This gene is transcribed first into a 1,300-nucleotide-long message from which a 34.5-kdal polypeptide is translated. This precursor is inserted into the thylakoid membrane and is processed to the 32-kdal form [Grebanier et al, 1978]. Isolated maize plastids can synthesize (or complete the synthesis of) the 34.5-kdal precursor and insert it into membranes. The protein product of photogene 32 has been identified as the azido-atrazine binding protein that is the determinant of the electron transport function of a bound plastid quinone molecule in the photosystem II complex [Steinback et al, 1981; Arntzen et al, 1982].

Current research is focused on DNA sequence analyses of photogene 32 and other photogenes to determine whether they have common sequences that may be involved in the regulation of their transcription.

Leaf Cell Type Genes

Maize is a C4 plant. Carbon dioxide is fixed first into oxaloacetate in the mesophyll cells that are located between the epidermis of the leaf and the vascular bundles (the xylem, phloem, and associated tissues). The oxaloacetate is converted to malate in the mesophyll cells and the malate is transported to the cells surrounding the vascular bundles, the bundle sheath cells, where it is decarboxylated and the carbon dioxide first fixed in the mesophyll cells is released. The CO_2 transported into the bundle sheath cells in this way serves, with ribulosebisphosphate, as a substrate for the enzyme Rubpcase. This enzyme forms two molecules of phosphoglycerate from CO_2 plus ribulosebisphosphate. The phosphoglycerate is reduced to phosphoglyceraldehyde using photosynthetically generated ATP and $NADPH_2$. The C4 carbon dioxide fixation system is important because it reduces photorespiration and permits plants such as maize to operate efficiently at high light intensities at which photosynthesis is proceeding rapidly and $_2$ production is high. Rubpcase can fix either CO_2 or oxygen into ribulosebisphosphate. When oxygen is fixed, one molecule of phosphoglycolate and one molecule of phosphoglycerate are formed. The phosphoglycolate is oxidized to CO_2 and water in a process called photorespiration that appears to be metabolically wasteful.

Bundle sheath cells contain Rubpcase but are relatively poor in the photosystem II activity responsible for oxygen production. These cells contain the enzyme necessary for decarboxylation of malate; chloroplasts of both cell types carry on photosynthetic phosphorylation. The availability of a cloned maize chloroplast DNA fragment containing *rcL* as well as part of *cf1B* (ie, part of the gene for the beta subunit of CF_1) permitted analyses of the levels of mRNAs for these two genes in bundle sheath and mesophyll cells [Link and Bogorad, 1980].

By treatment with pectinases and cellulases, mesophyll protoplasts were liberated from leaves and then were separated from strands of bundle sheath cells. The mesophyll protoplast preparations were contaminated with about 3% of the bundle sheath cells (assuming that the enzyme Rubpcase occurs only in bundle sheath cells). RNA prepared from these two cell types was labeled with ^{32}P and hybridized against restriction fragments from within the coding regions of *rcL* and *cf1B*. DNA fragments were separated electrophoretically in agarose gels and transferred to nitrocellulose for hybridization [Southern, 1975]. Transcripts of *cf1B* are abundant in both mesophyll and bundle sheath cells but transcripts of *rcL* are confined to bundle sheath cells (assuming that the rare transcripts in the mesophyll protoplast preparation are from contaminating bundle sheath cells). Thus, these two genes on the same DNA fragment are expressed differently in adjacent cell types of maize leaves [Link et al, 1978; Jolly et al, 1981].

The arrangement of these two genes in the maize chloroplast chromosome is seen in Figure 1. The two genes are transcribed from positions about 406 nucleotides apart and from opposite strands—they are transcribed divergently. Details of the sequences upstream of transcription initiation for these two genes is discussed below, but first it is interesting to compare DNA sequences upstream of transcription initiation in *rcL* of maize [McIntosh et al, 1980] with that of spinach [Zurawski et al, 1981]. Spinach is a C3 plant and Rubpcase is present in all of the photosynthetic cells of the leaf. As is indicated in Figure 6, transcription of maize *rcL* begins about 63 nucleotides

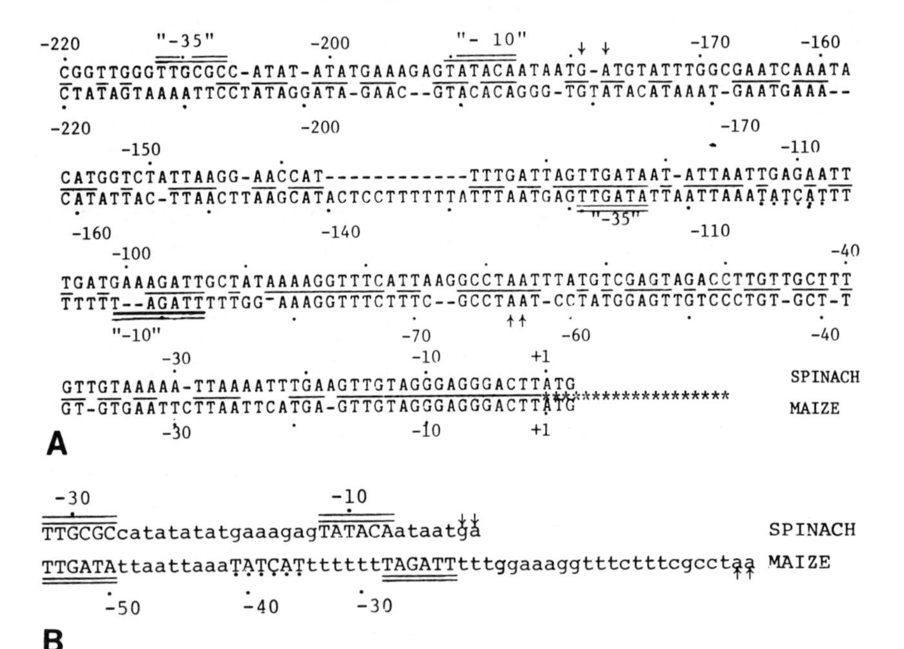

Fig. 6. A) DNA sequences upstream of translation initiation (position +1) of the spinach [Zurawski et al, 1981] and maize plastid [McIntosh et al, 1980] *rcL* genes that code for LS Rubpcase. (The portions of the two genes that code for amino acids are symbolized by a row of asterisks.) Nucleotides are numbered with reference to translation initiation, designated position +1. Transcription is initiated at approximately −63–64 in the maize gene (marked by two upward-facing arrows) and at about −178 to 179 in the spinach gene. Regions of homology between the two sequences are marked by lines between the two sequences. Sequences reminiscent of conserved −10 and −35 promoter regions of bacterial genes are indicated by double overlines for spinach and double underlines for maize. B) Nucleotide sequence upstream of the transcription initiation sites for the two genes. In this figure numbering is with respect to transcription, rather than translation, initiation.

upstream of the translation initiation site, whereas that of spinach begins at about −178. The most conspicuous difference is the presence of a sequence of 12 nucleotides (−129 to −140) in the maize DNA that is absent from spinach. Overall, the homology in these regions is extensive: Using the maize numbering of positions, from −1 to −63 the homology is 75%; from −65 to −179 it is 60%; and from −181 to −220 it is 54%. Long stretches of homologous sequences are seen in a few regions between the sites of transcription initiation of the maize *rcL* and the spinach *rcL*. It is not possible to judge now which, if any, of these sequences is significant with regard to control of expression, but it is interesting to note that the sequence of the "−35-like" region at positions −114 to −119 of the maize sequence is present at the corresponding place in the spinach sequence but the "−10-like" sequence of maize *rcL* is only partially matched in spinach. Conversely, the "−10-like" sequence of spinach is largely conserved in maize (−186 to −191), but there is virtually complete divergence at −207 to −212—positions corresponding to the "−35" region of spinach.

Nucleotide sequences immediately upstream of transcription initiation sites on maize *rcL* and *cf1BE*, the two adjacent plastid genes that are differently regulated, are aligned at their "−35-like" sequences in Figure 7. Focusing on the sequences that resemble procaryotic promoters, the first difference between these two genes is the distance from transcription initiation sites to "−10-like" sequences. A second difference is in the conserved sequences themselves—TAGATT for *rcL* and TAGTAT for *cf1BE* and a single nucleotide difference between the two "−35-like" sequences. Finally, the distances between the "−10-like" and "−35-like" sequences are different. The two conserved regions, as noted earlier, commonly have 17 to 18 nucleotides between them—as in the case of *cf1BE*—but this distance is slightly longer in *rcL*. These data and the comparisons just made emphasize the need for carrying on investigations to reveal which of the nucleotide sequences in and

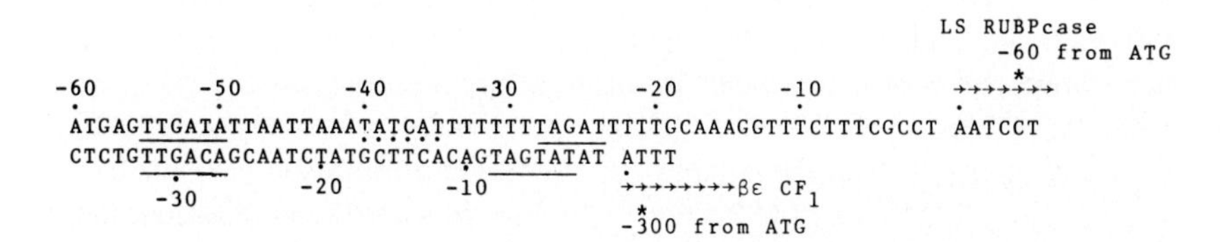

Fig. 7. Untranscribed DNA sequences lying upstream of the transcription initiation start sites for *rcL* (upper) and *cf1BE* (lower). Numbering is from the site of initiation of transcription. The two sequences are aligned at their presumptive "−35" sequences. Those sequences have very high homology to one another but the spacing between these and the much less conserved "10 sequences" as well as the distances from the "−10 regions" to the starts of transcription all vary. These two genes are under different expression controls in vivo and are transcribed to different levels in vitro.

around these genes may be part of the apparatus for regulating the relative rates of their transcription in cells where both genes are expressed and which may play roles in the differential expression of these two genes in mesophyll and bundle sheath cells of maize.

MAIZE PLASTID DNA-DEPENDENT RNA POLYMERASE

Maize plastids purified by density gradient centrifugation can synthesize RNA. The polymerase activity is associated with thylakoid membranes and is partially dependent on exogenous DNA, ie, the activity is increased by addition of exogenous DNA [Bogorad, 1967]. The DNA-dependent RNA polymerase can be solubilized by incubation of plastid fragments in the absence of magnesium ions or, even more effectively, in the presence of low concentration of ethylene diamine tetraacetate [Bottomley et al, 1971]. The enzyme prepared in this way has been further purified by glycerol density gradient centrifugation followed by chromatography on diethylaminoethyl (DEAE) cellulose succeeded by chromatography either on phosphocellulose or DNA-cellulose. The most highly purified preparations contain polypeptides of 180, 140, 100, 95, 85, and 40 kdal [Smith and Bogorad, 1974].

The maize plastid RNA polymerase can also be prepared from vigorously homogenized leaf preparations. After removal of particulate matter by high-speed centrifugation and fractionation by ammonium sulfate precipitation, the enzyme is further purified by chromatography on DEAE-cellulose followed by phosphocellulose [Kidd and Bogorad, 1980]. The enzyme preparation recovered with 0.15 M ammonium sulfate from the phosphocellulose column is composed of the following polypeptides: 180, 140, 120, 110, 100, 95, 85, 70, 42, 40, 38, and 27 kdal with smaller amounts of 75-, 55-, 42-, and 38-kdal polypeptides. This pattern matches quite well that obtained by analysis of the enzyme prepared from isolated plastids and purified by the same sequence of procedures. Maize nuclear RNA polymerase II [Strain et al, 1971; Mullinix et al, 1973; Kidd and Bogorad, 1980] is composed of 180-, 160-, 43-, 41-, 28-, and 22-kdal polypeptides that are present in about the same proportions to one another in different preparations. In addition, there are polypeptides of 66, 64, and 51 kdal whose molar ratios vary greatly with each preparation. Thus, the nuclear polymerase II preparation is distinctively different in polypeptide composition from that prepared from plastids; there are also marked differences in other properties. On the other hand, the presence in each enzyme of a 180-kdal polypeptide, the possibility of the 140-kdal plastid polypeptide being derived from the 160-kdal nuclear polymerase II polypeptide, the size similarity of the 42- and 43-kdal and the 27- and 28-kdal polypeptides of nuclear polymerase II and plastid polymerase respectively, all raise questions of whether the two polymerases may have some common polypeptides. Tryptic peptide maps of these pairs of polypeptides revealed that they are all different [Kidd and Bogorad, 1979].

Plastid RNA synthesis in dark-grown seedlings is stimulated by light. ^{32}P-phosphate supplied to leaves of dark-grown plants is incorporated into plastid rRNAs at a relatively slow rate as long as the plants are kept in the dark but incorporation is strongly accelerated by illumination of the leaves. The accelerated rRNA synthesis can be detected within 45 to 60 min and is marked after seedlings have been illuminated for 120 min [Bogorad, 1967]. When RNA polymerase activity of maize plastid membrane fragments, with which the enzyme is associated, is measured by titration with added calf thymus DNA, preparations from plants grown and maintained in darkness have much less activity than those from dark-grown plants illuminated for 120 min. The activity of solubilized maize plastid RNA polymerase rises three- to fourfold during 16 hr of illumination of dark-grown maize seedlings [Apel and Bogorad, 1976]. Is the synthesis of plastid polymerase stimulated by light? The polypeptides of RNA polymerase preparations obtained from plastids of dark-grown plants illuminated for 2 hr and for 16 hr have been analyzed and compared by polyacrylamide gel electrophoresis. The amounts of the two largest polypeptides of the polymerase (after purification of the enzyme by glycerol gradient centrifugation and phosphocellulose chromatography) are not significantly different in the two preparations. The failure to detect differences in the amounts of the large polypeptides, despite a three- to fourfold change in activity, suggests that there may be more subtle changes that occur during light-induced plastid maturation—such as addition of some minor polypeptides or changes in the DNA templates.

To pursue the understanding of mechanisms of developmentally regulated selective expression of genes in plastids we have turned to the use of cloned and characterized maize plastid DNA sequences as templates for the polymerase. An objective was the development of an in vitro system for the analysis of differential gene expression in vivo.

When RNA polymerase solubilized from sucrose density gradient-purified plastids is applied to a column of DEAE-cellulose, the bulk of the RNA polymerase activity can be eluted with approximately 0.2 M KCl. This enzyme is especially active with commercial calf thymus DNA, either in its "native" broken linear form or in a heat denatured form. The activity is almost six times greater with the "native" calf thymus DNA than when supercoiled DNA of the plasmid pZmc 134 was used. (The latter plasmid is a chimera of the bacterial vector pMB 9 and maize plastid DNA fragments Eco a and 1.) Maize plastid DNA is circular and some molecules—perhaps all—are supercoiled. An objective of the research was to reconstruct an enzyme system which has a higher capacity for transcribing plastid DNA in a form similar to its natural state than for transcribing linear fragments of heterologous DNA. We found that the fraction that is eluted from DEAE-cellulose with 0.5 M KCl, when added to the RNA polymerase eluting at 0.2 M KCl, stimulates the transcription of the chimeric plasmid about fivefold but has no effect on the rate of transcription of the calf thymus DNA. The active material in the 0.5 M KCl fraction proved to be a polypeptide of about 27.5 kdal, designated the S factor [Jolly and Bogorad, 1980]. In subsequent experiments, transcription of pZmc 134 DNA by the plastid RNA polymerase was stimulated up to 15-fold by the S factor.

Maize plastid fragment Eco 1 cloned in pMB 9 (pZmc 150) was used as a template for a quantitative assessment of the effect of the S factor on relative the transcription of cloning vehicle (pMB 9) DNA vs cloned chloroplast DNA sequence. In the absence of the S factor, the amount of transcription of pMB9 and maize plastid sequence Eco 1 DNA was within 10% of being proportional to the relative sizes of the two DNA segments. On the other hand, in the presence of the S factor, the maize chloroplast DNA sequence was transcribed about eight times more actively than the pMB 9 DNA. In these experiments, supercoiled cloned chimeric plasmid pZmc 150 was used as a template.

Relative transcription of maize plastid DNA sequence Bam 9 (which contains *rcL* and part of *cf1BE*) and the bacterial plasmid vector RSF 1030 by maize plastid polymerase was studied to extend the experiments with pZmc 150. In the absence of the S factor the chloroplast sequence was transcribed about 8.7 times more actively than the vector DNA, but in the presence of the S factor transcription of the chloroplast DNA sequence was considerably increased and transcription of the vehicle DNA was undetectable [Jolly et al, 1981]. These data showed that not only does the maize plastid polymerase system differentiate between the DNA sequences of the bacterial plasmid and the plastid, as in transcribing the pMB 9 and plastid sequence Eco 1 of pZmc 150, but that the relative utilization of chloroplast DNA and plasmid vehicle sequences also depends upon the particular DNA sequences of the bacterial plasmid vehicle. No experiment was designed and performed to test the point explicitly, but transcription of RSF 1030 DNA, relative to that of the plastid sequence Bam 9, was much below that of pMB9 DNA vs Eco 1. That is, transcription from RSF 1030 by the maize polymerase is relatively low and is suppressed below the level of detection by the S factor; transcription of pMB9 DNA is greater without S and is less completely suppressed. But, in a sense, the experiments we have described are competition experiments in which different DNA sequences that bind RNA polymerase are competing with one another for the enzyme, the stimulation of transcription of the chloroplast sequences reduces the amount of enzyme available for transcription of bacterial plasmid sequences. This question is answerable but the relevant experiments have not been done.

To examine the effect of the configuration of the DNA on enzyme activity and specificity, supercoiled pZmc 150 DNA (the cloning vector pMB 9 plus maize plastid DNA sequence Eco RI 1) was incubated with E coli omega protein (topoisomerase type I) until all of the DNA was relaxed, as assayed by gel electrophoresis. S accelerated the transcription of the enzymatically relaxed circular and supercoiled DNAs by the plastid RNA polymerase to about the same extent. However, quantitative hybridization showed that when enzymatically relaxed circular pZmc 150 DNA was used as the template by plastid polymerase plus S, the transcription ratio of fragment 1 to pMB 9 DNA was 2.5:1 compared with the 8:1 ratio when the template was the same DNA but in supercoiled form [Jolly and Bogorad, 1980]. It was also found that S influences the relative transcription of the two strands of Eco fragment 1 DNA [Jolly and Bogorad, 1980].

The 2,200-bp-long maize chloroplast chromosome fragment Eco 1 was found to contain a gene for tRNA^{His} *(tH-GUG)* and part of a gene for a 1.6-kb RNA that includes an open reading frame. These two genes overlap one another by at least a few nucleotides and are divergently transcribed, ie, they are on opposite strands. Comparison of the 5′ end of the in vitro transcript of *tH-GUG* with tRNA^{His} recovered from maize plastids, showed the same 5′ termini [Schwarz et al, 1981a]. Comparable "correct initiation" in vitro was later found in comparing the 5′ ends of the in vivo and in vitro transcripts of maize *rcL* [Jolly et al, 1981].

The preceding sets of experiments demonstrated that the maize plastid RNA polymerase plus S-factor system discriminates in vitro between maize plastid DNA sequences and E coli plasmid sequences as templates, preferring the former when the DNA was in a supercoiled form. The next question to be raised was whether this in vitro transcription system can discriminate between two maize plastid genes. We studied the transcription of the maize genes *rcL* and *cf1B* to determine whether they are transcribed to the same or different extents in vitro. For this purpose the template used was DNA from the plasmid pZmc 37–11, a chimera of RSF 1030 and a part of Bam HI fragment 9 of maize plastid DNA [Jolly et al, 1981]. The plastid DNA fragment contains all of *rcL* and approximately 700 bp of *cf1B* [Coen et al, 1977; Link and Bogorad, 1980; McIntosh et al, 1980; Krebbers et al, 1982]. These two genes are transcribed divergently (each from one of the two separate strands of DNA), and the initiation sites for their transcription are about 406 bp apart. In vivo, in young maize plants in which *rcL* transcription may occur in all cells, the ratio of transcription of these two genes is approximately 3:1 in favor of *rcL* [S. Rodermel and L. Bogorad, unpublished]. In vitro, with maize plastid polymerase but in the absence of the S factor, the ratio of transcription of *rcL* to *cf1B* was 1.1:1, but in the presence of the S factor the ratio was 3.2:1. The DNA sequences upstream of the transcripts of these two genes are shown and compared in Figure 7. It remains to be determined whether these or other DNA sequences associated with the two genes play roles in their differential transcription. An additional interesting question that remains to be answered is why *rcL* transcripts are absent from mesophyll cells of maize [Link et al, 1978]. It is not known whether control of *rcL* transcription in mesophyll cells is positive or negative: ie, whether a factor is required for transcription in bundle sheath cells, or whether there is a factor in mesophyll cells that suppresses transcription of this gene, or if both occur. Furthermore, regulation by selective destruction of mRNAs in vivo cannot be excluded.

Finally, we have taken advantage of knowledge of the DNA sequences of a number of maize plastid tRNA genes [Steinmetz et al, 1983] to construct chimeric plasmids containing two maize plastid tRNA genes with strikingly different sequences 5′ to their transcription initiation sites to use as templates in comparing the activities of E coli and maize plastid RNA polymerases. Preceding the transcription initiation site for *tH-GUG* (the gene for histidine tRNA in maize plastid fragment Eco 1) is the sequence TGAATG starting at −38; 18 nucleotides downstream is the sequence TTAGCT. Both of these are quite unlike the sequences present in the same regions near the serine tRNA gene, *tS-GGA*. The latter are, in the "−10 region," TAAGAT and, in the "− 35 region," TTCACT. Both of the latter sequences resemble comparable regions on E coli genes. In fact, the "−10" region sequence is identical to that of the phage $T7A_2$ promoter. A chimeric plasmid was made containing *tH-GUG* and *tS-GGA* cloned into pBR322. When supercoiled DNA was used as a template for E coli RNA polymerase, *tS* was transcribed vigorously, but tH was virtually untranscribed. On the other hand, the maize plastid polymerase plus S factor transcribed both of these genes strongly—in some experiments *tS* was transcribed more vigorously than *tH* [A. A. Steimetz, S. O. Jolly, and L. Bogorad, unpublished]. Although it is attractive to believe that these transcription differences are related to the "−10-like" and "−35-like" sequences, there is as yet no direct experimental evidence that this is so.

The availability of cloned sequenced plastid genes that are differentially expressed in various cell types (eg, *rcL* in mesophyll and bundle sheath cells of maize leaves, or at different stages of development, eg, photogene 32 in dark-grown versus illuminated plants) together with the in vitro system that differentially transcribes chloroplast genes should provide opportunities for analysis of mechanisms for developmental regulation of plastid gene expression.

DISCUSSION AND SUMMARY

The 139-kilobase (kb) pair circular maize plastid DNA bears two large inverted repeated sequences separated by two single-copy regions. This pattern is, so far, the most common among plastid DNA molecules. A few plastid chromosomes have been found which lack the inverted repeated sequences, and the chromosome of Euglena gracilis is unusual in having three small repeated segments. The large repeated segments in DNA of the maize type (found in plants from the alga Chlamydomonas to maize, tobacco, and spinach) carry genes for rRNAs, tRNAs, and proteins. The three smaller repeats of Euglena each also carries a set of genes for rRNAs.

Genes for all tRNAs of plastids, as exemplified by those in maize, appear to be coded in the plastid genome and to have a mixture of characteristics of eucaryotic and procaryotic tRNA genes. Four maize tRNA genes, those for alanine, isoleucine, valine, and leucine, have introns in their anticodon loops. The positions of the introns vary and the sizes of the introns range from 458 to more than 900 bp in length. The CAA 3′ terminal sequence of tRNAs are not encoded in the maize plastid genes and consequently must be added after transcription. These two features are "eucaryotic." In contrast, maize tRNAs for phenylanine and threonine are more than 75% homologous in their sequences to the corresponding tRNAs of E coli. The maize plastid genes have conserved sequences upstream of their transcription start sites, which, in some cases, bear close resemblances to those at the "−10" and "−35" regions of E coli genes.

Regardless of the origin of plastids, ie, exogenously as in the endosymbiont hypothesis or endogenously as in the cluster-clone hypothesis [Bogorad, 1975, 1982], plastid genes show the effect of living in a cell with nuclear genes. If plastid genomes originated from procaryotic endosymbionts that invaded cells in which nuclear gene structure evolved, the "intron-habit" might well have come from the nucleus. On the other hand, if intron processing is different in plastids than in the nuclear genome, we may simply be seeing tolerable introns—introns that could be removed when introduced and that did not destroy the capacity of the plastid to survive.

The maize plastid genes for proteins resemble most genes, save those of fungal and mammalian mitrochondria, in utilizing the universal code for amino acids. They resemble bacterial genes in the presence of nucleotide sequences preceding the translation initiation sites comparable to Shine-Dalgarno sequences of bacterial genes. Furthermore, conserved sequences upstream of transcription start sites resemble those of −10 and −35 regions of procaryotic gene promoters. In the case of maize *rcL*, these conserved sequences are extraordinarily distant from the transcription initiation start site even in comparison to the *rcL* genes of tobacco and spinach. It should be borne in mind that only a very few genes for plastid proteins have been sequenced to date and some of the "generalizations" may not survive.

The maize plastid RNA polymerase with the 27.5-kdal S factor preferentially transcribes cloned plastid DNA sequences over bacterial plasmid DNA from the same supercoiled plasmid. This system provides the possibility of studying the roles of DNA sequences of plastid genes in the control of transcription—as in differential expression of *rcL*, the gene for the large subunit of ribulose bisphosphate carboxylase in mesophyll and bundle sheath cells of maize leaves, or the light-induced expression of photogene 32.

ACKNOWLEDGMENTS

We are grateful to the National Science Foundation, the Competitive Research Grants Office of the US Department of Agriculture, and the National Institute of General Medical Sciences for research grant funds which, in part, permitted this work to be done. The research was also supported in part by the Maria Moors Cabot Foundation for Botanical Research of Harvard University. Karen M.T. Muskavitch and I.M. Larrinua were recipients of NRSA Postdoctoral Fellowships awarded by the National Institute of General Medical Sciences. Enno T. Krebbers was supported by a NIGMS training grant. A. Steinmetz is Charge de Recherche at CNRS, France and was the recipient of a NATO Research Grant.

REFERENCES

Apel K, Bogorad L (1976): Light-induced increase in the activity of maize plastid DNA-dependent RNA polymerase. Eur J Biochem 67:615–620.

Arntzen CJ, Darr SC, Mullet JE, Steinback KE, Pfister K (1982): Polypeptide determinants of plastoquinone function in photosystem II of chloroplasts. In Trumpower B (ed): "Function of Quinones in Energy Conserving Systems." New York: Academic Press (in press).

Barrell BG, Anderson S, Bankier AT, DeBruijn MHL, Chen E, Coulson AR, Drouin J, Eperon IC, Nierlich DP, Roe BA, Sanger F, Schreier PH, Smith AJH, Staden R, Young IG (1980): Different pattern of codon recognition by mammalian mitochondrial tRNAs. Proc Natl Acad Sci USA 77:3164–3166.

Bedbrook JR, Bogorad L (1976): Endonuclease recognition sites mapped on Zea mays chloroplast DNA. Proc Natl Acad Sci USA 73:4309–4313.

Bedbrook JR, Kolodner R, Bogorad L (1977): Zea mays chloroplast ribosomal RNA genes are part of a 22,000 base pair inverted repeat. Cell 11:739–749.

Bedbrook JR, Link G, Coen DM, Bogorad L, Rich A (1978): Maize plastid gene expressed during photoregulated development. Proc Natl Acad Sci USA 75:3060–3064.

Bedbrook JR, Coen DM, Beaton A, Bogorad L, Rich A (1979): Location of the single gene for the large subunit of ribulosebisphosphate carboxylase on the maize chloroplast chromosome. J Biol Chem 254:905–910.

Bogorad L (1967): Control mechanisms in plastid development. Dev Biol Suppl 1:1–31.

Bogorad L (1975): Evolution of organelles and eukaryotic genomes. Science 188:891–898.

Bogorad L (1981): Chloroplasts. J Cell Biol 91:256s–270s.

Bogorad L (1982): Regulation of intracellular gene flow in the evolution of eukaryotic genomes. In Schiff JA (ed): "On the Origins of Chloroplasts." Amsterdam: Elsevier North-Holland, pp 277–295.

Bogorad L, Jolly SO, Link G, McIntosh L, Schwarz Z, Steinmetz A (1980): Studies of the maize chloroplast chromosome. In Bucher T, Sebald W, Weiss H (eds): "Biological Chemistry of Organelle Formation." Berlin: Springer Verlag, pp 87–96.

Bonitz SG, Berlani R, Coruzzi G, Li M, Macino G, Nobrega FG, Nobrega MP, Thalenfeld

BE, Tzagoloff A (1980): Codon recognition rules in yeast mitochondria. Proc Natl Acad Sci USA 77:3167–3170.
Bottomley W, Smith HJ, Bogorad L (1971): RNA polymerases of maize: Partial purification and properties of the chloroplast enzyme. Proc Natl Acad Sci USA 26:2412–2416.
Coen DM, Bedbrook JR, Bogorad L, Rich A (1977): Maize chloroplast DNA fragment encoding the large subunit of ribulosebisphosphate carboxylase. Proc Natl Acad Sci USA 74:5487–5491.
Edwards K, Kossel H (1981): The rRNA operon from Zea mays chloroplast: Nucleotide sequence of 23S rDNA and its homology with E. coli 23S rDNA. Nucleic Acids Res 9:2853–2868.
Gray PW, Hallick RB (1978): Physical mapping of the Euglena gracilis chloroplast DNA and ribosomal RNA gene region. Biochemistry 18:284–290.
Grebanier AE, Coen DM, Rich A, Bogorad L (1978): Membrane proteins synthesized but not processed by isolated maize chloroplasts. J Cell Biol 78:734–746.
Grebanier AE, Steinback KE, Bogorad L (1979): Comparison of molecular weights of proteins synthesized by isolated chloroplasts with those which appear during greening. Plant Physiol 63:436–439.
Heckman JE, Sarnoff J, Alzner-DeWeerd B, Yin S, RajBhandary UL (1980): Novel features in the genetic code and codon reading patterns in Neurospora crassa mitochondria based on sequences of six mitochondrial tRNAs. Proc Natl Acad Sci USA 77:3159–3163.
Jenni B, Stutz E (1979): Mapping of a DNA sequence complementary to 16S rRNA outside of the three rRNA sets. FEBS Lett 102:95–99.
Jolly SO, Bogorad L (1980): Preferential transcription of cloned maize chloroplast DNA sequences by maize chloroplast RNA polymerase. Proc Natl Acad Sci USA 77:822–826.
Jolly SO, McIntosh L, Link G, Bogorad L (1981): Differential transcription in vivo and in vitro of two adjacent maize chloroplast genes: The large subunit of ribulosebisphosphate carboxylase and the 2.2-kilobase gene. Proc Natl Acad Sci USA 78:6821–6825.
Kidd GH, Bogorad L (1979): Peptide maps comparing subunits of maize chloroplast and type II nuclear DNA-dependent RNA polymerase. Proc Natl Acad Sci USA 76:4890–4892.
Kidd GH, Bogorad L (1980): A facile procedure for purifying maize chloroplast RNA polymerase from whole cell homogenates. Biochim Biophys Acta 609:14–30.
Koch W, Edwards K, Kossel H (1981): Sequencing in the 16S-23S spacer in a ribosomal RNA operon of Zea mays chloroplast DNA reveals two split tRNA genes. Cell 25:203–214.
Koller B, Delius H (1980): Vicia faba chloroplast DNA has only one set of ribosomal RNA genes as shown by partial denaturation mapping and R-loop analysis. Mol Gen Genet 178:261–269.
Krebbers ET, Larrinua IM, McIntosh L, Bogorad L (1982): The maize genes for the beta and epsilon subunits of the photosynthetic coupling factor CF_1 are fused. Nucleic Acids Res 10:4985–5002.
Link G, Bogorad L (1980): Sizes, locations and directions of transcription of two genes on a cloned maize chloroplast DNA sequence. Proc Natl Acad Sci USA 77:1832–1836.
Link G, Coen DM, Bogorad L (1978): Differential expression of the gene for the large subunit of ribulosebisphosphate carboxylase in maize leaf cell types. Cell 15:725–731.
McCarty RE (1979): Roles of a coupling factor for photophosphorylation in chloroplasts. Annu Rev Plant Physiol 30:79–104.
McIntosh L, Poulsen C, Bogorad L (1980): Chloroplast gene sequence for the large subunit of ribulosebisphosphate carboxylase of maize. Nature 288:556–560.
Mullinix KP, Strain GC, Bogorad L (1973): RNA polymerases of maize. Purification and molecular structure of DNA-dependent RNA polymerase II. Proc Natl Acad Sci USA 70:2386–2390.
Palmer JD, Thompson WF (1981): Rearrangements in the chloroplast genomes of mung bean

and pea. Proc Natl Acad Sci USA 78:5533–5537.

Rawson JRY, Kushner SD, Vapnek D, Alton VNK, Boerma CL (1978): Chloroplast ribosomal RNA genes in Euglena gracilis exist as three clustered tandem repeats. Gene 3:191–209.

Saraste M, Gay NJ, Eberle A, Runswick MJ, Walker JE (1981): The atp operon: Nucleotide sequence of the genes for the alpha, beta and epsilon subunits of Escherichia coli ATP synthase. Nucleic Acids Res 9:5287–5296.

Schwarz Z, Kossel H (1980): The primary structure of 16S rDNA from Zea mays chloroplast is homologous to E. coli 16S rRNA. Nature 283:739–742.

Schwarz Z, Jolly SO, Steinmetz AA, Bogorad L (1981a): Overlapping divergent genes in the maize chloroplast chromosome and in vitro transcription of the gene for $tRNA^{His}$. Proc Natl Acad Sci USA 78:3423–3427.

Schwarz Z, Kossel H, Schwarz E, Bogorad L (1981b): A gene coding for $tRNA^{Val}$ is located near 5′ terminus of 16S rRNA gene in Zea mays chloroplast genome. Proc Natl Acad Sci USA 78:4748–4752.

Shine J, Dalgarno L (1974): The 3′-terminal sequence of Escherichia coli 16S Ribosomal RNA: Complementarity to nonsense triplets and ribosome binding sites. Proc Natl Acad Sci USA 71:1342–1346.

Smith HJ, Bogorad L (1974): The polypeptide subunit structure of the DNA-dependent RNA polymerase of Zea mays chloroplasts. Proc Natl Acad Sci USA 71:4839–4842.

Southern EM (1975): Detection of specific sequences among DNA fragments separated by gel electrophoresis. J Mol Biol 98:503–517.

Steitz J, Jakes K (1975): How ribosomes select initiator regions in mRNA: Base pair formation between the 3′ terminus synthesis in Escherichia coli. Proc Natl Acad Sci USA 72:4734–4738.

Steinback KE, McIntosh L, Bogorad L, Arntzen CJ (1981): Identification of the triazene receptor protein as a chloroplast gene product. Proc Natl Acad Sci USA 78:7463–7467.

Steinmetz AA, Gubbins EJ, Bogorad L (1982): The anticodon of maize chloroplast gene for $tRNA^{Leu}$UAA is split by a large intron. Nucleic Acids Res 10:3027–3037.

Steinmetz AA, Krebbers ET, Schwarz Z, Gubbins EJ, Bogorad L (1983): Nucleotide sequences of five maize chloroplast transfer RNA genes and their flanking regions. J Biol Chem (in press).

Strain GC, Mullinix KP, Bogorad L (1971): RNA polymerases of maize: Nuclear RNA polymerases. Proc Natl Acad Sci USA 68:2647–2651.

Walker JE, Saraste M, Runswick MJ, Gay NJ (1982): Distantly related sequences in the alpha- and beta-subunits of ATP synthase, myosin, kinases and other ATP-requiring enzymes and a common nucleotide binding fold. EMBO J 1:945–951.

Zurawski G, Perrot B, Bottomley W, Whitfeld PR (1981): The structure of the gene for the large subunit of ribulose 1,5-bisphosphate carboxylase from spinach chloroplasts. Nucleic Acids Res 9:3251–3280.

Gene Structure and Regulation in Development, pages 33–61

Structure, Evolution, and Developmental Expression of the Chorion Multigene Families in Silkmoths and *Drosophila*

Fotis C. Kafatos
Department of Cellular and Developmental Biology, The Biological Laboratories, Harvard University, Cambridge, Massachusetts 02138, and Department of Biology, University of Crete, Iraclion, Crete, Greece

INTRODUCTION

Insects are particularly tractable systems for studying development, including programed gene expression, regulated in space and time. An important reason is that insects undergo the exquisite phenomenon of hormonally triggered postembryonic metamorphosis: Accessible and well defined tissues and organs can be studied as they undergo predictable and profound developmental changes within a brief period of a few hours or days during each metamorphic molt. A second reason has to do with the insect body plan: Compared to vertebrates, insect tissues and organs are of relatively simple design, and thus reasonably homogenous populations of developing cells can often be isolated for study. A third reason is the sophisticated genetics of Drosophila melanogaster, which is yielding unprecedented insights into developmental mechanisms, as it is combined with the powerful molecular analysis now made possible by recombinant DNA technology.

Insects are also favorable for the study of evolution: Numerous examples ranging from natural history to population genetics immediately come to mind. Important reasons in this respect are the short (for higher eucaryotes) life cycles of many insects; their evolutionary success (recorded in numerous radiations, which have resulted in a huge number of species, and in numerous strains); and, again, the sophisticated genetics and cytogenetics of Drosophila. Recombinant DNA technology, permitting rapid, definitive and detailed studies of genomic structure, has already profoundly accelerated the molecular analysis of evolution in insects.

My laboratory's research is centered around these two central (and interrelated) biological processes, evolution and development, with emphasis on cellular and molecular approaches. We use the formation of the eggshell (chorion) in insects as a model system for studying development and molecular evolution. In this essay I will briefly summarize our findings, referring to the work of other laboratories as appropriate. This essay is an updated version of a recently published summary [Kafatos, 1981]; some of the new material is treated more fully in a forthcoming detailed review [Regier and Kafatos, 983].

STRUCTURE AND MORPHOGENESIS OF THE INSECT CHORION

The insect chorion is fascinating in the complexity of its structure and morphogenesis. Furthermore, it is widely variable in different orders and even families [Kafatos et al, 1977; Hinton, 1969, 1981]. The chorion has been studied extensively by a combination of morphological and biochemical procedures in only two groups: the silkmoths (Order: Lepidoptera) and Drosophila (Order: Diptera).

The silkmoth chorion (Fig. 1) is a largely (over 96%) proteinaceous, extracellular structure that surrounds the egg. Both in construction and in physiological function, its complexity rivals that of cuticle, with which it is in many ways analogous [Hinton, 1969, 1981; Smith et al, 1971; Kafatos et al, 1977; Mazur et al, 1980, 1982; Regier et al, 1980, 1982]. As an eggshell, it must protect the developing embryo after fertilization and oviposition, prevent it from drying out, and yet facilitate its respiration. The chorion is formed toward the end of oogenesis, by a monolayer of follicular epithelium that surrounds the single oocyte of each follicle [Paul et al, 1972]. More than 100 structural proteins have been resolved in the mature chorion of the saturniid moth Antheraea polyphemus [Regier et al, 1980]. These proteins are secreted inward by the approximately 10,000 large, polyploid follicular cells, onto the surface of the oocyte, where they assemble into the chorion.

The thin, innermost part of the chorion (closest to the oocyte) is made of vertical columns separated by air spaces (trabecular layer). The rest (lamellar chorion), like cuticle, has the basic structure of "universal plywood": layers of fibrils, parallel to each other and to the oocyte surface, are found on top of each other, but are progressively rotated in terms of fibril direction, each by a small, constant angle relative to the preceding layer. Thus, the entire structure is built of helicoidally rotated fibril layers, and is a biological analogue of a cholesteric liquid crystal [Bouligand, 1972; Mazur et al, 1982]. Numerous details, superimposed on this basic structural plan, are thought to serve specific physiological functions [Mazur et al, 1982]. For example, five radial parts can be distinguished in the polyphemus chorion on the basis of the pitch of helicoidal rotation, the evenness or disruption of the fibril layers, the existence of radial or horizontal empty spaces, etc. (Fig. 1a). These variations of helicoidal order, and its interruptions and "defects," are presumably important in dissipating impact strains on the chorion, and in creating an elaborate internal system of air spaces to convey respiratory gases between the environment and the trabecular layer apposed to the oocyte surface. The surface outlets of this internal air space are the *aeropyles*, which in certain parts of the chorion are equipped with chimney-like projections, the aeropyle crowns [Regier et al, 1980]. This structural and functional complexity of the mature chorion is one explanation for its biochemical complexity, ie, the high number of component proteins. Both the ultrastructure and the chorion protein electrophoretic patterns differ substantially between silkmoth species [Kafatos et al, 1977], perhaps reflecting differences in physiology. For example, the embryos of Bombyx mori undergo diapause and thus require a tighter, more impermeable chorion than the rapidly developing embryos of saturniid moths. Bombyx chorion (Fig. 1b) uniquely includes a class of specialized proteins that are high in cysteine (Hc-proteins, ca 30 molar percent cysteine); these proteins are produced at the end of choriogenesis and in part permeate and cement the chorion, and in part form a tight, outermost chorion layer lacking in saturniids [Kafatos et al, 1977; Mazur et al, 1982].

A second, related explanation of the biochemical complexity is found in the requirements for chorion morphogenesis [Regier et al, 1982; Mazur et al, 1982]. The lamellar chorion is first formed as a thin framework, with wide angles between fibril layers. The framework then expands by intercalation of additional layers between preexisting ones. Densification follows, as the fibrils thicken and ultimately become indistinguishable by the addition of more material [Smith et al, 1971; Mazur, Regier, and Kafatos, in preparation]. Finally, the surface sculpturings, including the aeropyle crowns, are deposited [Mazur et al, 1980]. Corresponding to, and making possible, this morphogenetic program is a program of differential gene expression [Paul and Kafatos, 1975]. The chorion proteins are not all produced in parallel, as a single set: Subsets are synthesized in overlapping succession, during characteristic developmental periods, ranging from a few hours to more than half of the total 51-hr period of choriogenesis (Fig. 2). In broad terms, four major developmental groups of proteins have been recognized: early, middle, late, and very late [Sim et al, 1979]. A reasonable hypothesis is that early proteins are associated with framework formation [Regier et al, 1982], middle proteins with framework expansion, late proteins with densification, and very late proteins with formation of surface sculpturings. Because of the temporal overlap of expansion and densification, the functions of middle and late proteins are the most difficult to establish rigorously. There is little question of the function of early proteins, since their synthesis and formation of the framework are brief and coincident processes [Regier et al, 1982]. The

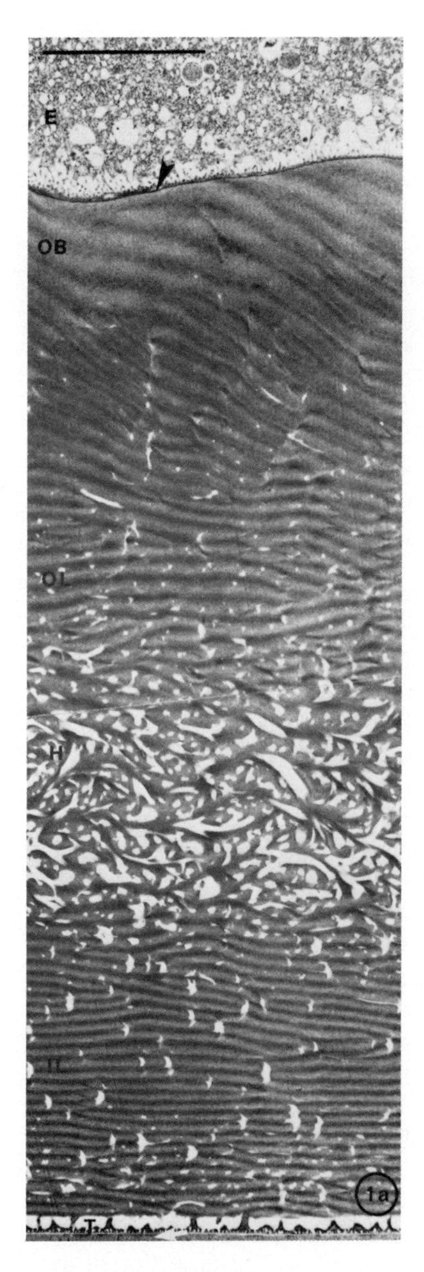

Figure 1a.

Fig. 1. a) Transmission electron micrograph of a mature (synthetic stage Xd) chorion of A polyphemus. At the bottom of the micrograph, apposed to the vitelline membrane (arrow), is the trabecular layer (T) consisting of pillars surrounded by air spaces. Within the bulk of the lamellar chorion, four types can be distinguished: thin lamellae of the inner lamellar layer (IL), thick distorted spongy lamellae of the holey layer (H), lamellae of the outer lamellary layer (OL) and, lying at an angle to the rest, the thick lamellae of the oblique layer (OB). Interposed between the microvilli of the apical surface of the follicular epithelial cell (E), and the lamellar chorion, is the thin electron-dense sieve layer (arrowhead). Scale marker = 5 μm (courtesy Dr G.D. Mazur; from Regier et al [1982] and Mazur et al [1982]). b) Transmission electron micrograph of the chorion in B mori. Adjacent to the follicular epithelial cell (e) at upper right is the highly electron dense outer crust (c), which consists of high-cysteine proteins arranged in thin, even lamellae. The bulk of the lamellar chorion (L) shows much fewer air spaces than in A polyphemus (cf Fig. 1a). The trabecular layer is indicated by (t). Scale marker = 5 μm (Courtesy Dr G.D. Mazur; from Kafatos et al [1977] and Mazur et al [1982]). c) (see page 38) Transmission electron micrograph of a mature (stage 14) Drosophila follicle. The eggshell can be seen interposed between the oocyte (bottom) and the follicular cells (top, E). The eggshell layers are, from innermost to outermost, the vitelline membrane (I), the innermost chorionic layer (L), the endochorion complex, and the fibrous exochorion (III), which has just been secreted by the follicular epithelial cells. The endochorion consists of a fenestrated inner endochorion (F), an outer endochorion, or roof (R), pillars (P), and protrusions forming the roof network (arrow). A loose material is found within the cavities of the endochorion (C). A different type of wispy material, found between vitelline membrane and innermost chorionic layer, corresponds to the "wax layer." Note the gross similarity between endochorion and the trabecular layer of silkmoths (Fig. 1a,b). Scale = 1 μm (courtesy Dr L.H. Margaritis; from Margaritis et al [1980]).

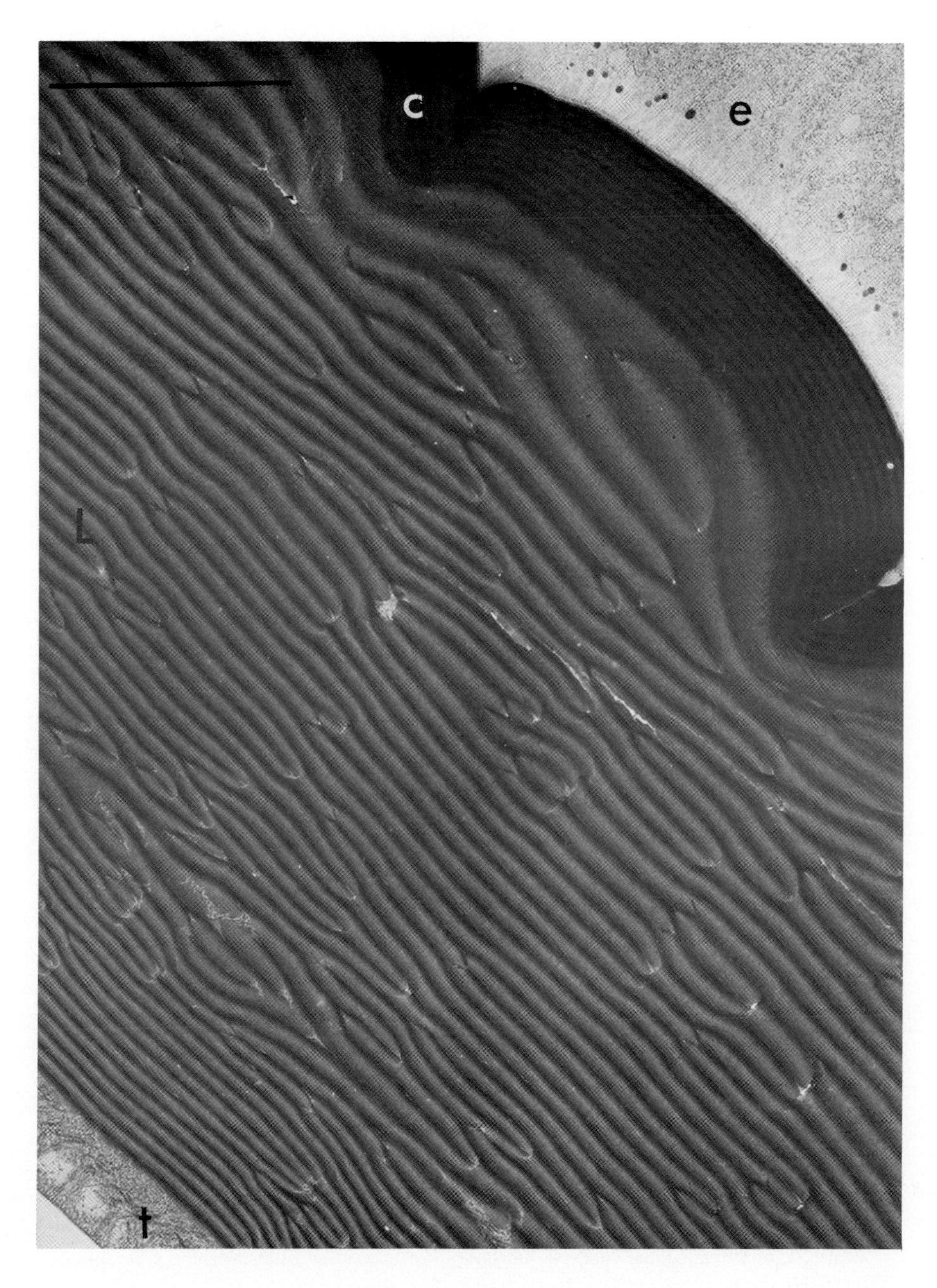

Figure 1b.

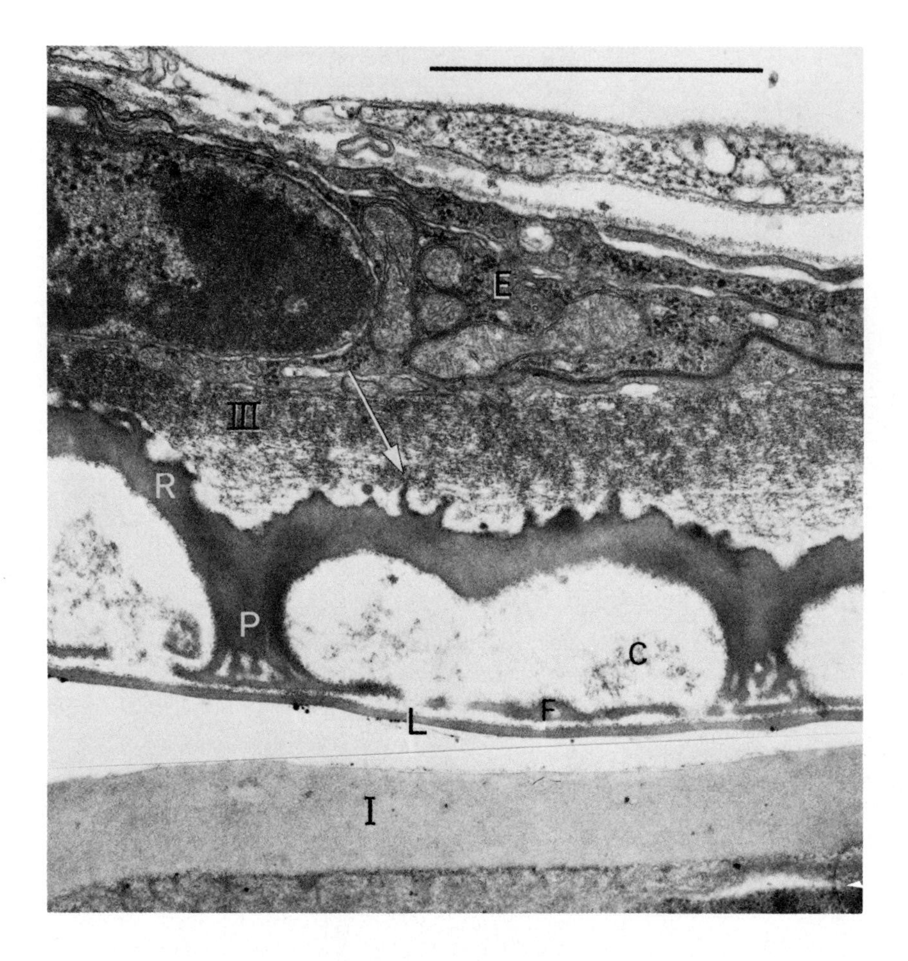

Figure 1c. Legend on page 36.

Fig. 2 (opposite page). Changing patterns of silkmoth chorion protein synthesis during development. Protein-synthetic profiles are presented, corresponding to 17 stages of choriogenesis (Ia to Xd; see Paul and Kafatos [1975]), plus the last prechoriogenic stage (O). A single ovariole of A polyphemus was labeled for 30 min in culture with ^{3}H-leucine. Sequential follicles were then deyolked, dissolved in sample buffer, and analyzed by sodium dodecyl sulfate (SDS)-polyacrylamide slab gel electrophoresis. The fluorogram shows the newly synthesized proteins at each developmental stage. The A, B, C, D, and E size classes are identified, as are individual bands or subclasses (1–6) (courtesy M.D. Koehler: from Sim et al [1979]).

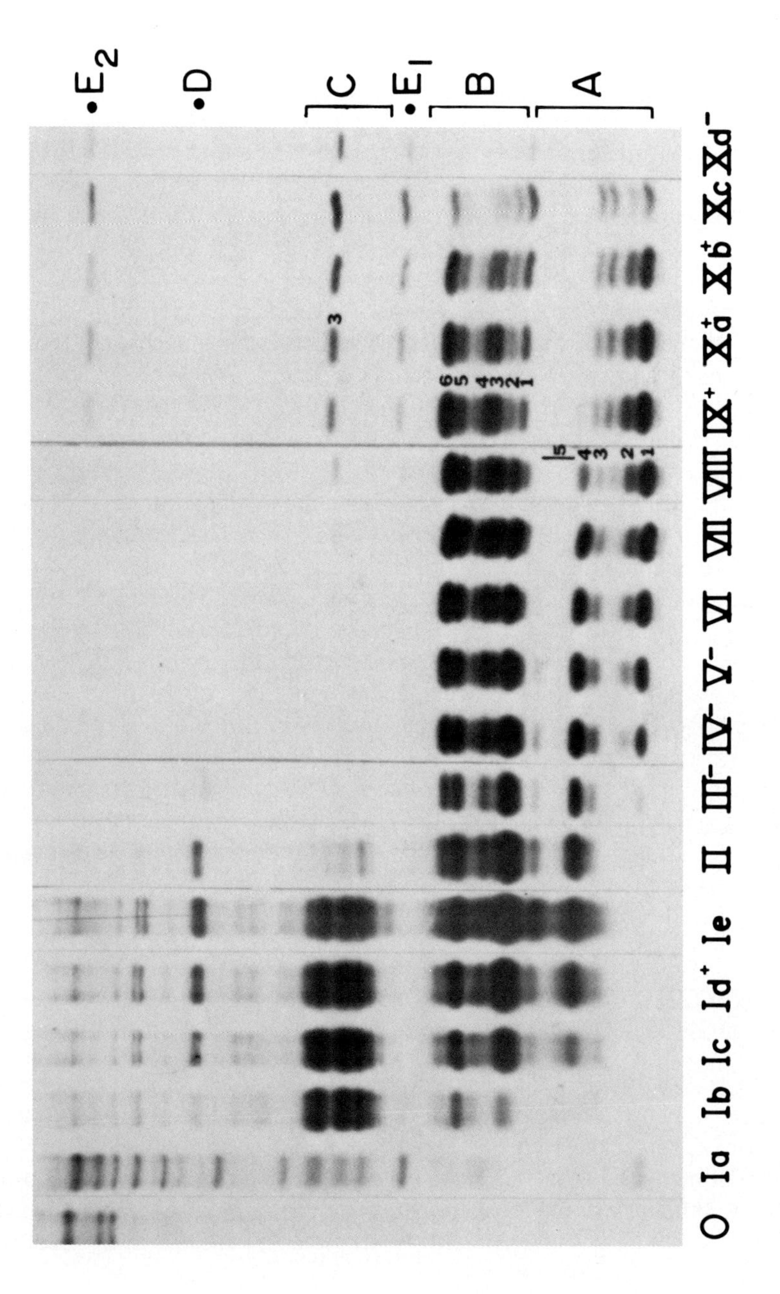

Figure 2.

functions of the 23 very late polypeptides of A polyphemus are also clear [Regier et al, 1980]: They are only produced in specialized regions of the chorion by distinct cell populations, and are associated with aeropyle crown formation [Mazur et al, 1980]. The remaining, early to late proteins, accounting for 95% of the chorion mass, are apparently synthesized by all follicular cells and form the main body of the chorion [Kafatos et al, 1977; Regier et al, 1980].

A third explanation of the biochemical complexity of the silkmoth chorion concerns its evolution. Many chorion genes are multicopy, as will be discussed below, and parts of their sequence are evolving relatively rapidly, presumably permitting rapid evolution of chorion structure and physiology. Some of the complexity of chorion proteins may be functionally neutral, representing early steps in the evolution of identical gene copies into functionally distinct members of a multigene family.

The structure of the Drosophila eggshell was originally described by King and collaborators (for a review, see King [1970]). More recently, detailed structural analysis was performed by Margaritis and collaborators [Margaritis et al, 1976, 1979, 1980; Margaritis, 1983). Lamellar chorion of the type described in silkmoths is missing. Instead, the eggshell consists of five radial layers, the most prominent of which, the endochorion, is somewhat reminiscent of the trabecular layer of silkmoths (Fig. 1c). Regional complexity is even more pronounced than in silkmoths. As in the moths, the eggshell is formed by a monolayer of follicular cells (in this case approximately 1,000). Ten subpopulations of follicular cells have been distinguished, responsible for formation of the main body of the eggshell and its specialized regions. Chief among the latter are two long respiratory appendages, the operculum area (a trapdoor through which the larva hatches), the micropyle through which fertilization occurs, and the fenestrated posterior pole. The prominent structural differences between fly and moth eggshells are correlated with the major differences in their component proteins (see below).

SILKMOTH CHORION PROTEINS ARE PRODUCTS OF MULTIGENE FAMILIES

High-resolution two-dimensional gels reveal not only an amazingly large number of chorion proteins, but also substantial clustering of these proteins in terms of molecular weight. Three major size classes of proteins have thus been defined [Paul et al, 1972]: A (average molecular weight ca 10,000), B (ca 13,000), and C (ca 18,000). An additional, higher molecular weight class, D, is complex but quantitatively very minor; a fifth class, E, is defined by unusual solubility properties.

The A and B classes together account for approximately 85% of the chorion mass [Kafatos et al, 1977]. Partial protein sequences have been determined for eleven A proteins and seven B proteins of A polyphemus, as well as three A and two B proteins of A pernyi [Regier et al, 1978; Rodakis, 1978; Moschonas, 1980; Rodakis et al, 1982, 1983]. The results clearly show that these proteins are encoded by distinct DNA sequences: The proteins differ by amino acid replacements or insertions/deletions. At the same time, it is clear that within each size class the amino acid sequences are extensively similar, thus establishing that the proteins are encoded by two respective families of evolutionarily related genes (multigene families).

Most members of the C class are early proteins and are thought to correspond to the chorion framework [Regier et al, 1982]. Although C proteins have not been sequenced, their amino acid compositions clearly show that they are related, and can be assigned to three subgroups, one of which resembles A and B components in composition, whereas the other two are more distinct [Regier et al, 1983]. Rare A and B proteins of early developmental specificity also exist. Abundant members of the A and B classes are either middle or late proteins, and thus are presumably involved in framework expansion and densification. Among the very late proteins of A polyphemus, the eight members of the E class have been identified as an architectural "filler," which helps mold the aeropyle crowns [Mazur et al, 1980]; their amino acid composition is distinct, but nothing is known about their sequences. The remaining very late proteins (members of the C, A, and B classes) presumably constitute the lamellae of the surface structures. In B mori the very late proteins are high-cysteine (Hc) components belonging to two multigene families (see below).

In summary, the silkmoth chorion is encoded by a large number of developmentally regulated genes, belonging to several multigene families. Each family must have arisen by several cycles of reduplication of an ancestral gene, followed by sequence divergence of the repeated copies to form distinct genes. At least two families, A and B, include members expressed at different developmental periods. Clearly, at least some of the reduplicated gene copies must have changed their developmental specificity, ie, their regulation, as they evolved into distinct genes. Study of the silkmoth chorion should help provide answers to two fundamental questions concerning eucaryotic genomes:

1. How do reduplicated copies of a gene evolve into distinct genes?
2. How do genes change their developmental specificity—how does developmental regulation evolve?

CHARACTERIZATION OF SILKMOTH CHORION GENES VIA DNA LIBRARIES

To obtain probes corresponding to individual chorion genes of A polyphemus, a population of double-stranded DNA copies (ds-cDNA) was synthesized using as template total poly $(A)^+$ follicular RNA (which is largely chorion mRNA), and was cloned in bacteria using a plasmid vector [Sim et al, 1979]. By the very nature of this procedure, each transformed bacterial colony contained hybrid plasmid derived from a single molecule, ie, represented in homogeneity a single chorion mRNA species. Approximately 20 different chorion cDNA clones have been analyzed in detail by hybridization [Sim et al, 1979; Kafatos et al, 1979], and many have also been characterized by sequencing [Jones et al, 1979; Tsitilou et al, 1980; Jones and Kafatos, 1982; Regier et al, 1983]. Corresponding clones of chromosomal DNA have also been obtained and similarly characterized (see below; Jones and Kafatos [1982]). The results have given us a detailed picture of the primary structure of chorion genes and of the corresponding proteins.

Many clones that were easily distinguishable by restriction endonuclease analysis proved to cross-hybridize significantly. Hybrid-selected translation [Ricciardi et al, 1979], ie, hybridization of total mRNA to cloned DNA, recovery of the specifically hybridized RNA, and translation in the wheat germ system [Thireos and Kafatos, 1980], proved that one group of cross-hybridizing clones corresponded to the B family, and other groups corresponded to the A family. By performing hybridization experiments or melts under conditions of progressively higher stringency, it was possible to estimate the extent of sequence homology between different sequences [Sim et al, 1979]. These results were confirmed and extended by complete sequencing of selected clones [Jones and Kafatos, 1982]. Within a single family, mRNA sequences ranged from highly homologous (ca 1% mismatching) to highly divergent (ca 50% mismatching, no detectable cross-hybridization under normal conditions). Accordingly, sequences were classified as being (nonidentical) "copies" of the same gene type (eg, the cDNA clone pc401 and the genomic clone sequences 401a and 401b, less than 5% mismatching); different gene types but members of the same "subfamily" (eg, 408 and 10, 10 to 20% mismatching); or members of different subfamilies but of the same family (eg, 401 and 10, 30 to 50% mismatching). Interestingly, in both the A and the B families, two respective subfamilies were distinguished, one of which is expressed during the middle period of development and one during the late period (Figs. 3, 4).

As a result of the sequencing studies, we now know a great deal about the evolutionary history of these families and the structural features of the corresponding proteins (Figs. 3, 4). Computer predictions [Hamodrakas et al, 1982a] indicate that both A and B proteins are characterized by extensive β-pleated sheet conformation—in agreement with laser Raman spectroscopic evidence [Hamodrakas et al, 1982b] and x-ray diffraction [Hamodrakas et al, 1983]. In both families, the most highly structured portion appears to be a central "domain," which is also remarkably conserved within the family in

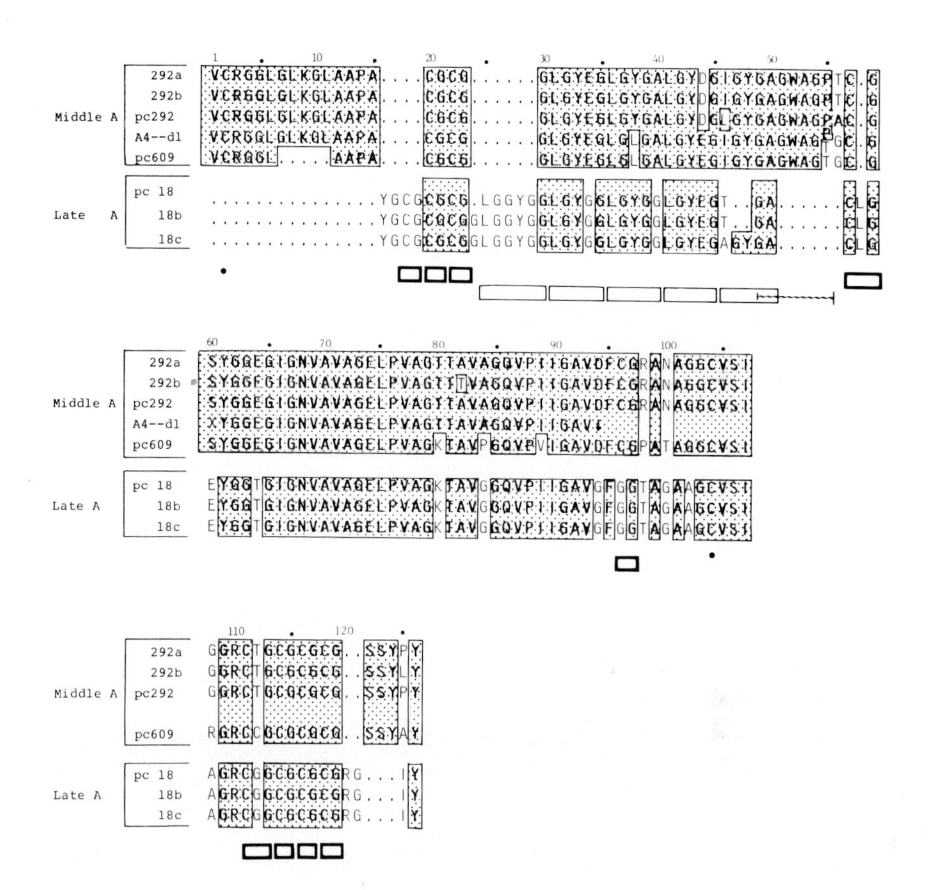

Fig. 3. Summary of mature chorion protein sequences of the A family in A polyphemus. Each sequence is presented in the one-letter code and is read from left to right, sequentially in three panels: left-arm (top), central domain (middle), right-arm (bottom). Sequences derived directly from proteins are identified by a double code (eg, A4--d1) and those derived from DNA are identified by numbers as explained in the text (eg, 292a or pc292). Sequences are clustered by developmental subfamily (middle or late). Identical residues are stippled, replacements are shown in plain letters, and gaps are indicated by dots. Symbols are used to identify repetitive peptides or other sequence features: heavy boxes, CG repeats or variants; light boxes, LGYGG repeats or variants; heavy dots, cysteines not part of CG repeats; wavy line, W-containing peptide which may be a variant of the LGYGG repeats (see Rodakis et al [1982]). Sequence A4--dl was obtained by J.C. Regier [unpublished], pc292 by S.G. Tsitilou [Tsitilou et al, 1980] and the rest by C.W. Jones [Jones and Kafatos, 1982]. (From Regier et al [1983].)

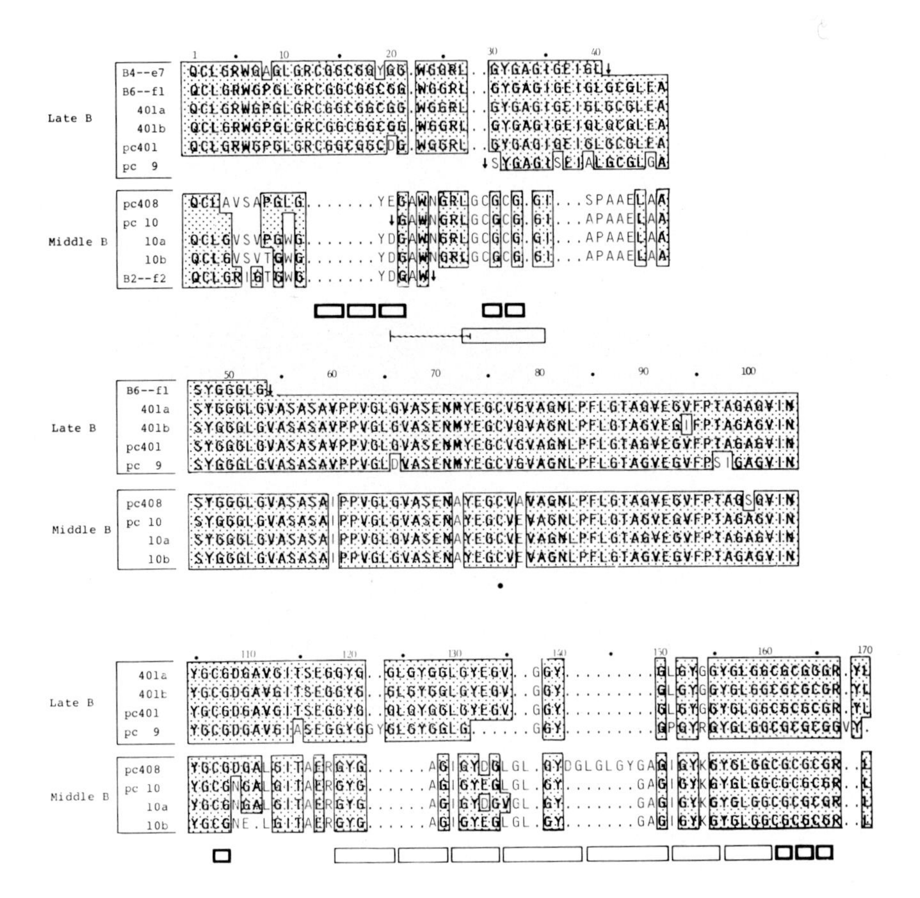

Fig. 4. Summary of mature chorion protein sequences of the B family in A. polyphemus. Conventions as in Figure 3. Sequences B4--e7, B6--f1 and B2--f2 were determined by G.C. Rodakis [Rodakis et al, 1983], pc9 by J. Pustell [unpublished], and the rest by C.W. Jones [Jones and Kafatos, 1982]. (From Regier et al [1983].)

primary structure. The rest of the molecule consists of a "left-arm" (amino-terminal) and a "right-arm" (carboxyterminal), which are more variable in sequence and less consistent in structure. We presume that the central, conservative domain serves family-wide functions, such as fibril formation, whereas the more variable flanking arms are responsible for more specific functions, such as those that differentiate between middle and late proteins. One C family sequence has been determined from cloned cDNA and shown to have the same tripartite structure [Regier et al, 1983].

Sequence information from B mori (a member of a different moth family, Bombycidae) is not extensive as yet. However, cross-hybridization analysis between cDNA clones clearly established the existence of multigene families in that species, as well [Iatrou et al, 1982]. Several B mori clones have been sequenced and shown to encode proteins unmistakably belonging to the A or B families, but to subfamilies distinct from those known in A polyphemus (Fig. 5) [Rodakis et al, 1982; Tsitilou et al, 1983; Lecanidou et al, 1983].

Although A and B sequences appear quite distinct, close comparisons, in combination with a limited amount of sequence information available for members of the C and Hc classes, reveal a remarkable fact: Different chorion gene families are related, constituting a superfamily. Thus, the one C sequence known [Regier et al, 1983] shows unmistakable homology to the B sequences in the central domain (ca 63% and 57% at the nucleotide and amino acid levels, respectively) and in immediately flanking portions of the arms; for most of their length, however, the arms are essentially unrelated to the arms of the B family (Fig. 5). Similarly, Hc sequences have turned out to belong to two families, one related to the A and one to the B family (Hc-A and Hc-B, respectively). In each case, the homology is obvious in the central domain, but the arms have evolved radically through large deletions and reduplications, to become polypeptide segments consisting almost exclusively of cysteine and glycine [Rodakis and Kafatos, 1982; G.C. Rodakis, K. Iatrou, and S.G. Tsitilou, unpublished; see Fig. 5]. A and B sequences themselves show limited similarities, in both the central domain and in the arms, suggesting a possible ancient homology [Jones et al, 1979; Tsitilou et al, 1980]. In summary, silkmoth chorion genes apparently constitute a superfamily with two major branches, one A-like (A and Hc-A families) and one B-like (B, C, and Hc-B families).

THE CHORION PROTEIN SYNTHETIC PROGRAM REFLECTS A PROGRAM OF mRNA PRODUCTION

Characterization of the cloned chorion sequences has also permitted us to establish conditions sufficiently stringent to detect in hybridization assays specific sequences (or small groups of highly related sequences, such as copies of the same gene type), rather than the full range of sequences in a family or subfamily (see the next section below).

Using rather stringent hybridization conditions, we determined the developmental stages during which various cloned cDNA sequences are represented in total cellular RNA [Sim et al, 1979]. These results permitted us to classify the sequences as developmentally early, middle, or late. Since the encoded proteins were also known, from hybrid-selected translations, it was possible to show that each RNA sequence is present in the cell in significant

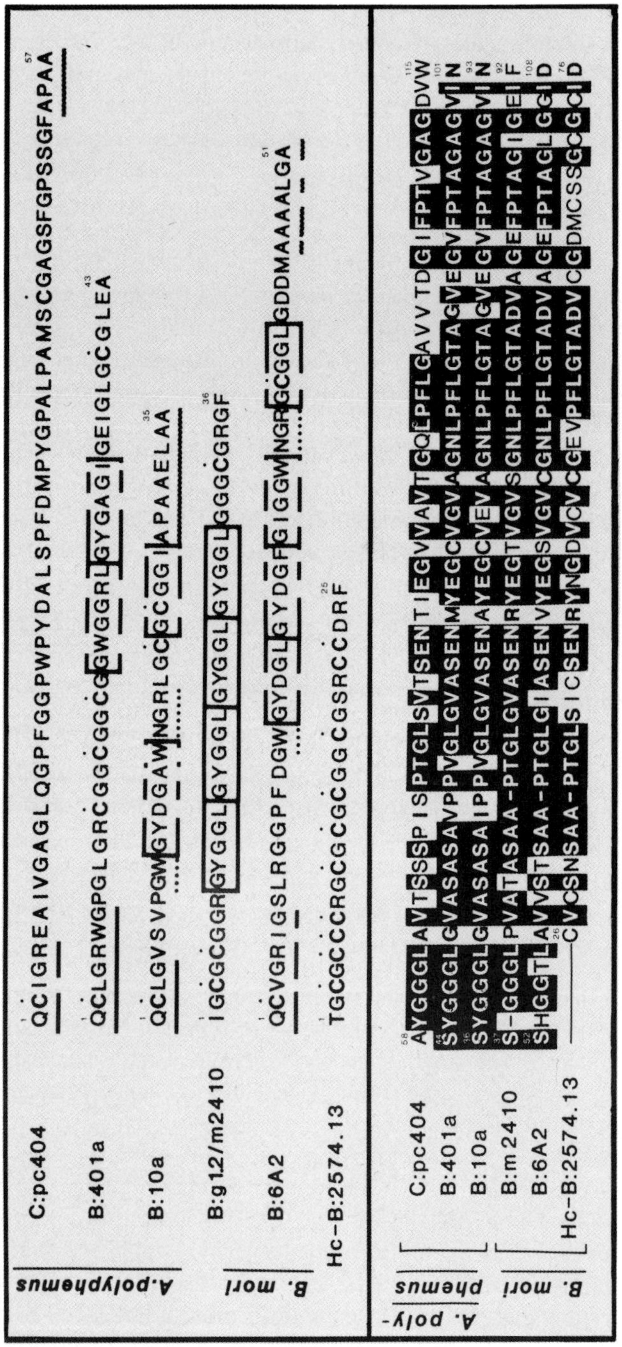

A.polyphemus

C:pc404 QCIGREAIVGAGLQGPFGGPWPYDALSPFDMPYGPALPAMSCGAGSFGPSSGFAPAA 57

B:401a QCLGRWGPGLGRCGGCGGCGGWGGRLGYGAGIGEIGLGCGLEA 43

B:10a QCLGVSVPGWGYDGAWNGRLGCGCGGIAPAAELAA 35

B. mori

B:g12/m2410 IGCGCGGRGYGGLGYGGLGYGGLGYGGLGGGCGRGF 36

B:6A2 QCVGRIGSLRGGPFDGWGYDGLGYDGFGIGGWNGRGCGGLGDDMAAAALGA 51

Hc-B:2574.13 TGCGCCCRGCGCGCGGCGSRCCDRF 25

A. polyphemus

C:pc404 AYGGGLAVTSSSPISPTGLSVTSENTIEGVVAVTGQLPFLGAVVTDGIFPTVGAGDVW 115

B:401a SYGGGLGVASASAVPPVGLGVASENMYEGCVGVAGNLPFLGTAGVEGVFPTAGAGVIN 101

B:10a SYGGGLGVASASAIPPVGLGVASENAYEGCVEVAGNLPFLGTAGVEGVFPTAGAGVIN 93

B. mori

B:m2410 S-GGGLPVATASAA-PTGLGVASENRYEGTVGVSGNLPFLGTADVAGEFPTAGIGEIF 92

B:6A2 SHGGTLAVVSTSAA-PTGLGIASENVYEGSVGVCGNLPFLGTADVAGEFPTAGLGGID 108

Hc-B:2574.13 ————CVCSNSAA-PTGLSICSENRYNGDVCVCGEVPFLGTADVCGDMCSSGCGCID 76

Figure 5.

amounts only at stages when the corresponding protein is synthesized in vivo [Thireos and Kafatos, 1980]. Similarly, a temporal program of accumulation and disappearance of specific chorion mRNA sequences has been established in B mori [Bock et al, 1983; Lecanidou et al, 1983].

CHROMOSOMAL ORGANIZATION OF SILKMOTH CHORION GENES

Overall Features

Members of the chorion multigene families are clustered in the silkmoth genome. In Bombyx mori, genetic analysis by M. Goldsmith and co-workers has established that the majority, if not all, of the chorion structural genes are clustered within 3.7 map units at one end of chromosome number 2 [n = 28; Goldsmith and Basehoar, 1978; Goldsmith and Clermont-Rattner, 1979, 1980]. We have undertaken a detailed structural analysis of the silkmoth chorion locus, and of its developmental and evolutionary properties, using recombinant DNA procedures. Our main findings are that the fundamental unit of chorion gene organization is a compact, divergently oriented, and coordinately expressed gene pair; that such pairs are embedded in longer

---

Fig. 5. Summary comparisons of various families and subfamilies of B-like chorion proteins from A polyphemus and B mori. Proteins are presented in the one-letter code and are numbered sequentially from the aminoterminus of the mature protein; signal peptide sequences are omitted. *pc404* is the only available C family sequence [Regier et al, 1983]; the first two residues, not represented in the incomplete pc404 cDNa clone, are derived from a corresponding genomic sequence [F. Toneguzzo and F.C. Kafatos, unpublished]. The genomic *401a* and *10a* sequences typify the middle and late B subfamilies, respectively, in A polyphemus (see Fig. 4). *gl.2/m2410* is typical of the known B sequences in B mori, and is a combination of sequences derived from protein gl.2 and from the cDNA clone, m2410 [S.G. Tsitilou et al, 1983]. 6A2 is an early B protein sequence that represents a distinct subfamily [Lecanidou et al, 1983]. *2574.13* typifies the two available Hc-B sequences [G.C. Rodakis, K. Iatrou, and S.G. Tsitilou, unpublished]. (Top panel) Sequences of the left arm are compared. Because of their substantial variability, no alignment is attempted. Landmark features are highlighted, with variations indicated by interruptions in highlighting. Boxes outline peptide repeats of consensus GYGGL [Regier et al, 1978]. Solid underlining indicates a recurrent aminoterminal sequence, wavy underlining an alanine-rich sequence, and dotted underlining some additional residues shared between the 10a subfamily and 6A2. Filled dots draw attention to cysteine residues, which are thought to be important for cross-linking of chorion proteins [Regier et al, 1978; Hamodrakas et al, 1982a], and are especially abundant in Hc-B proteins: (Bottom panel) Sequences of the conservative central domain are compared. Residues shared by three or more prototypical sequences are shown in white letters on black background. Deletions are indicated by dashes. The right arm (not shown) is also variable (cf Fig. 4; Regier et al [1983]). (From Lecanidou et al, [1983].)

chromosomal DNA units, which are quite variable; and that neighboring pairs tend to be similar both in sequence and in developmental control.

A library of A polyphemus chromosomal DNA was constructed using the charon 4 derivative of phage λ as vector [Maniatis et al, 1978]. A number of genomic clones proved to hybridize with two distinct cloned cDNA probes, indicating close linkage of the respective genes [Jones and Kafatos, 1980a]. Figure 6 diagrams the organization of two of these genomic clones, which have been studied in detail [Jones and Kafatos, 1980b]. Clone APc110 contains gene pairs consisting of two types of genes that are expressed during the late period of development: gene 18, which belongs to the A family; and gene 401, which belongs to the B family. Similarly, clone APc173 contains AB pairs (consisting of genes 292 and 10, respectively)—in this case expressed during the middle period of development. Thus, A and B genes are paired according to their developmental specificity (time of expression). The clones contain two or more copies of the respective gene pairs.

The AB gene pairs are divergently transcribed [Jones and Kafatos, 1980a,b]. All gene pairs and immediately adjacent flanking sequences in APc110 and APc173 have been sequenced [Jones and Kafatos, 1980b]. The positions of the 5′ ends of the genes were identified by comparison with the 5′ terminal sequences (cap sites) of the corresponding mRNAs. Remarkably, the 5′ flanking DNA, which separates the two genes in each pair, is very short: 325 base pairs (bp) for the 401/18 pairs and 264 bp for the 10/292 pairs. This DNA has structural features indicating that it contains paired, divergently operating promoters for the two genes.

Each gene is accompanied by a canonical Hogness-Goldberg TATA$^{T}_{A}$AA box in the appropriate strand, and at the correct distance from the 5′ end (centered at positions −24 to −26). In addition, each gene is accompanied by short sequences that are centered approximately at position −87 and show developmental rather than family specificity: For the paired late A and B genes (18 and 401) the sequences are T$^{A}_{T}$CGTGAA$^{G}_{C}$TT$^{T}_{C}$TAT, whereas for the middle A and B genes (292 and 10) they are TG$^{A}_{T}$TA. The generality of such specific upstream sequences remains to be established by analysis of additional gene pairs, and their significance can only be determined by functional tests (in vitro mutagenesis and testing for stage-specific transcriptional activation, in vitro or in vivo). However, the short length of the 5′ flanking region and its symmetrically located, short conserved sequences suggest that coordinate expression of the gene pair may be accomplished by some type of interaction between paired stage-specific promoters or associated regulatory elements.

On the other hand, more global analysis has suggested that neighboring chorion-gene pairs also tend to be expressed coordinately. This is seen most clearly in B mori, where chromosomal "walking" has permitted analysis of

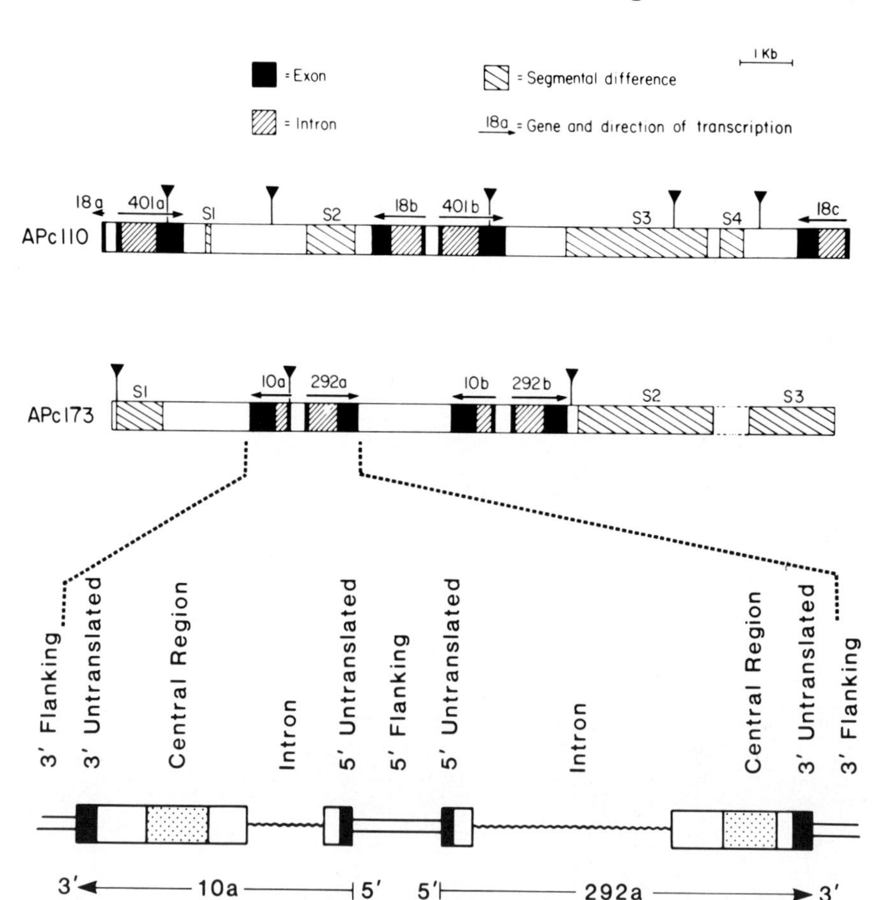

Fig. 6. Diagrams of A polyphemus genomic clones bearing multiple chorion genes (top) and expanded diagram showing the organization of a typical chorion gene pair (bottom). Each chorion gene is interrupted by a single intron, invariably within the signal peptide-coding sequence. The 5′-flanking DNA between the divergent genes of each pair is short, whereas the 3′-flanking DNA between gene pairs is long and variable because of the presence of variable inserted DNA elements (S1 to S3). (From Regier and Kafatos [1983] modified from Jones and Kafatos [1980, 1982].)

a 270-kilobase (kb) region of DNA, probably representing one-third to one-fifth of the chorion locus [Eickbush and Kafatos, 1982]. By Southern analyses of this region, using as probes broadly staged "middle" and "late" follicular RNAs, it has been shown that the central, ca 130-kb region contains at least 34 "late" genes, and the flanking, ca 100-kb and ca 40-kb regions contain a total of at least 36 "middle" genes (Fig. 7). The central region was studied

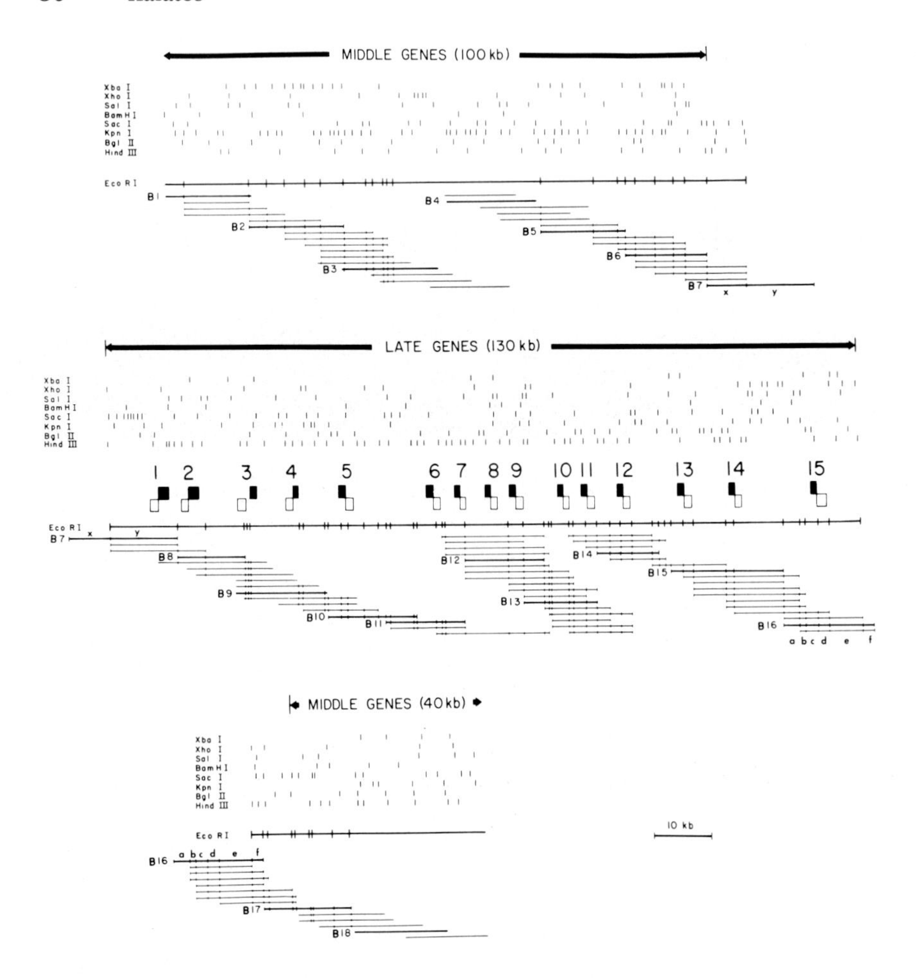

Fig. 7. Molecular map of a 270-kb portion of the chorion locus in B mori. The continuous, 270-kb map was derived from overlapping charon 4 genomic clones. It is shown in three segments, corresponding to developmentally specific clusters. The central, 130-kb region includes 15 pairs of genes, Hc-A (black boxes)/Hc-B (white boxes); four additional gene-containing segments exist in the late region (not shown). (Courtesy Dr T. Eickbush; from Eickbush and Kafatos [1982].)

in greater detail and proved to consist of 15 pairs of Hc-A/Hc-B genes, actually expressed during the very late developmental period; at least four additional, less-well characterized genes of similar developmental specificity exist within the same region. The flanking 100-kb and 40-kb regions are populated by AB gene pairs [T. Eickbush, S.G. Tsitilou and N. Spoerel, unpublished]. These results establish that the organization of chorion genes into coordinately expressed pairs is of wide generality, and that the pairs themselves tend to be clustered according to the period of their expression. However, since developmental characterization has been based to date on coarsely staged RNA, the precision of that clustering remains to be tested by further work: whereas the Hc-A and Hc-B genes are all expressed during the very late period of choriogenesis [Nadel and Kafatos, 1980; Bock et al, 1983; Lecanidou et al, 1983], typical A and B genes are known to be expressed during several distinguishable periods [Nadel and Kafatos, 1980], which were not assayed separately in the initial characterization of the 270 kb of DNA from the chorion locus.

Limited developmental clustering of gene pairs has also been documented in A polyphemus [Jones and Kafatos, 1981; G.A. Beltz and F.C. Kafatos, in preparation]. Approximately 15 copies of the 401/18 gene pair exist per genome. Chromosomal walking has localized these pairs into three noncontiguous arrays (three to five pairs each). Flanking each array are other chorion gene pairs or genes, of types different than 401 and 18. Several of these genes are expressed at other developmental periods and, interestingly, are found in inverted orientation relative to the 401/18 pairs. This analysis was only made possible by our detailed knowledge of the 401 and 18 sequences; their 3′ portions, encompassing the coding sequences for the respective right arms plus the 3′ untranslated sequences, are known to be rather divergent from the corresponding portions of other members of the same family. Therefore, 3′-specific chorion gene fragments can be used as probes under appropriately stringent conditions of hybridization [Beltz et al, 1983], to detect copies of the same sequence during the chromosomal walk, and to document the temporal specificity of their transcription [Beltz and Kafatos, in preparation].

Although the extent of developmental clustering remains to be established, its existence is intriguing from the perspective of both development and evolution: Are developmental clusters merely the consequence of relatively recent reduplications? Do they permit, or are they required for, mutual correction of the structural genes during evolution, and to what degree? Do they function in developmental regulation, for example, by permitting activation of a whole battery of genes in chromatin structure? Further structural and functional analysis is necessary to answer such questions.

A question of special interest, for both evolution and development, is how chorion gene pairs change their developmental specificity. We know that such changes have occurred in evolution since, say, the late- and middle-401/18 and 10/292 gene pairs are homologous, as are middle- and late-AB pairs and the very late Hc-A/Hc-B pairs in B mori. Yet changes in developmental specificity are apparently infrequent: Comparison of the timing of expression of 12 different cloned cDNA sequences in A polyphemus and A pernyi showed no differences between species [Moschonas, 1980]. Comparisons of 5′ flanking sequences and gene organization, both within and between species, should provide clues as to whether temporal regulation is exerted through regulatory DNA elements in the short 5′ flanking sequences, through global organization of the locus, or both.

Detailed Sequence Organization

Complete sequence analysis of the gene-bearing segments of APc110 and APc173 revealed that the gene copies found on each clone are not identical [Jones and Kafatos, 1980b, 1982]. Their DNA sequences differ by both base substitutions and small insertions or deletions (usually reduplications or deletions involving short direct repeats; Jones and Kafatos [1982]). The intercopy nucleotide sequence divergence for genes 401 and 18 is only 0.9%, whereas copies of genes 10 and 292 differ by 5.5%. This finding might suggest that the duplication (or correction) of genes 10 and 292 is a more ancient evolutionary event than the duplication (or correction) of genes 401 and 18. Other interpretations are possible, however: Limited sequence analysis of several 401/18 genes in the three arrays discussed earlier suggests that the copies of this gene pair are surprisingly conserved. In any case, in APc110 and APc173 the copies of each gene differ by at least one base substitution that results in an amino acid replacement. Thus, by the criterion of differences in the respective proteins, these copies have already taken the first step toward becoming different genes.

All chorion genes that have been sequenced to date, fully or in part, consist of two exons and one intron [Jones and Kafatos, 1980b; K. Iatrou, G.C. Rodakis, S.G. Tsitilou and N. Spoerel, unpublished]. Like other secretory proteins, chorion proteins are translated into protein precursors bearing an aminoterminal signal peptide [Thireos and Kafatos, 1980]. In each gene the intron interrupts the signal peptide-encoding region between codons -4 and -5 (counting from the mature protein NH_2-terminus). The introns for the different gene copies are quite similar both in sequence and length, although on the average they are clearly more divergent than the copies of exon sequences. No introns are found between the domains of the mature protein. Thus, the domain-specific evolutionary diversification of the chorion genes, involving slow evolution of the central domain and rapid or even radical evolution of the arms, has not been based on introns.

By contrast to the short 5′-flanking DNA that connects the two genes in each pair, the 3′-flanking DNA that separates gene pairs is long and variable (1.8 to > 5.5 kb for the clones shown in Fig. 6). Nevertheless, heteroduplex and restriction endonuclease analysis has established that much of the 3′-flanking DNA is homologous between gene pair copies [Jones and Kafatos,

1980b]. Length differences are due to the presence of insertion/deletion elements (labeled S1 through S4 in Fig. 6), which vary in restriction map, location, and length (0.065–2.9 kb). The adjacent copies of a gene pair, as well as the homologous portions of the 3′-flanking DNAs between these copies, are found in the same orientation (Fig. 6). Thus, the fundamental unit of organization of A and B genes appears to be a long, tandem DNA repeat, which contains one divergently transcribed AB gene pair, and varies by multiple inserts of nonhomologous DNA. The three arrays of 401/18 copies differ in the degree of diversification of their 3′-flanking sequences [Jones and Kafatos, 1981; G.A. Beltz and F.C. Kafatos, in preparation].

We suspect that the 3′-flanking inserts play an important role in the evolution of multicopy chorion genes into multigene families [Jones and Kafatos, 1980b]. Since chorion genes exist in tandem arrays, we would expect such arrays to undergo mispairing leading to gene correction, or unequal crossing-over and, therefore, expansion and contraction. Both of these processes tend to "homogenize" the sequences, ie, to reduce mutational differences between gene copies [Smith, 1973, 1976; Slightom et al, 1980; Ohta, 1980]. Variable inserts in the 3′-flanking DNA would tend to "dilute" the direct sequence repetition of the fundamental units, perhaps reducing the frequency of gene correction or unequal crossing-over, and thereby permitting accumulation of sequence differences between adjacent copies of the array. The differences may be useful for rapid evolution of the chorion genes, leading to rapid evolution of chorion structure and physiology. Clearly, comparative analysis of multiple copies of gene types that are rather well conserved (eg, 401/18), and others that appear to be more variable (eg 10/292), should be fruitful, when combined with analysis of their 3′-flanking DNA elements. Questions of interest include whether the inserts in 3′-flanking DNA are nomadic elements, and if so of what type and what reiteration frequency. Are the same elements found outside the chorion locus, and how frequently? Is the chorion locus in fact an exceptionally favorable acceptor of such elements, and if so how?

CHROMOSOMAL ORGANIZATION AND EXPRESSION OF DROSOPHILA CHORION GENES

Oogenesis in Drosophila has been divided into 14 stages by morphological criteria [King, 1970]. The chorion is produced during stages 11 to 14, lasting approximately 5 hr. Although a number of major and minor chorion protein components have been identified [Petri et al, 1976; Waring and Mahowald, 1979; Margaritis et al, 1980], attention has largely focused on the most abundant ones, belonging to six molecular weight size groups (named s15, s16, s18, s19, s36, and s38, by reference to approximate M_r in kilodaltons).

Each group is produced at a characteristic period during choriogenesis [Petri et al, 1976; Waring and Mahowald, 1979]; s36 and s38 are synthesized "early" (primarily stages 11 and 12); whereas s15, s16, s18, and s19 are synthesized at various "late" periods (primarily stages 13 and 14). Size classes of mRNAs encoding the major chorion proteins have been identified and characterized [Spradling and Mahowald, 1979; Thireos et al, 1979]. cDNA clones for most of the chorion mRNAs have been generated and used to identify the chromosomal location of the genes [Griffin-Shea et al, 1980; Spradling et al, 1980]. Genetic locations of several components were independently established using strains with protein isoelectric focusing variants [Yannoni and Petri, 1980, 1981]. Thus two chorion gene clusters were identified: Genes coding for the s36 and s38 proteins are clustered on the X chromosome, in the region of band 7F1, whereas those for the s15 through s19 proteins are on the third chromosome in chromosomal region 66D [Spradling, 1981; Griffin-Shea et al, 1982]. The X-chromosome cluster actually includes several additional, as yet unidentified chorion genes [Spradling, 1981].

Each of the identified major chorion genes is present in a single copy in the chromosomal regions cloned; furthermore, whole genomic Southerns at the commonly used criterion of stringency indicate that these genes are single-copy in the genome as a whole—in marked contrast to the situation in the silkmoths. However, sequences that are distantly homologous with chorion genes exist at various locations in the genome: By using very permissive criteria, K. Jacobs [personal communication] has identified four different unlinked clones homologous with s15. It is not known yet whether these sequences represent chorion components. Although the identified major chorion genes do not cross-hybridize at commonly used criteria, they are probably distantly homologous. Partial proteolysis of isolated chorion proteins followed by fractionation on SDS gels shows similarities in the digestion patterns, particularly between s18 and s19, s36, and s38 [Spradling et al, 1980]; sequence homologies between s15 and s18 have also been detected at the nucleotide level [Y. -C. Wong, personal communication]. Sequence analysis of chorion DNAs and homologous sequences is proceeding and will shortly reveal in more detail the features of this gene family or families. The results to date do not show homology to the silkmoth chorion gene families—as might be expected, since the ultrastructures and protein compositions of the chorion are so different between flies and moths (Fig. 1) [Petri et al, 1976; Kafatos et al, 1977]. In summary, although gene families are also involved in formation of the chorion in Drosophila, they are apparently different than those used in moths, and they show substantially different properties: Each gene type is present in a single copy per genome, the gene types are only distantly homologous, and they are clustered in at least two and probably several chromosomal loci.

In contrast to the lack of multiple, closely related germ-line genes, the quantitative demands of choriogenesis would appear to be substantially higher in Drosophila than in moths: Choriogenesis is completed ten times faster, within approximately 5 rather than 50 hr. As an alternative strategy to the redundancy characteristic of the silkmoth chorion locus, Drosophila chorion genes undergo specific amplification during development [Spradling and Mahowald, 1980]. The entire cluster of genes in each major locus seems to amplify as a unit, with amplification probably beginning from a single origin and extending over 50 to 100 kb [Spradling and Mahowald, 1981]. The importance of this amplification was established by analysis of *ocelliless* [Spradling et al, 1979], a small inversion with a breakpoint very near the s36 gene. Rearrangement of the DNA by the inversion leads to elimination of amplification at one end (presumably the end separated from a putative origin of replication), reduction in amplification of the other end, and spreading of amplification into the new DNA, brought into proximity with the origin as a consequence of the inversion [Spradling, 1981]. The consequence is *cis*-limited underproduction of proteins encoded by the affected chorion genes [Spradling et al, 1979], with attendant disruption of the chorion structure and female sterility [Johnson and King, 1974]. In summary, chorion gene clustering in Drosophila would seem to be related, at least in part, to coordinate amplification.

Careful analysis of the third chromosome cluster by the Northern procedure has established that its four genes are subject to some gene-specific controls, in addition to being coordinately and equally amplified [Griffin-Shea et al, 1982]. As shown in Figure 8, despite their equal gene dosage, the four genes are expressed into mRNA in amounts that differ severalfold; furthermore, although all four genes are predominantly expressed late in choriogenesis (stages 13 and 14), they differ slightly in their developmental kinetics of mRNA accumulation and disappearance, as would be expected from the synthetic kinetics of the respective proteins. These temporal differences may not be unrelated to the structural organization of the genes: Unlike the situation in moths, the Drosophila genes are organized in direct orientation, and are not found in pairs (Fig. 8).

The developmental Northerns, in conjunction with developmental Southerns that were designed to determine the time-course of amplification, led to a curious and unexpected finding [Thireos et al, 1980; Griffin-Shea et al, 1982]: s15 (but not the other three genes in the cluster) is prematurely transcribed at stages 8 to 10, at approximately the same time as the onset of amplification, even though the corresponding protein is not synthesized until

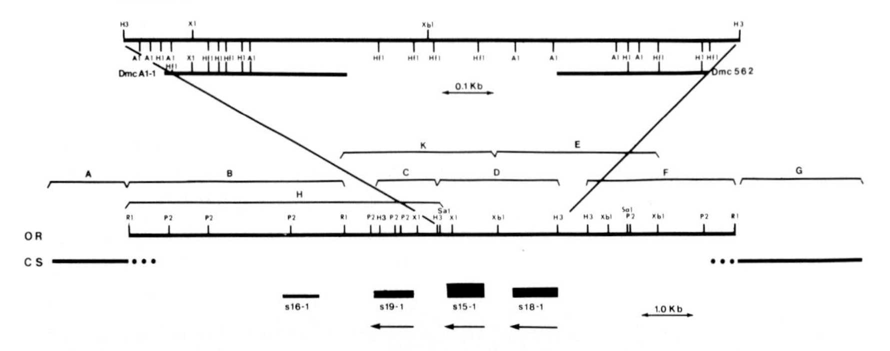

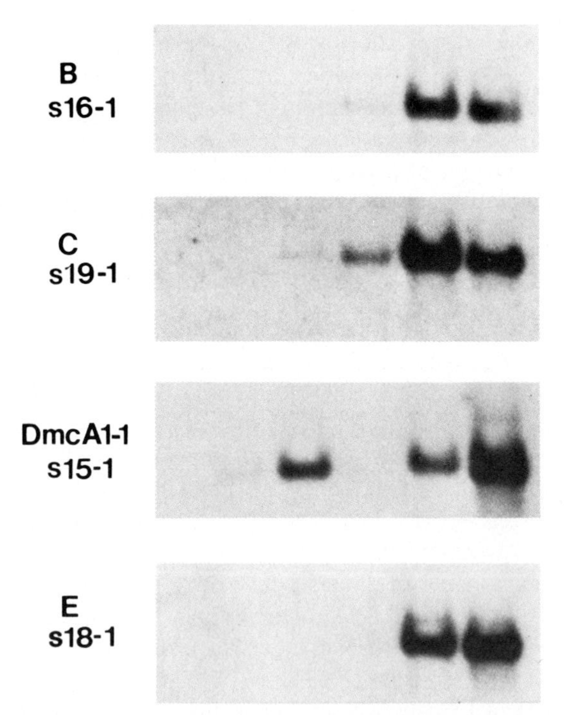

Fig. 8. Diagram of the third chromosome cluster of chorion genes in Drosophila (top), and determination of their developmental specificity by Northern analysis (bottom). The diagram includes a restriction map of the cluster; a magnified portion aligned with the cDNA clones, DmcA1-1 and Dmc5G2; restriction fragments (A-G) used as probes in various experiments; and a bar indicating location of each gene and relative abundance of its RNA in steady-state ovarian RNA. For Northern blots, total RNA was extracted from 40 follicles each of stages 8 + 9, 10, 11 + 12, 13, and 14, and from embryos (E), electrophoresed on methyl mercury gels, and blotted on DBM paper. The paper was sequentially hybridized, following dehybridization, with ^{32}P-nick-translated fragments, each specific for one gene (B, C, DmcA1-1, E). (Courtesy Drs. R. Griffin-Shea and G. Thireos; from Griffin-Shea et al [1982].)

many hours later (Fig. 8). The premature transcript is not stored for subsequent activation: It enters the cytoplasm but does not associate with polysomes, and is subsequently degraded, only to be replaced by new (and translated) transcripts at the appropriate (late) period of choriogenesis. These results indicate the existence of some type of translational control, preventing in a "fail-safe" manner the premature expression into protein of the prematurely transcribed gene. The function of this premature transcription is a mystery. We speculate that it may somehow be related to the amplification mechanism.

For critical analysis of developmental regulation, Drosophila has two major advantages: Sophisticated genetics, and the possibility of germline transformation using cloned DNA [Rubin and Spradling, 1982]. We are currently using both of these approaches. Analysis of chorion mutants (preselected as female-steriles; Komitopoulou et al [1983]) has already proven fruitful: Two such mutants, which are recessives, lead to underproduction of all major chorion proteins through strong underamplification of the two gene clusters [W. Orr and K. Komitopoulou, personal communication]. These mutants are located at band regions 5D5–6 to 6C12, and 12A6–7 to 12D3 on the X-chromosome, ie, are not adjacent to either chorion gene cluster. In addition to confirming the importance of amplification for normal chorion production, these mutants have provided an unexpected opening into the study of regulation of developmentally specific gene amplification by *trans*-acting factors.

SYNOPSIS

In concluding, I will not attempt to summarize this essay, which is already a summary. I merely wish to point out that control of choriogenesis operates at many levels, and these are different in moths and Drosophila. In Drosophila, an important control is amplification, which operates coordinately on several clustered, single-copy genes; but the genes are not all clustered within a giant locus. They are oriented in tandem, and they are additionally regulated by gene-specific mechanisms, some of them presumably acting through DNA elements flanking the individual genes. In the case of s15, some type of translational control apparently ensures that the premature transcription (possibly related to amplification) does not result in premature translation. By contrast, in silkmoths chorion genes are not ontogenetically amplified [Jones and Kafatos, 1981]. The germ-line genes are multicopy and are clustered within a single giant locus spanning 3.7 centimorgans [Goldsmith and Clermont-Rattner, 1979]. The moth chorion genes are found in tightly organized, divergent, and coordinately expressed pairs, and form arrays that evolve rather rapidly into a large number of new genes, while retaining features (such as the central domain) that are characteristic of the entire gene family. There is no indication of translational control, in that accumulation and disappearance of specific mRNAs parallel the time-course of corresponding protein synthesis in vivo. Finally, the structural gene families themselves are apparently different in moths and Drosophila. For biochemists oriented toward universal mechanisms, this picture may appear bewildering; for biologists comfortable with the complexities of development and evolution, it should be exhilarating.

ACKNOWLEDGMENTS

I thank the former and present members of my laboratory for their participation in the enjoyable and productive team effort summarized in this essay. In particular, I wish to thank the following persons for the use of figures and unpublished data or ideas: G.D. Mazur, J.C. Regier, C.W. Jones, T. Eickbush, K. Iatrou, S.G. Tsitilou, G.C. Rodakis, R. Lecanidou, N.K. Moschonas, S.J. Hamodrakas, J. Pustell, G.A. Beltz, N. Spoerel, F. Toneguzzo, and M. Alexopoulou (Silkmoth chorion project); and K. Komitopoulou, W. Orr, L.H. Margaritis, K. Jacobs, Y-C. Wong, R. Griffin-Shea, and G. Thireos (Drosophila chorion project). Our work was supported by the NSF, the NIH, the American Cancer Society, and the National Foundation—March of Dimes.

REFERENCES

Beltz GA, Jacobs KA, Eickbush TH, Cherbas P, Kafatos FC (1983): Isolation of multigene families and determination of homologies by filter hybridization methods. In Wu R, Grossman L, Moldave K (eds): "Methods in Enzymology. Recombinant DNA, Part B." Amsterdam: Elsevier North-Holland, Inc (in press).

Bock SC, Tiemeir DC, Mester K, Goldsmith MR (1983): Differential patterns in the temporal expression of Bombyx mori chorion genes. Wilhelm Roux Arch Dev Biol (in press).

Bouligand Y (1972): Twisted fibrous arrangements in biological materials and cholesteric mesophases. Tissue Cell Res 4:189–217.

Eickbush TH, Kafatos FC (1982): A walk in the chorion locus of Bombyx mori. Cell 29:633–643.

Goldsmith MR, Basehoar G (1978): Organization of the chorion genes of Bombyx mori, a multigene family. I. Evidence for linkage to chromosome 2. Genetics 90:291–310.

Goldsmith MR, Clermont-Rattner E (1979): Organization of the chorion genes of Bombyx mori, a multigene family. II. Partial localization of three gene clusters. Genetics 92:1173–1185.

Goldsmith MR, Clermont-Rattner E (1980): Organization of the chorion genes of Bombyx mori, a multigene family. III. Detailed marker composition of three gene clusters. Genetics 96:201–212.

Griffin-Shea R, Thireos G, Kafatos RC, Petri WH, Villa-Komaroff L (1980): Chorion cDNA clones of D. melanogaster and their use in studies of sequence homology and chromosomal location of chorion genes. Cell 19:915–922.

Griffin-Shea R, Thireos G, Kafatos FC (1982): Organization of a cluster of four chorion genes in Drosophila and its relationship to developmental expression and amplification. Dev Biol 91:325–336.

Hamodrakas SJ, Jones CW, Kafatos FC (1982a): Secondary structure predictions for silkmoth chorion proteins. Biochim Biophys Acta 700:42–51.

Hamodrakas SJ, Asher SA, Mazur GD, Regier JC, Kafatos FC (1982b): Laser Raman studies of protein conformation in the silkmoth chorion. Biochim Biophys Acta 703:216–222.

Hamodrakas SJ, Paulson JR, Rodakis GC, Kafatos FC (1983): X-ray diffraction studies of a silkmoth chorion. Int J Bio Macromol (in press).

Hinton H (1969): Respiratory systems of insect eggshells. Annu Rev Entomol 14:343–368.

Hinton H (1981): "Biology of Insect Eggs." Oxford: Pergamon Press.

Iatrou K, Tsitilou SG, Kafatos FC (1982): Developmental classes and homologous families of chorion genes in Bombyx mori. J Mol Biol 157:417–434.

Johnson DD, King RC (1974): Oogenesis in the ocelliless mutant of Drosophila melanogaster Meigen (Dipter:Drosophilidae). Int J Insect Morphol and Embryol 3(3/4):385.

Jones CW, Rosenthal N, Rodakis GC, Kafatos FC (1979): Evolution of two major chorion multigene families as inferred from cloned cDNA and protein sequences. Cell 18:1317–1332.

Jones CW, Kafatos FC (1980a): Coordinately expressed members of two chorion multigene families are clustered, alternating and divergently orientated. Nature 284:635–638.

Jones CW, Kafatos FC (1980b): Structure, organization and evolution of developmentally regulated chorion genes in a silkmoth. Cell 22:855–867.

Jones CW, Kafatos FC (1981): Linkage and evolutionary diversification of developmentally regulated multigene families: Tandem arrays of the 401/18 chorion gene pair in a silkmoth. Mol Cell Biol 1:814–828.

Jones CW, Kafatos FC (1982): Accepted mutations in a gene family: Evolutionary diversification of duplicated DNA. J Mol Evol 19:87–103.

Kafatos FC (1981): Structure, evolution and developmental expression of the silkmoth chorion multigene families. Am Zool 21:707–714.

Kafatos FC, Jones CW, Efstratiadis A (1979): Determination of nucleic acid sequence homologies and relative concentrations by a dot hybridization procedure. Nucleic Acids Res 7:1541–1552.

Kafatos FC, Regier JC, Mazur GD, Nadel MR, Blau HM, Petri WH, Wyman AR, Gelinas GE, Moore PB, Paul M, Efstratiadis A, Vournakis JN, Goldsmith MR, Hunsley JR, Baker B, Nardi J, Koehler M (1977): The eggshell of insects: Differentiation-specific proteins and the control of their synthesis and accumulation during development. In Beermann W (ed): "Results and Problems in Cell Differentiation," Vol 8. Berlin: Springer-Verlag, pp 45–145.

King RC (1970): "Ovarian Development in Drosophila melanogaster." New York: Academic Press.

Komitopoulou K, Gans M, Margaritis LH, Kafatos FC, Masson M (1983): Isolation and characterization of sex-linked female-sterile mutants in Drosophila melanogaster with special attention to eggshell mutants. Genetics (in press).

Lecanidou R, Eickbush TH, Rodakis GC, Kafatos FC (1983): Novel B family sequence from an early chorion cDNA library of Bombyx mori. Proc Natl Acad Sci USA (in press).

Maniatis T, Hardison RC, Lacy E, Lauer J, O'Connell C, Quon D, Sim GK, Efstratiadis A (1978): The isolation of structural genes from libraries of eucaryotic cDNA. Cell 15:687–701.

Margaritis LH (1983): Structure and physiology of the egg shell. In Kerkut GA, Gilbert LI (eds): "Embryogenesis and Reproduction: Comprehensive Insect Physiology, Biochemistry, and Pharmacology," Vol I (in press).

Margaritis LH, Petri WH, Kafatos FC (1976): Three dimensional structure of the endochorion in wild-type Drosophila melanogaster. J Exp Zool 198:403–408.
Margaritis LH, Petri WH, Wyman AR (1979): Structural and image analysis of a crystalline layer from a dipteran eggshell. Cell Biol Int Rep 3:61.
Margaritis LH, Kafatos FC, Petri WH (1980): The eggshell of Drosophila melanogaster I. Fine structure of the layers and regions of the wild-type eggshell. J Cell Sci 43:1–35.
Mazur GD, Regier JC, Kafatos FC (1980): The silkmoth chorion: Morphogenesis of surface structures and its relation to synthesis of specific proteins. Dev Biol 76:305–321.
Mazur GD, Regier JC, Kafatos FC (1982): Order and defects in the silkmoth chorion, a biological analogue of a cholesteric liquid crystal. In Akai H, King RC (eds): "Insect Ultrastructure," Vol 1. New York: Plenum Press, pp 150–185.
Moschonas N (1980): Evolutionary comparisons of chorion structural and regulatory genes in two wild silkworm species. PhD thesis. University of Athens, Greece.
Nadel MR, Kafatos FC (1980): Specific protein synthesis in cellular differentiation IV. The chorion proteins of Bombyx mori and their program of synthesis. Dev Biol 75:26–40.
Ohta T (1980): Evolution and variation of multigene families. In Levin S (ed): "Lectures in Biomathematics," vol 37. Heidelberg: Springer-Verlag.
Paul M, Goldsmith MR, Hunsley JR, Kafatos FC (1972): Specific protein synthesis in cellular differentiation I. Production of eggshell proteins by silkmoth follicular cells. J Cell Biol 55:653–680.
Paul M, Kafatos FC (1975): Specific protein synthesis in cellular differentiation II. The program of protein synthetic changes during chorion formation by silkmoth follicles, and its implementation in organ culture. Dev Biol 42:141–159.
Petri WH, Wyman AR, Kafatos FC (1976): Specific protein synthesis in cellular differentiation III. The eggshell proteins of Drosophila melanogaster and their program of synthesis. Dev Biol 49:185–199.
Regier JC, Kafatos FC, Goodfliesh R, Hood L (1978): Silkmoth chorion proteins: Sequence analysis of the products of a multigene family. Proc Natl Acad Sci USA 75:390–394.
Regier JC, Mazur GD, Kafatos FC (1980): The silkmoth chorion: Morphological and biochemical characterization of four surface regions. Dev Biol 76:286–304.
Regier JC, Mazur GD, Kafatos FC, Paul M (1982): Morphogenesis of silkmoth chorion: Initial framework formation and its relation to synthesis of specific proteins. Dev Biol 92:159–174.
Regier JC, Kafatos FC (1983): Molecular aspects of chorion formation. In Kerkut GA, Gilbert LI (eds): "Embryogenesis and Reproduction: Comprehensive Insect Physiology, Biochemistry, and Pharmacology," Vol 1 (in press).
Regier JC, Kafatos FC, Hamodrakas, SJ (1983): Silkmoth chorion multigene families constitute a superfamily: Comparison of C and B family sequences. Proc Natl Acad Sci USA 80:1043–1047.
Ricciardi RP, Miller JS, Roberts BE (1979): Purification and mapping of specific mRNAs by hybridization-selection and cell-free translation. Proc Natl Acad Sci USA 76:4927–4931.
Rodakis GC (1978): The chorion of the lepidopteran Antheraea polyphemus: A model system for the study of molecular evolution. PhD thesis, University of Athens, Greece.
Rodakis GC, Kafatos FC (1982): Origin of evolutionary novelty in proteins: How a high-cysteine chorion protein has evolved. Proc Natl Acad Sci USA 79:3551–3555.
Rodakis GC, Moschonas NK, Kafatos FC (1982): Evolution of a multigene family of chorion proteins in silkmoths. Mol Cell Biol 2:554–563.
Rodakis GC, Moschonas NK, Regier JC, Kafatos FC (1983): The B multigene family of chorion proteins in saturniid silk moths. J Mol Evol (submitted).

Rubin GM, Spradling AC (1982): Genetic transformation of Drosophila with transposable element vectors. Science 218:348–353.

Slightom JL, Blechl AE, Smithies O (1980): Human fetal $^{G}\gamma$- and $^{A}\gamma$-globin genes: Complete nucleotide sequences suggest that DNA can be exchanged between these duplicated genes. Cell 21:627–638.

Sim GK, Kafatos FC, Jones CW, Koehler MD, Efstratiadis A, Maniatis T (1979): Use of a cDNA library for studies on evolution and developmental expression of the chorion multigene families. Cell 18:1303–1316.

Smith GP (1973): Unequal crossover and the evolution of multigene families. Cold Spring Harbor Symp Quant Biol 38:507–513.

Smith GP (1976): Evolution of repeated DNA sequences by unequal crossover. Science 191:528–535.

Smith DS, Telfer WH, Neville AC (1971): Fine structure of the chorion of a moth, Hyalophora cecropia. Tissue Cell Res 3:477–498.

Spradling AC (1981): The organization and amplification of two chromosomal domains containing Drosophila chorion genes. Cell 27:193–201.

Spradling AC, Mahowald AP (1979): Identification and genetic localization of mRNAs from ovarian follicle cells of Drosophila melanogaster. Cell 16:589–598.

Spradling AC, Mahowald AP (1980): Amplification of genes for chorion proteins during oogenesis in Drosophila melanogaster. Proc Natl Acad Sci USA 77:1096–1100.

Spradling AC, Mahowald AP (1981): A chromosome inversion alters the pattern of specific DNA replication of Drosophila follicle cells. Cell 27:203–209.

Spradling AC, Waring GL, Mahowald AP (1979): Drosophila bearing the ocelliless mutation underproduce two major chorion proteins both of which map near this gene. Cell 16:609–616.

Spradling AC, Digan ME, Mahowald AP, Scott M, Craig EA (1980): Two clusters of genes for major chorion proteins of Drosophila melanogaster. Cell 19:905–914.

Thireos G, Griffin-Shea R, Kafatos FC (1979): Identification of chorion protein precursors and the mRNAs that encode them in Drosophila melanogaster. Proc Natl Acad Sci USA 76:6279–6283.

Thireos G, Griffin-Shea R, Kafatos FC (1980): Untranslated mRNA for a chorion protein of Drosophila melanogaster accumulates transiently at the onset of specific gene amplification. Proc Natl Acad Sci USA 77:5789–5793.

Thireos G, Kafatos FC (1980): Cell-free translation of silkmoth chorion mRNAs: Identification of protein precursors and characterization of cloned DNAs by hybrid-selected translation. Dev Biol 78:36–46.

Tsitilou SG, Regier JC, Kafatos FC (1980): Selection and sequence of a cDNA clone encoding a known chorion protein of the A family. Nucleic Acids Res 8:1987–1997.

Tsitilou SG, Rodakis GC, Alexopolou M, Kafatos FC, Iatrou K (1983): The B family of chorion proteins in Bombyx mori. EMBO J (submitted for publication)

Waring GL, Mahowald AP (1979): Identification and time of synthesis of chorion proteins in Drosophila melanogaster. Cell 16:599–607.

Yannoni CZ, Petri WH (1980): Characterization by isoelectric focusing of chorion protein variants in Drosophila melanogaster and their use in developmental and linkage analysis. Wilhelm Roux Arch 189:17–24.

Yannoni CZ, Petri WH (1981): Drosophila melanogaster eggshell development: Localization of the S19 chorion gene. Wilhelm Roux Arch 190:301–303.

II. Control of Functional Cell States

Gene Structure and Regulation in Development, pages 65–75

Determination of Yeast Cell Type

Ira Herskowitz
Department of Biochemistry and Biophysics, University of California, San Francisco, California 94143

INTRODUCTION

One important question that must be answered for a molecular understanding of development concerns the mechanism by which differentiated cells produce a specific set of specialized proteins. Liver-specific proteins, brain-specific proteins, etc, are each encoded by a set of genes that must be transcribed in the appropriate cell type. Recent analysis indicates that in many cases the genes for these specialized proteins are transcribed *only* in the specialized cells [Derman et al, 1981]. Having defined a set of specialized genes, the next question in a molecular understanding of development is "How is this set of genes activated?" which gives rise to the question "How does this set of genes become activated in the course of development?" In this article, I shall try to put these questions into focus and to provide some answers through descriptions of gene control circuitry in yeast and in phage λ.

For phage λ, the "specialized cell types" are the mutually exclusive outcomes of λ infection, lysis or lysogenization of the host. The lytic outcome is reflected by its specialized set of proteins, including those for particle formation and for cell lysis and the lysogenic outcome is reflected by its specialized set of proteins, notably the λ repressor (coded by the cI gene) and integration protein, integrase (coded by the int gene). How are all of the genes of the lytic set activated; all of the genes of the lysogenic set activated? Two mechanistically different solutions, with a common theme, are used by λ. The λ late genes (the two dozen genes concerned with lysis and particle formation) are coordinately regulated: their expression requires the activator protein coded by the λ Q gene [Forbes and Herskowitz, 1982; Grayhack and Roberts, 1982]. The arrangement of these late genes similarly plays a role in control: all are contiguous and all are expressed from a single promoter.

The int and cI genes are also coordinately regulated: their expression requires the λ activator protein cII [see Miller et al, 1981]. In this case, the int and cI genes are not adjacent but are preceded by a cII recognition site. Expression of lytic and lysogenic gene (and protein) sets thus is coordinately regulated by action of appropriate regulatory proteins, Q and cII, that recognize specific DNA sequences. Interestingly, the mechanisms by which these regulatory proteins act are different: cII stimulates transcription by aiding RNA polymerase binding, whereas Q stimulates transcription by allowing already initiated transcription to continue past a termination site and into the late genes.

THE SPECIALIZED CELL TYPES OF YEAST

Although yeast is a unicellular eucaryote, it has three different types of cells [reviewed in Herskowitz and Oshima, 1981]. Two of the cells are the mating types **a** and α, and they are haploid. The third cell type, the **a**/α cell, is a diploid that is formed by mating between **a** and α haploid cells. The key macroscopic differences between these cell types are that **a** and α cells are specialized for mating, and the **a**/α cell is specialized to respond to nutritional signals (starvation), which induces sporulation (that is, meiosis to yield four haploid nuclei, each of which is then encapsidated into separate spores). As might be expected, these macroscopic behaviors are reflected in specializations at a "microscopic," biochemical, level. The mating types each produce an appropriate oligopeptide pheromone (α-factor by α cells and **a**-factor by **a** cells) that causes cell cycle arrest in the opposite cell type as a prelude to mating. **a**/α cells, which do not mate, produce neither pheromone and respond to neither pheromone. A further specialization for mating is agglutination: **a** and α cells agglutinate with each other, but neither to an **a**/α cell. Thus one type of major specialization is the **a** cell versus the α cell versus the **a**/α cell. In addition, there are differences between cells that can be termed haploid (whether **a** or α) versus diploid (**a**/α): **a** and α haploids exhibit mating type interconversion, whereas **a**/α cells do not. Likewise, haploids exhibit a pattern of budding that differs from that of **a**/α cells. Thus there are quite a few differences among the three cell types—not a vast number, but some very distinctive differences of great importance to the life cycle of yeast.

However many differences between **a** and α cells in cell behavior (cell phenotype), there is only one genetic difference. A single genetic locus, the mating type locus (MAT) determines all of the differences among **a**, α, and **a**/α cells. Cells with the **a** allele (genotypically MAT**a**) exhibit the **a** cell type, cells with the α allele (MATα) exhibit the α cell type, and those with both alleles (MAT**a**/MATα) are of course **a**/α cells. There is a moderately long and interesting history on thoughts as to the relationship between these alleles, which is discussed elsewhere [Hawthorne, 1963; Hicks and Herskowitz, 1977; Sprague et al, 1981a]. The bottom line is that MAT**a** and MATα differ from each other in containing sizable nonhomologous regions: MAT**a** contains a segment of DNA (termed Y**a**) of 642 base pairs (bp) that is absent from MATα, and MATα contains a segment of 747 bp (Yα) that is absent from MAT**a** [Astell et al, 1981].

How does a single genetic locus determine all of the different properties of the yeast cell types? MacKay and Manney [1974] made the proposal that the alleles of the mating-type locus code for regulators that govern expression of unlinked genes concerned with mating, a reasonable hypothesis and one that has proved to be correct. We shall next consider the identification of these genes that are thought to be regulated by the mating-type locus and then, further down the line, discuss the specific hypothesis (the α1-α2 hypothesis) by which the products of the mating-type locus alleles carry out this task.

IDENTIFYING OTHER GENES REQUIRED FOR MATING

One approach to studying cell-type determination that is being productively employed in several eucaryotes is to identify genes that are expressed in one cell type and not in another by differential hybridization methods [eg, Benton and Davis 1977]. This type of analysis and its variations are currently being undertaken in yeast as well. Identifying segments that are expressed in one cell type and not in another is, of course, of great potential utility in documenting just how many of these types of differences exist between two cell types. Furthermore, in yeast, one has the opportunity to use the cloned segment to produce mutations and thus assess the functional role of the DNA segment that has been cloned.

The powerful genetic opportunities in yeast make it possible to identify cell-type differences in a way that does not presuppose the level of regulation. The method for yeast is simply the classical method of genetics—identification of genes that are important to a process by isolation of mutants defective in that process. With one important additional wrinkle. After identifying a mutant that is defective in mating, it is then possible to glean some very important and useful information about the gene so identified—whether it is **a**-specific, α-specific, or not specific to mating type—in a single cross [MacKay and Manney, 1974].

To demonstrate this method, let us consider a yeast mutant derived from an α cell that is now defective in mating, due to a mutation somewhere in its genome. This mutant strain obviously carries a mutation in a gene that is necessary for mating by an α cell. The question is whether an **a** cell carrying this mutation will also be defective in mating. In practice, the question amounts to constructing a recombinant that carries this mutation, which we shall call X^- for the purpose of discussion, and the **a** allele of MAT. This feat is accomplished by mating the original mating-deficient strain to an **a** strain (selecting for the rare matings between the two types of cells by selection for cells that can grow under the appropriate nutritional conditions). The resulting diploids are then induced to sporulate and the meiotic products grown up and analyzed. If the X^- mutation does not affect mating by MAT**a** cells, then *all* tetrads will contain two segregants that mate as **a**. However, if the X^- mutation does affect mating by **a** cells, there will be tetrads that do *not* contain two **a** segregants per tetrad. (If we assume that the X^- mutation and MAT**a** are unlinked to each other—a reasonable assumption for yeast, which has 16 chromosomes and a high rate of recombination—the ratio of tetrads with two **a**, one **a**, and no **a** segregants would be 1:4:1 in this case).

Mutations that are necessary for mating only in α cells define α-specific genes (αsg), and those necessary for mating only in **a** cells define **a**-specific genes (**a***sg*). By this type of analysis, four α-specific genes have been identified: STE3, STE13, TUP1, and KEX2. At least three **a**-specific genes have likewise been identified: STE2, STE6, and STE14. There are six genes that are not mating-type specific: STE4, STE5, STE7, STE11, STE12, and STE15.

Although it is premature to make a firm statement, it is likely that there are not many more genes of this type to be identified. The best information on this point is provided by work in progress by Kathy Wilson, who is carrying out an extensive search for **a**-specific mutations. She has identified many new alleles of STE2, 6, and 14, as well as alleles of perhaps one or two additional **a**-specific genes. Thus there will probably be only three or four or so **a***sg*. As noted below, it is likely that there are other genes whose products are specific to **a** or α cells but which have not yet been identified because they are redundant and thus not readily identifiable by mutation.

The genes distinct from the mating-type locus are natural candidates for genes that might be under control by the mating-type locus. The specific hypothesis by which the mating-type locus regulates these genes is derived from consideration of properties of mutations of the mating-type locus itself. The experiments, observations, and arguments are described elsewhere and will not be repeated here, except to note some basic and striking findings and to make some simple points:

1. There are two complementation groups in MATα: These are denoted $\alpha 1$ and $\alpha 2$.

2. Mutants defective in the MAT$\alpha 2$ gene exhibit several properties of **a** cells. We thus argue that the role of the wild-type $\alpha 2$ product is to keep the **a**-specific functions quiet.

3. Mutants deficient in $\alpha 1$ don't do much of anything (that is, they mate neither as α nor **a**). We thus argue that the role of the wild-type $\alpha 1$ product is to induce expression or activity of the α-specific genes.

4. The only mutations known in MAT**a** affect sporulation and not mating: Mutants defective in this function (mat**a**1) mate fine as **a** but are unable to form sporulating diploids (that is, mat**a**1/MATα diploids don't sporulate). Mutants defective in the MAT$\alpha 2$ gene also are unable to sporulate after mating with **a** cells (to form MAT**a**/mat$\alpha 2$ diploids).

THE $\alpha 1$-$\alpha 2$ HYPOTHESIS

As indicated above, the $\alpha 1$-$\alpha 2$ hypothesis (Fig. 1) [Strathern et al, 1981] was proposed on the basis of genetic arguments. It has been (and is being) tested genetically [Strathern et al, 1981; Sprague et al, 1981b; Sprague and

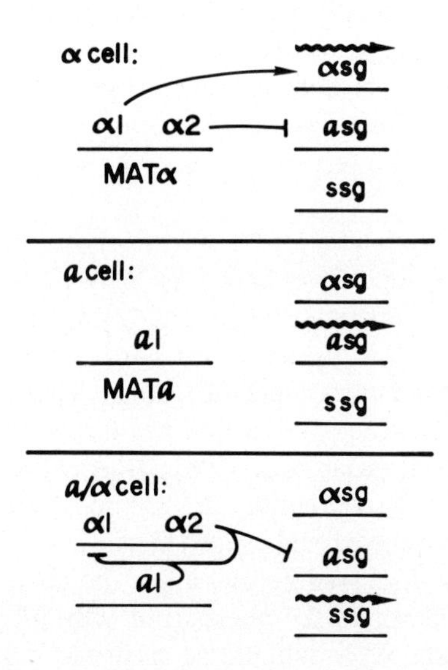

Fig. 1. The $\alpha 1$-$\alpha 2$ model for control of yeast cell type. The mating type locus is drawn to the left and other genes are drawn to the right for an α cell (top), an **a** cell (middle), and an **a**/α cell (bottom). *αsg* (α-specific genes) and ***a**sg* (**a**-specific genes) are described in the text; *ssg* are sporulation-specific genes and other genes whose activity is specific to **a**/α cells. Wavy line indicates expression (RNA production) of the indicated locus. Line with pointed arrowhead indicates activation of indicated gene or gene product; line with blunt arrowhead indicates inhibition of indicated gene or gene product. Other information is given in the text (reprinted with permission from Cold Spring Harbor Laboratory Press, pp 181–209, 1981).

Herskowitz, 1981] and biochemically [Sprague et al, 1983; Kathy Wilson, unpublished]: first, the hypothesis stated as if it dropped from a tree. Then a few words on its status.

The α Cell Type

MATα codes for two products, α1 and α2. α1 is a positive regulator of at least some α-specific genes; α2 is a negative regulator of at least some **a**-specific genes. Consequently, α cells express *αsg* but not **a***sg*. (Why α cells and **a** cells do not sporulate is discussed below.)

The a Cell Type

MAT**a** does not contain analogues to α1 and α2 (that is, one product to activate expression of **a**-specific genes and another to negatively regulate expression of α-specific genes). Rather, **a**-specific genes are simply expressed constitutively. α-Specific genes are not expressed due to the absence of the α1 product; **a**-specific genes are free to be expressed due to the absence of the α2 product. (This view gives rise to the prediction that a double mutant defective in both α1 and α2 genes would have the mating behavior of an **a** cell. This turns out to be true.)

The a/α Cell Type

There are two key questions here: (1) Why *don't* these cells mate (or exhibit any of the associated processes); and (2) why *do* these cells sporulate? We argue that α2 does in **a**/α cells just what it does in α cells—turn off expression of **a***sg*. And the α-specific genes? There are several different possibilities, but a sufficient explanation has turned out to be that **a**/α cells don't produce α1 RNA [Klar et al, 1981; Nasmyth et al, 1981]; hence, as hypothesized from our first premise, *αsg* are not expressed. This turn-off of α1 RNA requires both the **a**1 and α2 products.

As will be briefly discussed below, all of the above points are on rather firm footing. The answer to the question of why **a**/α cells sporulate is on softer footing, although we would not be surprised if our explanation turned out to be correct. In this case, we propose [Rine et al, 1981] that haploid cells produce a product (coded by the RME gene [Kassir and Simchen, 1976]) that inhibits sporulation in some manner. Perhaps it represses expression of a gene that is essential for sporulation or inhibits the activity of a protein essential for sporulation. According to our hypothesis, this inhibitor is not produced in **a**/α cells, because its synthesis is turned off by the action of **a**1 and α2.

TESTING THE HYPOTHESIS

Genetic and physiologic tests (for example, the behavior of the matα1 matα2 double mutant) have given us a lot of confidence that this formal framework for understanding cell type determination in yeast is correct. But this type of analysis does not indicate the level at which α1, α2, and **a**1-α2 act. We have therefore cloned several genes and tested directly whether synthesis of their RNA product is observed.

α Cells and α-Specific Genes

Of the four α-specific genes, expression of two (TUP1 and KEX2) is clearly not limited to α cells: MAT**a** kex2$^-$ and MAT**a** tup1$^-$ cells as well as MAT**a**/MATα cells carrying these mutations all exhibit a variety of mutant phenotypes. The answer for the other α-specific genes has come from cloning these genes by complementation and using these cloned genes to assay their corresponding RNA species. Interestingly, these two genes are regulated differently.

Expression of the STE3 gene is limited to α cells [Sprague et al, 1983]: Its RNA is not found in **a** and **a**/α cells. It was further found that production of STE3 RNA requires the α1 product of MATα (but not the α2 product). We see that the proposed role of α1 as a positive regulator of α-specific genes (at least for one of them) is borne out in this case: α1 acts to stimulate RNA synthesis.

The situation for STE13 is different [George Sprague, Jr, and Rob Jensen, unpublished observations]: Its RNA is produced in all three cell types. In α cells, the STE13 gene product is required for mating and for α-factor synthesis [Rine, 1979; Sprague et al, 1981b]. Its role in **a** and in **a**/α cells is not known.

Thus one of the four α-specific genes is known to be regulated by the mating-type locus. (We note that without RNA analysis, as has been done for STE3 and STE13, we cannot know whether α cells exhibit greater expression of KEX2 and TUP1 than the other two cell types.)

Are there other genes regulated like STE3? Are there other α-specific genes yet to be identified? α-Factor of course is an α-specific function, which ought to be encoded by an α-specific gene and which is an obvious candidate for a gene that might be regulated like STE3. None of the known mutants defective in mating define the α-factor structural gene. We have, however, been able to clone a gene for α-factor and have begun to learn something about its structure and regulation. The structure of the α-factor gene (MFα) has allowed us some inferences about the pathway of processing of α-factor; though interesting, this is not germane to the present discussion, and the reader is referred elsewhere [Kurjan and Herskowitz, 1982]. From preliminary RNA analysis, it has recently been observed that RNA species homologous to the α-factor gene (presumably the α-factor RNA itself) are regulated in a manner like STE3: α-limited and α1-dependent [Rob Jensen, Kathy Wilson, and Stan Fields, unpublished observations]. Southern hybridization analysis indicates that there are additional sequences that are homologous to the α-factor gene that we have cloned. Hence there may be multiple α-factor genes in the yeast genome. The existence of multiple α-factor genes would certainly provide an explanation for why mutants defective in the α-factor structural gene have not yet been identified by standard genetic methods (making of course the additional, reasonable assumption that an α-factor defective mutant would be defective in mating).

a Cells and a-Specific Genes

The **a**-specific genes STE2, STE6, and STE14 have no known roles in α and **a**/α cells; MATα and MAT**a**/MATα strains defective in these genes exhibit no known phenotypes. Thus expression of these genes may be limited to **a** cells. Cloning of these genes and analysis of their RNAs is underway. An exciting preliminary result [Kathy Wilson, personal communication] has been obtained with a plasmid that complements a ste6 mutation and thus that is presumed to code for STE6. Its corresponding RNA is produced by **a** cells but not by α or **a**/α cells. Furthermore, it is produced by mutants defective in the MATα2 gene (but not by mutants defective in MATα1). These preliminary studies thus point the way toward STE6 being a gene whose expression is limited to **a** cells and toward α2 being a negative regulator of RNA synthesis.

a/α Cells and Various Genes

Genes that are required for mating by both **a** and α cells are candidates for differential regulation as well: haploid (**a** or α) versus diploid (**a**/α). Analysis of the STE5 gene indicates exactly this type of control: the STE5 transcript is produced in **a** and in α cells but not in **a**/α cells [Vivian MacKay, Jeremy Thorner, and Kim Nasmyth, personal communication]. In addition, we have recently shown for the HO gene (which controls the switching of yeast cassettes) that it too exhibits "haploid-diploid" regulation, being expressed in **a** and α cells but not in **a**/α diploids [Jensen et al, 1983]. Another example of this type of haploid-diploid regulation is seen for the major RNA species coded by the Ty1 element and by mutants for loci whose expression is driven by Ty1 (see Errede et al [1980]; Elder et al [1981]).

In all of these haploid-diploid types of regulation, diploidy is monitored by the genetic composition of the mating-type locus. Each mating-type locus contributes a product (**a**1 from MAT**a** and α2 from MATα) that is necessary for inhibiting expression in the diploid cell. Thus MAT**a**/mat$\alpha 2^-$, mata1^-/

MATα, MAT**a**/MAT**a**, and MATα/MATα cells all behave as their corresponding haploid cell types.

The original example of **a**1-α2 regulation documented at the RNA level, as noted above, was the inhibition of α1 RNA in **a**/α cell. The list of different genes whose expression is inhibited in **a**/α cells is quite large and is expected to get larger. In fact, we think that **a**1-α2 inhibition of yet another gene (RME) is responsible for sporulation being turned on in **a**/α cells [Rine et al, 1981].

CONCLUDING COMMENTS

We see that yeast cell type is determined by a single genetic locus, with two alleles—MAT**a** and MATα. These two alleles conspire to make it possible for yeast cells to have three different types of regulatory activities: α1, to turn on α-specific genes; α2, to turn off **a**-specific genes; and **a**1-α2, to turn off a variety of interesting genes. These activities are responsible for the three different types of yeast cell. The form of the answer to our question of how cell type is determined in yeast, and how a single genetic locus can be responsible for so many differences, is thus in hand: Regulatory proteins activate and inhibit expression of genetic information that is scattered throughout the yeast genome. The form of the answer is therefore similar to the solution that phage λ employs in regulating expression of its gene sets, the use of wide-ranging regulatory proteins as agents for coordinate regulation.

Having identified certain yeast genes as coding for regulatory proteins and likewise identified potential targets for their action, we are in a position to begin determining the mechanism by which these regulatory proteins actually work. Do we think that eucaryotic activator proteins will be like those of λ? This is a loaded question because λ itself uses several different mechanisms for gene activation: cI and cII are RNA polymerase helpers, N and Q are transcription termination antagonists, and cIII may be a protease inhibitor (reviewed by Herskowitz and Hagen [1980]).

The logic of the analysis in λ and in yeast has been to identify the genes and proteins that constitute a specialized set and then to determine the mechanism for coordinate regulation. It is now possible to return to the biological questions that motivated the original work: What determines λ's choice between lytic and lysogenic growth? How do cells become differentiated during development? Answers to these questions will require learning how the regulatory proteins are themselves regulated. For example, understanding the lysis-lysogeny decision of phage λ boils down to understanding what factors control the level of activity of the cII protein. Interestingly, one way in which its level is controlled is by proteolysis. What is responsible for the orderly appearance of specialized cell types in development of multicellular eucaryotes? In yeast, the mating-type locus is a "master regulatory locus," the locus that determines its cell type. It would seem reasonable to invoke analogous master regulatory loci that determine the specialized cell types of higher eucaryotes. These loci would code for regulatory products that govern expression of the genes for specialized proteins. One additional requirement for such a master regulatory locus to play an important role in development is that it be itself regulated. The mating-type locus offers exactly this situation: Expression of the mating-type locus is governed both by genetic rearrangement (cassette switching) as well as by more "conventional" means (inhibition of $\alpha 1$ by **a**1-$\alpha 2$).

ACKNOWLEDGMENTS

I thank Kathy Wilson, Rob Jensen, and Stan Fields for allowing me to cite their unpublished results and Marilyn Hersh for preparation of the manuscript. I thank the National Institute of Allergy and Infectious Diseases of the US Public Health Service for support.

REFERENCES

Astell CR, Ahlstrom-Jonasson L, Smith M, Tatchell K, Nasmyth KA, Hall BD (1981): The sequence of the DNAs coding for the mating-type loci of Saccharomyces cerevisiae. Cell 27:15–23.

Benton WD, Davis RW (1977): Screening λ*gt* recombinant clones by hybridization to single plaques in situ. Science 196:180–182.

Derman E, Krauter K, Walling L, Weinberger C, Ray M, Darnell JE Jr (1981): Transcriptional control in the production of liver-specific mRNAs. Cell 23:731–739.

Elder RT, St. John TP, Stinchcomb DT, Davis RW (1981): Studies on the transposable element Ty1 of yeast. I. RNA homologous to Ty1. Cold Spring Harbor Symp Quant Biol 45:581–584.

Errede B, Cardillo TS, Sherman F, Dubois E, Deschamps J, Wiame JM (1980): Mating signals control expression of mutations resulting from insertion of a transposable repetitive element adjacent to diverse yeast genes. Cell 22:427–436.

Forbes D, Herskowitz I (1982): Polarity suppression by the Q gene product of bacteriophage λ. J Mol Biol 160:549–569.

Grayhack EJ, Roberts JW (1982): The phage λ Q gene product: Activity of a transcription antiterminator in vitro. Cell 30:637–648.

Hawthorne DC (1963): A deletion in yeast and its bearing on the structure of the mating type locus. Genetics 48:1727–1729.

Herskowitz I, Hagen D (1980): The lysis-lysogeny decision of pahge λ: Explicit programming and responsiveness. Annu Rev Genet 14:399–445.

Herskowitz I, Oshima Y (1981): Control of cell type in Saccharomyces cerevisiae: Mating type and mating-type interconversion. In Strathern JN, Jones EW, Broach JR (eds): "Molecular Biology of the Yeast Saccharomyces: Life Cycle and Inheritance." Cold Spring Harbor: Cold Spring Harbor Laboratory Press, pp 181–209.

Hicks JB, Herskowitz I (1977): Interconversion of yeast mating types. II. Restoration of mating ability to sterile mutants in homothallic and heterothallic strains. Genetics 85:373–393.

Jensen R, Sprague GF Jr, Herskowitz I (1983): Regulation of yeast mating type interconversion: Feedback control of HO gene expression by the mating type locus. Proc Natl Acad Sci USA (in press).

Kassir Y, Simchen G (1976): Regulation of mating and meiosis in yeast by the mating-type region. Genetics 82:187–206.

Klar AJS, Strathern JN, Broach JR, Hicks JB (1981): Regulation of transcription in expressed and unexpressed mating type cassettes of yeast. Nature 289:239–244.

Kurjan J, Herskowitz I (1982): Structure of a yeast pheromone gene (MFα): A putative α-factor precursor contains four tandem copies of mature α-factor. Cell 30:933–943.

MacKay VL, Manney TR (1974): Mutations affecting sexual conjugation and related processes in Saccharomyces cerevisiae. II. Genetic analysis of nonmating mutants. Genetics 76:273–288.

Miller HI, Abraham J, Benedik M, Campbell A, Court D, Echols H, Fischer R, Galindo JM, Guarneros G, Hernandez T, Mascarenhas D, Schindler D, Schmeissner U, Sosa L (1981): Regulation of the integration-excision reaction by bacteriophage λ. Cold Spring Harbor Symp Quant Biol 45:439–445.

Nasmyth KA, Tatchell K, Hall BD, Astell C, Smith M (1981): A position effect in the control of transcription at yeast mating type loci. Nature 289:244–250.

Rine JD (1979) Regulation and transposition of cryptic mating type genes in Saccharomyces cerevisiae. PhD thesis (Biology). University of Oregon, Eugene, Oregon.

Rine J, Sprague GF Jr, Herskowitz I (1981): rme1 mutation of Saccharomyces cerevisiae: Map position and bypass of mating type locus control of sporulation. Mol Cell Biol 1:958–960.

Sprague GF Jr, Herskowitz I (1981): Control of yeast cell type by the mating type locus. I. Identification and control of expression of the **a**-specific gene, BAR1. J Mol Biol 153:305–321.

Sprague GF Jr, Rine J, Herskowitz I (1981a): Homology and non-homology at the yeast mating type locus. Nature 289:250–252.

Sprague GF Jr, Rine J, Herskowitz I (1981b): Control of yeast cell type by the mating type locus. II. Genetic interactions between MATα and unlinked α-specific STE genes. J Mol Biol 153:323–335.

Sprague GF Jr, Jensen R, Herskowitz I (1983): Control of yeast cell type by the mating type locus: Positive regulation of the α-specific STE3 gene by the MATα1 product. Cell 32:409–415.

Strathern J, Hicks J, Herskowitz I (1981): Control of cell type in yeast by the mating type locus: The α1-α2 hypothesis. J Mol Biol 147:357–372.

Gene Structure and Regulation in Development, pages 77–93

Protein Kinases Encoded by Avian Sarcoma Viruses and Some Related Normal Cellular Proteins

R.L. Erikson, Tona M. Gilmer, Eleanor Erikson, and J.G. Foulkes

Department of Pathology, University of Colorado School of Medicine, Denver, Colorado 80262 (R.L.E., T.M.G., E.E.) and Massachusetts Institute of Technology Center for Cancer Research, Cambridge, Massachusetts 02139 (J.G.F.)

INTRODUCTION

In this paper we will review some of the progress that has been made over the past several years in the understanding of viral transformation of cells and particularly transformation caused by RNA tumor viruses. Individual RNA tumor viruses have been isolated that cause specific types of neoplastic disease, specific types of leukemia, and/or fibrosarcomas. The first RNA tumor virus to be isolated and identified as such was the Rous sarcoma virus (RSV), which causes transformation of cells in culture as well as fibrosarcomas when injected into suitable host animals (for reviews, see Hanafusa [1977]; Bishop [1978]). The genome of RSV has been intensively studied for the past 10–12 years, and the organization of the genes along the RNA of the RSV genome is shown below. The transforming gene is denoted *src* for sarcoma and indicates that the virus itself carries a gene responsible for malignant transformation. This gene was identified genetically about 12 years ago through the isolation of temperature-sensitive mutants that were conditional for cell transformation in culture. These mutants were able to transform cells at a temperature of 35°C, the permissive temperature, but were unable to transform cells morphologically at 41°C, the nonpermissive temperature. The existence of these temperature-sensitive mutants implied that the product of the viral transforming gene, in the case of RSV, was a protein [Kawai and Hanafusa, 1971]. In addition to temperature-sensitive mutants, nonconditional mutants were isolated that had deletions of the *src* gene. These mutants are unable to transform cells in culture or to cause fibrosarcomas under most conditions.

About 4 years ago, the product of the *src* gene was identified as a phosphoprotein of M_r = 60,000; this protein was denoted $pp60^{src}$ [Purchio et al, 1978]. Very soon after the identification of the protein we and others ascribed to this molecule the function of protein phosphorylation [Collett and Erikson, 1978; Levinson et al, 1978]. However, there was considerable concern on our part, that the enzymatic activity observed could potentially be due to contamination of preparations of the viral *src* gene product by one of the many protein kinases encoded by the cells used for infection and transformation by RSV, the starting material for our preparations of $pp60^{src}$. Thus, a great deal of our effort over the past several years has been devoted to attempting to determine unambiguously the source of the enzymatic activity observed, because in order to understand cell transformation by RNA tumor viruses, it is important to define the exact nature of the functions that are viral as opposed to cellular. The RSV genome and the expression of the *src* gene is illustrated below:

Rous Sarcoma Virus Genome

```
5′ — gag — pol — env — src — 3′        39S RNA
                         ↓
                  5′ — src — 3′        21S
                                       mRNA
                         ↓
                    P         P
                    |         |
              N — ser — tyr — C        pp60src
```

RESULTS AND DISCUSSION

Expression of the RSV src Gene in Escherichia coli

One approach to determining whether or not $pp60^{src}$ was indeed a protein kinase was to obtain expression of the RSV *src* gene in a prokaryote, such as E coli. These organisms are believed to produce few, if any, protein kinases that carry out reversible protein phosphorylation, and thus if the product of the *src* gene expressed in E coli exhibited phosphotransferase activity, it would be strong evidence that the enzymatic activity observed was actually encoded by the *src* gene. The construction of plasmids for the generation of $p60^{src}$ in E coli is shown in Figure 1 [Gilmer et al, 1982]. In this case the RSV *src* gene was placed under the control of the *lac* promoter-operator in the hope of obtaining efficient transcription and, presumably, translation. Bacteria that expressed $p60^{src}$ were detected by immunoprecipitation of ^{35}S-labeled proteins and those E coli that were expressing $p60^{src}$ were used for

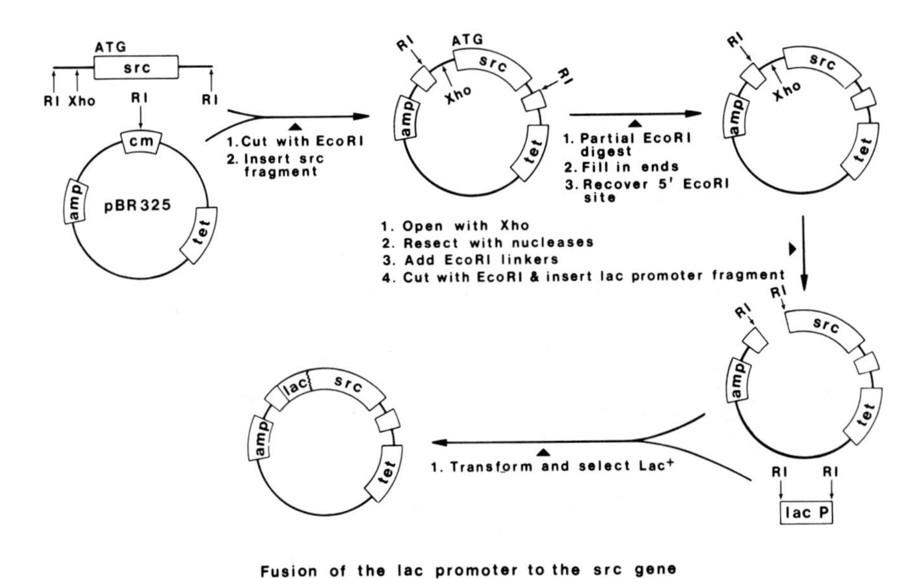

Fig. 1. Plasmid construction. The *Eco*RI fragment containing the *src* gene was isolated from a plasmid containing the entire Prague A-Rous sarcoma virus (RSV) genome and inserted into the *Eco*RI site of pBR325. The recombinant plasmid was subjected to partial *Eco*RI digestion, treatment with DNA polymerase I to fill in the *Eco*RI ends, and ligation with T4 DNA ligase to recircularize the molecule. After transformation of HB101, a plasmid that retained only the 5′ *Eco*RI site was recovered. This plasmid was digested with *Xho* I and resected with *Exo* III and S1 nuclease. *Eco*RI linkers were added and the mixture was digested with *Eco*RI. The *Eco*RI fragment, containing the UV5 *lac* promoter and the first seven amino acids of β-galactosidase, was inserted by ligation. HB101 was transformed with the ligation mixture and plated on 5-chloro-4-bromo-3-indolyl β-D-galactoside plates. Colonies containing the *lac* promoter fragment were identified by their blue color.

purification of the *src* gene product. The purification of p60^{src} from an expressor culture of E coli and a similar preparation from a control culture carrying only the *lac* promoter-operator but not the *src* gene is shown in Figure 2. When these preparations were tested for protein kinase activity, as shown in Figure 3, we found that, indeed, the protein produced in E coli did display protein kinase activity. It had the capacity to phosphorylate casein, α- and β-tubulin, and anti-pp60^{src} IgG. Thus, it had many characteristics identical to the protein, produced in eukaryotic host cells, that we had characterized previously. Furthermore, in each of the cases shown in this figure, the amino acid phosphorylated in the protein substrates was tyrosine [Gilmer and Erikson, 1981; Hunter and Sefton, 1980; Collett et al, 1980]. Thus, a number of characteristics suggest that the previously observed phos-

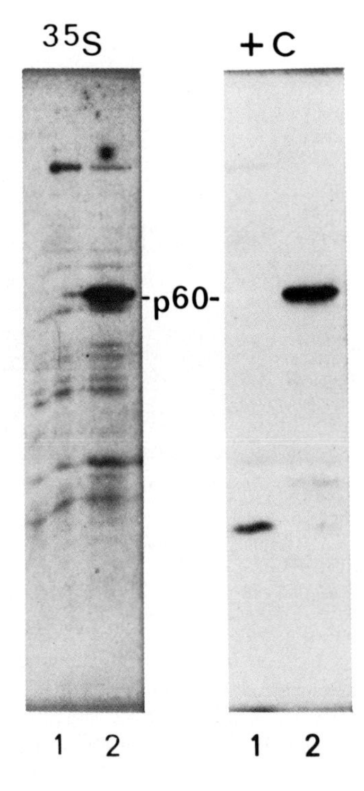

Fig. 2. Polyacrylamide gel electrophoretic analysis of bacterial extracts after immunoaffinity chromatography. Escherichia coli expressing p60 was detected by immunoprecipitation of ^{35}S-labeled extracts from cells containing the *lac* UV5 promoter-operator, and one of these cultures was used for the studies described here. Escherichia coli carrying the *lac* UV5 promoter in pBR325 in the same orientation as in the *src*-containing clone served as a control for the expression of p60. Lysates were prepared and subjected to immunoaffinity chromatography. ^{35}S-labeled proteins were visualized by fluorography and ^{32}P-labeled proteins by autoradiography with the aid of Dupont Lightning Plus intensifying screens. Left panel) Proteins purified by immunoaffinity chromatography of lysates prepared from bacteria labeled in culture with $^{35}SO_4^{2-}$. Track 1, *lac* pBR325; track 2, *lac-src* pBR325. Right panel) The preparations depicted in the left panel were incubated with CAT (+C) in the presence of 5 mM $mgCl_2$ and 1 μM [γ-^{32}P]ATP before analysis. Track 1, *lac* pBR325; track 2, *lac-src* pBR325. The catalytic subunit of cyclic AMP-dependent protein kinase (CAT) was a gift of James L. Maller.

photransferase activity closely associated with $pp60^{src}$ isolated from eukaryotic host cells was, in fact, an intrinsic property of the molecule itself. These data, taken together with those previously published by our laboratory, and others, lead to the near certain conclusion that $pp60^{src}$ encodes a protein kinase.

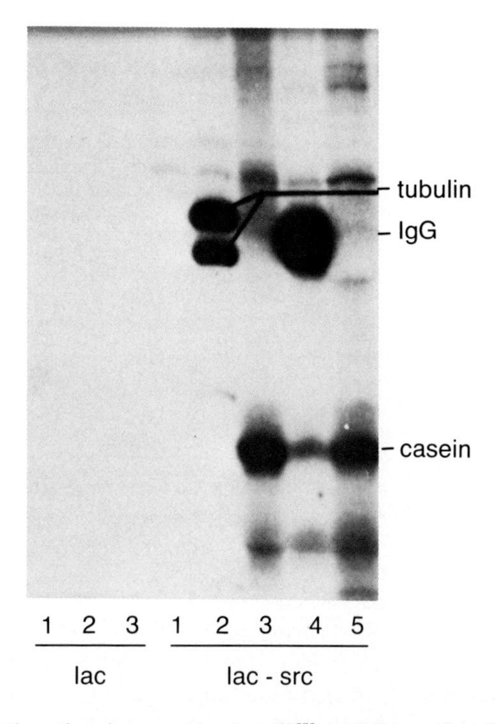

Fig. 3. Phosphorylation of various proteins by p60^{src} partially purified from Escherichia coli. 10 μl (10 ng) of the preparations described in Figure 2 were incubated at room temperature for 10 min with the indicated additions in a total volume of 25 μl. $MgCl_2$ and [γ-^{32}P]ATP (1,000–6,000 Ci mmol) were added to a final concentration of 5 mM and 1 μM, respectively, and incubation was continued for 30 min at room temperature. Proteins were then resolved by polyacrylamide gel electrophoresis and autoradiography. Left side (*lac*)) Track 1, *lac* pBR325 alone; track 2, plus α- and β-tubulin; track 3, plus casein. Right side (*lac-src*)) Track 1, *lac-src* pBR325 alone; track 2, plus α- and β-tubulin; track 3, plus casein; track 4, plus casein and anti-pp60^{src} IgG; track 5, plus casein and nonimmune IgG. Casein was present at 1 mg/ml, the IgGs at 350 μg/ml, and tubulin at 100 μg/ml.

A number of other avian sarcoma viruses (ASVs) have been now identified and characterized in the laboratories of Professor Hanafusa at the Rockefeller Institute, and Professor Peter Vogt at the University of Southern California. To date, there are at least three other classes of ASV that have transforming genes distinct from that of RSV and that encode distinct transforming gene products. Although they are antigenically distinct from pp60^{src}, these transforming gene products apparently have a functional similarity to pp60^{src} in that they are able to phosphorylate protein substrates at tyrosine residues. As illustrated in Table I, all four classes of ASVs have associated with them a

TABLE I. Protein Kinase Activity Associated With the Transforming Gene Products of Avian Sarcoma Viruses

Class	Avian sarcoma virus	Transforming gene product	Immunoprecipitated by		Phosphorylates TBR IgG on Tyrosine
I	Rous	$pp60^{src}$	TBR	+	+
			anti-gag	−	
II	Fujinami	$P140^{gag\text{-}fps}$	TBR	−	+
	PRCII	$P105^{gag\text{-}fps}$	anti-gag	+	
III	Y73	$P90^{gag\text{-}yes}$	TBR	−	+
	Esh	$P80^{gag\text{-}yes}$	anti-gag	+	
IV	UR2	$P68^{gag\text{-}ros}$	TBR	−	+
			anti-gag	+	

For references see Breitman et al [1981]; Feldman et al [1980, 1982]; Ghysdael et al [1981a, 1981b]; Neil et al [1981]; Pawson et al [1980]; Kawai et al [1980]; Lee et al [1980].

protein kinase activity that, when present in an immune complex with anti-*src* IgG from a RSV-tumor-bearing rabbit, results in the efficient phosphorylation of the heavy chain of IgG; thus, anti-$pp60^{src}$ IgG is a good substrate for the protein kinases encoded by other classes of ASV. There is no clear-cut explanation as to why this IgG is such a good substrate, but the result does demonstrate that it is likely that all classes of ASVs studied to date encode a protein kinase specific for tyrosine residues. In addition, the transforming proteins themselves become phosphorylated in the reaction, and tyrosine appears to be the sole phosphorylated residue.

Autophosphorylation of $pp60^{src}$

As shown previously [Collett et al, 1980], $pp60^{src}$ purified from eukaryotic sources is able to undergo self-phosphorylation or autophosphorylation when [γ-^{32}P]ATP is added to the purified protein. In addition, in the immune complex reaction, all the transformation-specific proteins indicated in Table I also undergo self-phosphorylation on tyrosine residues. However, there is no evidence, to date, that $p60^{src}$ produced in E coli, the prokaryotic host carrying the recombinant *src*-containing plasmid, is able to self-phosphorylate. Since we believe that the phosphorylation of $pp60^{src}$ itself on a tyrosine residue is likely to be important in the regulation of the enzymatic activity of the protein, we have attempted to elucidate the source of the enzyme responsible for the phosphorylation of the tyrosine residue of $pp60^{src}$.

As shown in Figure 4, the autophosphorylation reaction of enzyme purified from temperature-sensitive transformation mutant-infected cells is more

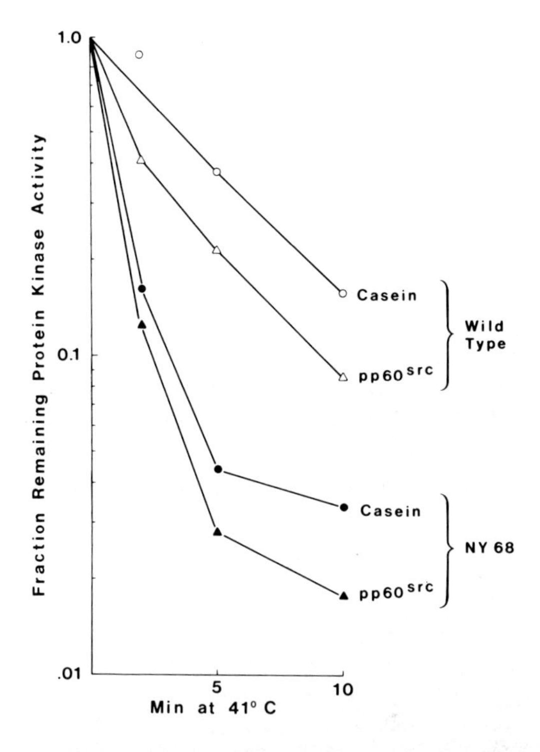

Fig. 4. Thermolability of pp60^{src} protein kinase activity. Chicken embryo fibroblasts were transformed with either wild-type Schmidt-Ruppin strain of RSV or the ts mutant of this strain, NY68. The cells were maintained at the permissive temperature, 35 °C. Immunoaffinity chromatography was used to prepare pp60^{src} from these cells. These preparations were heat-inactivated at 41 °C for the indicated times and then phosphotransferase reactions were carried out in the presence of 5 mM Mg^{-2} and 1 μM [γ-^{32}P]ATP (350 Ci/mmol) in the absence of any exogenously added substrates or in the presence of casein (1 mg/ml). After polyacrylamide gel electrophoresis, the phosphorylated proteins were localized by autoradiography, the bands were excised, radioactivity was determined, and percentage activity was calculated from the amount of radioactivity in the bands.

thermolabile than that from wild-type infected cells. These results could be interpreted to mean that pp60^{src} is responsible for its own phosphorylation. Unfortunately, an alternative explanation is that the molecule, when denatured, is no longer a good substrate for a putative pp60^{src}-specific kinase. One way to resolve this issue would be to isolate a kinase that is specific for phosphorylating the correct tyrosine residue in pp60^{src}. However, to date, although we have identified other tyrosine-specific protein kinases in eukaryotic host cells infected by RSV, we have been unable to show with certainty that any of these protein kinases are able to phosphorylate $pp60^{src}$ and are, in fact, the $pp60^{src}$-specific kinase. Thus this issue is still unresolved, although it would appear, from the study of the eukaryotic enzyme, that, most likely, $pp60^{src}$ is, in fact, capable of self-phosphorylation.

Production of Antibody Against $pp60^{src}$ Using Antigen Produced in Prokaryotes

Escherichia coli expressing $pp60^{src}$, under the control of the *lac* promoter-operator, produce a substantial amount of protein, thus making it relatively easy to purify large amounts of antigen in a denatured form for subsequent injection and production of antibody. This exercise was carried out and adult rabbits, injected with denatured $pp60^{src}$, produced antibody that was able to recognize not only native $pp60^{src}$ but also denatured $pp60^{src}$, unlike tumor-bearing rabbit serum that had been available to us previously. This result is shown in Figure 5. Normal rabbit serum (NRS) and tumor-bearing rabbit serum (TBR) are unable to precipitate denatured $pp60^{src}$, whereas the serum taken from rabbits injected with the denatured E coli protein readily immunoprecipitates denatured $pp60^{src}$ (αp60).

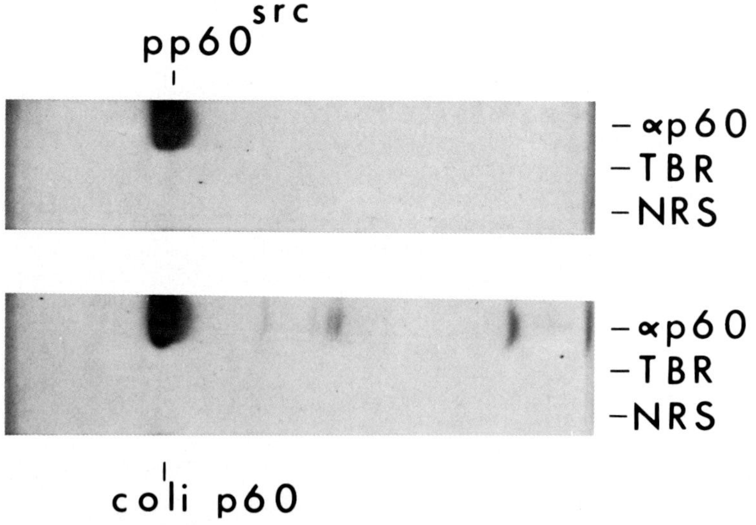

Fig. 5. Immunoprecipitation of denatured p60 from eukaryotic and prokaryotic cells. A preparation of $pp60^{src}$ from eukaryotic cells was phosphorylated by the addition of [γ-^{32}P]ATP as described in the legend of Figure 4, and $p60^{src}$ from E coli was phosphorylated as described in the legend to Figure 2. Radiolabeled protein from both sources was denatured by heating in 0.1% sodium dodecyl sulfate (SDS), and after the addition of NP40 to 1% and sodium deoxycholate to 0.5%, the preparations were immunoprecipitated as indicated for each track.

Normal Cellular Homologues of Viral-Transforming Gene Products

Some rabbits bearing RSV-induced tumors produce antibody that also recognizes, as shown in Figure 6, a phosphoprotein of M_r = 60,000 from normal uninfected cells. This protein, which has been denoted $pp60^{c\text{-}src}$, is presumed to be the product of the nucleic acid sequences homologous to the RSV *src* gene that are present in normal uninfected cells [see Collett et al, 1978], although no direct firm genetic evidence is yet available regarding this question. This molecule, the pp60 homologue of viral *src* in normal, uninfected cells, has also been purified in our laboratory and has been shown to have, at least qualitatively, all of the properties of the viral-transforming gene product. The only major difference is quantative. We find that virus-infected cells produce 40–50-fold more $pp60^{src}$ than would normally be expressed in these cells if only the cellular gene were functioning. Thus, one might imagine that transformation occurs because of a dosage phenomenon, because of the increased level of a protein that is very much like a normal

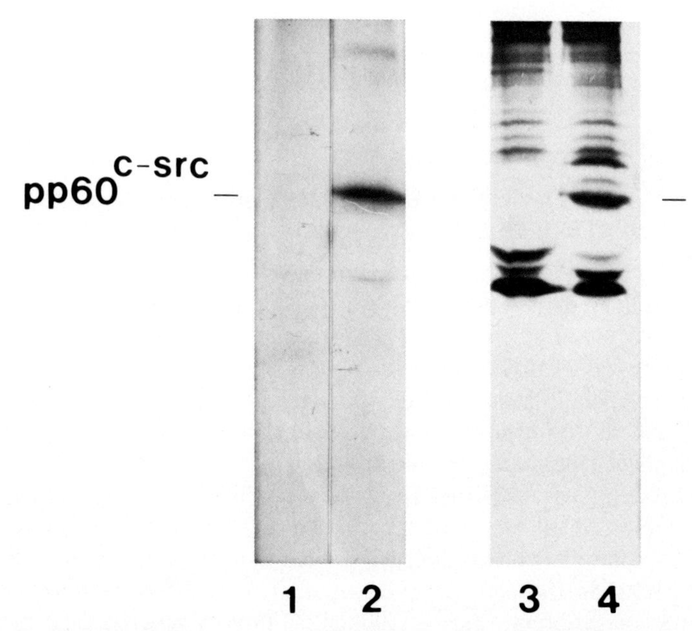

Fig. 6. Immunoprecipitation of the normal cellular homologue of $pp60^{src}$. Chicken embryo fibroblasts were labeled with either [^{32}P] (lanes 1 and 2) or [^{35}S]methionine (lanes 3 and 4). After immunoprecipitation of extracts with either normal serum (lanes 1 and 3) or cross-reacting TBR serum (lanes 2 and 4), the samples were electrophoresed in SDS-polyacrylamide gels and autoradiographed.

cell protein. The *c-src* protein is very highly conserved, being very similar in both avian and mammalian cells and, because of this high degree of conservation, it is presumed to play some crucial or essential role in normal cell metabolism [Collett et al, 1978]. Recently, a normal cellular protein has been identified that is related to the Fujinami sarcoma virus-transforming protein [Mathey-Prevot et al, 1982].

A summary of our understanding of the structure of $pp60^{c\text{-}src}$ and $pp60^{v\text{-}src}$ is shown below. Both molecules have a serine residue that is phosphorylated by a cyclic AMP-dependent protein kinase present in all host cells of RSV. We presume, but do not know for sure, that the phosphorylation of this serine residue has important implications concerning the function of $pp60^{src}$. To date, however, no strong quantitative data are available to suggest whether phosphorylation at the serine residue increases or decreases the enzymatic activity of the molecule. Furthermore, both molecules contain a phosphorylated tyrosine residue located somewhere in the carboxy terminus.

```
                P                        P
                |                        |
N      —       ser       ———————        tyr       —       COOH
                ↑                        ↑
         cAMP-dependent          cAMP-independent
```

Epidermal Growth Factor (EGF)-Stimulated Phosphorylation and Comparison to ASV-Induced Phosphorylation

One of the proteins that we and others have studied that appears to be directly phosphorylated by the activity of $pp60^{src}$ in the infected cell, or in vitro, is a molecule of M_r = 34,000 [Erikson and Erikson, 1980; Radke and Martin, 1979]. It is a fairly abundant protein in cultured fibroblasts. Because of the tyrosine-specific nature of EGF-stimulated phosphorylation [Cohen et al, 1980], and the tyrosine-specific phosphorylation of proteins by ASV-transforming gene products, we attempted to determine the specificity of these two protein kinase activities [Erikson et al, 1981]. As shown in Figure 7, when EGF is added to growing cells, the 34,000-dalton protein shows increased phosphorylation. Further, we wanted to know whether the sites of phosphorylation, in the case of EGF-stimulation or ASV-transformation, were similar or different. The 34,000-dalton protein was purified in the unphosphorylated form from the same cells that were used for the studies shown in Figure 7. Then purified $pp60^{v\text{-}src}$ or $pp60^{c\text{-}src}$ were used to phosphorylate the protein in vitro, using [γ-^{32}P]ATP. Phosphopeptide maps were

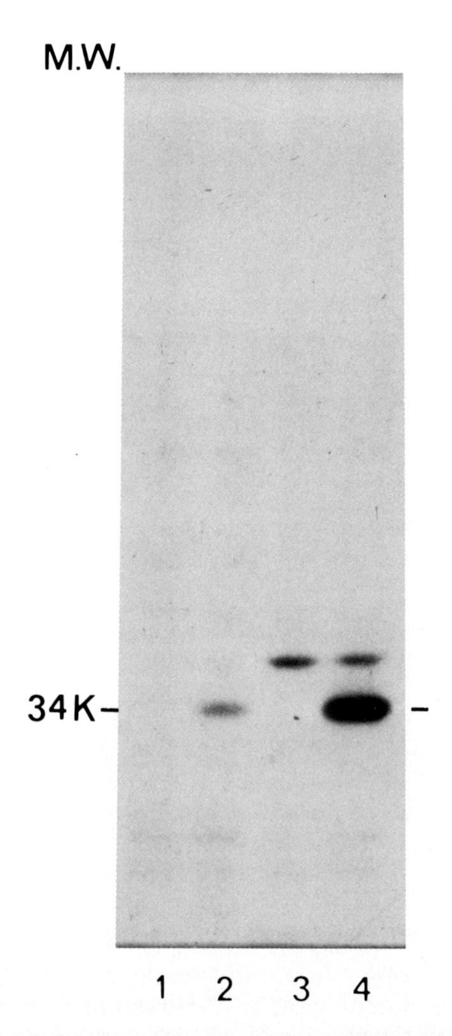

Fig. 7. Stimulation of phosphorylation of the M_r = 34,000 protein in epidermal growth factor (EGF)-treated cells. The medium was removed from cultures of A-431 cells that had just reached confluency. [^{32}P]orthophosphate-containing medium was added, and after 30 min EGF was added to one culture to a final concentration of 400 ng/ml. Two hours later the cells were harvested, the lysates were passed through DEAE-Sephacel, and samples of the flow-through fractions were immunoprecipitated with preimmune or anti-34K serum. The immunoprecipitated proteins were resolved by SDS-polyacrylamide gel electrophoresis and visualized by autoradiography. Lanes 1 and 2, cells radiolabeled in the absence of EGF; lanes 3 and 4, cells radiolabeled in the presence of EGF; lanes 1 and 3, preimmune serum; lanes 2 and 4, anti-34K serum.

prepared and compared to the phosphopeptide maps of the protein that had been phosphorylated after the addition of EGF to growing cells. As shown in Figure 8, we found two major phosphopeptides after tryptic digestion, and, indeed, in all three cases, these peptides were the same. Thus, although this experiment does not identify the enzyme responsible for phosphorylation of the 34,000-dalton protein in EGF-stimulated cells, it does show that the kinase responsible for this phosphorylation has the same specificity as $pp60^{src}$. This illustrates the very similar nature of the two types of phosphorylation and suggests that the 34,000-dalton protein is, perhaps, a common substrate for the different types of kinases under study.

Clearly, in the identification of substrates for any of the growth-regulated phosphorylations that we are discussing here, one must show some functional change associated with the phosphorylation observed. In order to understand the functional significance of the phosphorylation, one must, of course, first understand the function of the particular substrate under study. To date, we have no clear understanding of the function of the 34,000-dalton substrate described in the previous two figures, and without an understanding of that function it is difficult to assign any significance to its phosphorylation. Our best evidence, to date, is that in fibroblasts, the 34,000-dalton protein is a strong RNA-binding protein, seemingly associated with messenger RNA in the cytoplasm of growing fibroblasts. Another way of assessing the significance of a substrate for the protein kinase activities observed, is to determine the distribution of the particular protein in various tissues of the host that would normally be infected by a virus, such as RSV. When we looked at the distribution of the 34,000-dalton protein in a variety of tissues of the adult bird, we found that there is a vast difference in the expression of the protein. For instance, in red blood cells, as shown in Figure 9, it is undetectable, and there is only a low level of this protein in brain cells. On the other hand, there are intermediate levels of the protein in other tissues that were examined. This fact, of course, raises an obvious point that should be made when considering the significance of transforming proteins and their substrates, which is, that in order for viruses to be able to carry out cell transformation, one must not only have expression of the viral-transforming gene product, but must also have the presence of a suitable substrate for the activity of that product in the susceptible host cell.

The phosphorylation state of a particular protein will be a balance of the protein kinase activities and the protein phosphatase activities present in a particular cell (for review, see Krebs and Beavo [1979]). The phosphorylation state, and presumably the functional behavior, of a particular protein will be greatly influenced by these two levels, and, in order to understand the circuits involved in phosphorylation-dephosphorylation of proteins on tyrosine residues, one must be concerned not only about the specific kinases

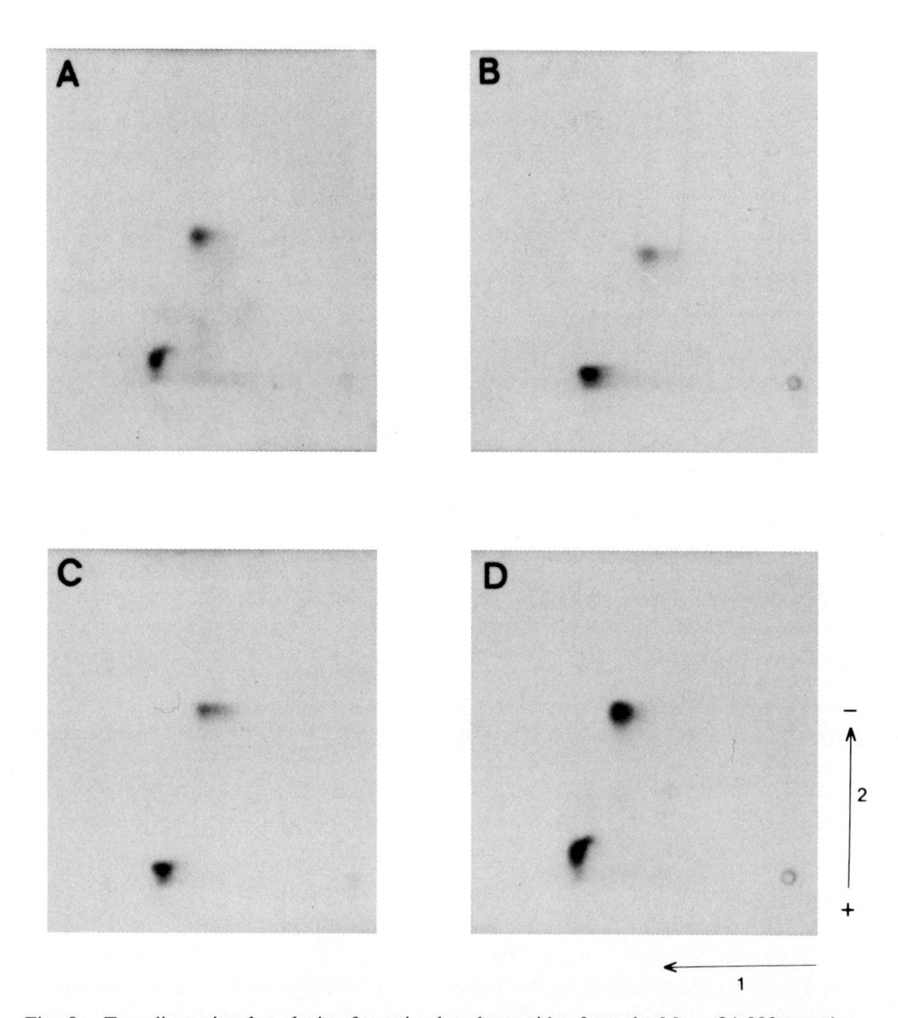

Fig. 8. Two-dimensional analysis of tryptic phosphopeptides from the M_r = 34,000 protein radiolabeled in culture in EGF-stimulated cells or phosphorylated in vitro by A-431 $pp60^{c\text{-}src}$ or by RSV $pp60^{src}$. Radiolabeled M_r = 34,000 protein was isolated from EGF-treated A-431 cells by immunoprecipitation and SDS-polyacrylamide gel electrophoresis as shown in Figure 7, lane 4. In addition, the M_r = 34,000 protein purified from A-431 cells was radiolabeled in vitro by A-431 $pp60^{c\text{-}src}$ or by RSV $pp60^{src}$ and then resolved by SDS-polyacrylamide gel electrophoresis. Tryptic digests were prepared and fractionated by ascending chromatography and electrophoresis at pH 3.5. A) M_r = 34,000 protein radiolabeled in culture in EGF-stimulated cells. B and C) M_r = 34,000 protein phosphorylated in vitro by A-431 $pp60^{c\text{-}src}$ or by RSV $pp60^{src}$, respectively. D) equal counts of A and B were mixed and fractionated together.

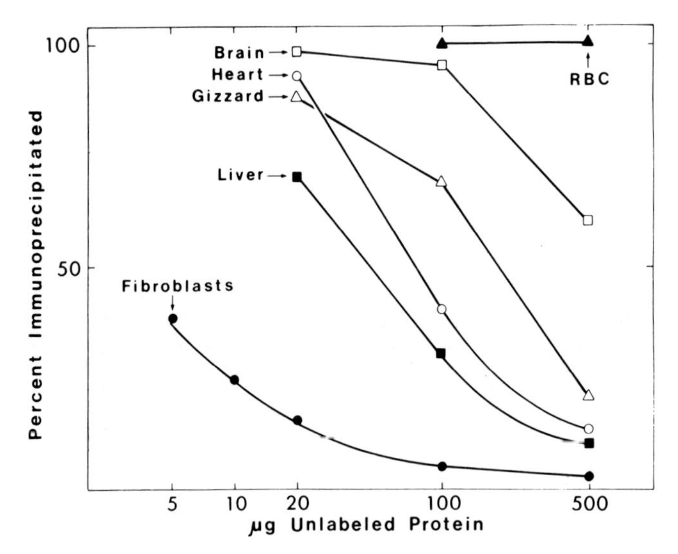

Fig. 9. Distribution of the M_r = 34,000 protein in various tissues. The indicated organs were removed from a 6-week-old chicken. Lysates were prepared from these tissues and from cultured chicken embryo fibroblasts, and samples containing from 5 to 500 μg of soluble protein were incubated from 1 hr with 1 μl of anti-34K serum. Then aliquots of a lysate prepared from ^{35}S-methionine-labeled chicken embryo fibroblasts were added and the immunoprecipitated proteins were analyzed by polyacrylamide gel electrophoresis. The 34K bands were excised and the radioactivity determined by liquid scintillation spectrometry. The 34K content of the various lysates is reflected by the ability of that lysate to block the immunoprecipitation of the radiolabeled protein.

responsible for the phosphorylation of these proteins, but also about the phosphoprotein phosphatases that are involved in their dephosphorylation.

Figure 10 illustrates results recently obtained by Gordon Foulkes, working in Denver, who studied the distribution of tyrosine-specific phosphoprotein phosphatases taken from chicken brain and fractionated on diethylaminoethyl (DEAE) cellulose. These results indicate that the phosphotyrosine-specific phosphatases are probably unique and quite different from the phosphoserine-specific phosphatases previously described.

Search for Other Substrates

The 34,000-dalton protein was identified by work in Denver, as well as by others, as a hyperphosphorylated normal cellular protein in RSV-transformed cells. However, when one examines phosphorylation patterns in transformed versus untransformed cells, many quantitative changes are ob-

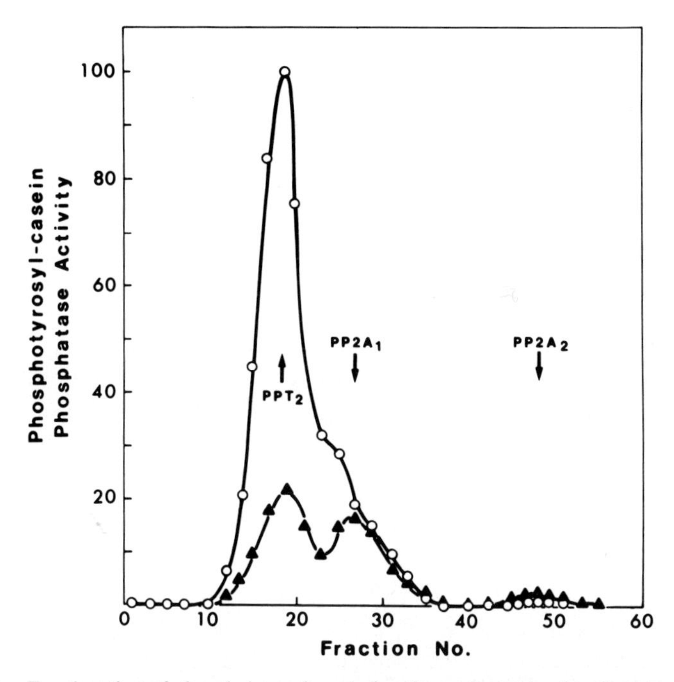

Fig. 10. Fractionation of phosphotyrosyl- and phosphoseryl-protein phosphatases on DEAE cellulose. An extract prepared from chicken brain was loaded onto DEAE cellulose and eluted with a linear gradient of 50–450 mM NaCl in 10 mM Tris, pH 7.0; 0.1 mM EDTA; 30 mM β-mercaptoethanol. Fractions were assayed for phosphatase activity using ^{32}P-phosphotyrosyl-casein as substrate. Assays were carried out in the presence of either 1 mM EDTA (○—○) or 1 mM Mn^{2+} (▲—▲). The positions of elution of the phosphoseryl-protein phosphatases $PP2A_1$ and $PP2A_2$ are indicated.

served, and most of these occur not at tyrosine residues but rather at serine residues. One dramatic example of that is the work of Decker [1981], which shows that the ribosomal protein S6 is heavily phosphorylated on serine residues in RSV-transformed cells. Even under conditions of serum starvation the phosphorylation of S6 seems to be under the control of the *src* gene product. In order to further assess the significance of this phenomenon, in collaboration with Jim Maller in the Department of Pharmacology in Denver, we undertook the microinjection of partially purified $pp60^{src}$ into Xenopus oocytes. As an assay for the activity of $pp60^{src}$ in oocytes, we chose the phosphorylation of the ribosomal protein S6. Our preliminary results show that 12 hr after the microinjection of either $pp60^{src}$, $pp60^{src}$ that had been inactivated by treatment at 60°C for 2 min, or progesterone, a hormone known to stimulate S6 phosphorylation, unheated $pp60^{src}$ is also able to stimulate the phosphorylation of S6. This result suggests other mechanisms for amplifying the activity of a viral-transforming gene product by its capacity to phosphorylate other proteins. One might imagine in this case that $pp60^{src}$ activates a protein kinase specific for serine, and that the increased activity of this serine-specific protein kinase results in phosphorylation of S6 which, in turn, leads to an increased efficiency of protein synthesis. Alternatively, $pp60^{src}$ could phosphorylate a phosphoprotein phosphatase specific for S6 dephosphorylation, inactivating it so that the protein kinases that are normally present now are able to cause an increased level of phosphorylation of S6. Either of these two mechanisms is possible at the moment, but they suggest obvious approaches to further study the process of transformation by viral-transforming gene products that encode protein kinases. Clearly, one of the most important areas of investigation in the near future will be directed at a biochemical description of the pathways that lead to neoplasia initiated by these and related viruses.

REFERENCES

Bishop JM (1978): Retroviruses. Annu Rev Biochem 47:35–88.

Breitman ML, Neil JC, Moscovici C, Vogt PK (1981): The pathogenicity and defectiveness of PRCII: A new type of avian sarcoma virus. Virology 108:1–12.

Cohen S, Carpenter G, King L Jr (1980): Epidermal growth factor-receptor-protein kinase interactions. J Biol Chem 255:4834–4842.

Collett MS, Erikson RL (1978): Protein kinase activity associated with the avian sarcoma virus *src* gene product. Proc Natl Acad Sci USA 75:2021–2024.

Collett MS, Brugge JS, Erikson RL (1978): Characterization of a normal avian cell protein related to the avian sarcoma virus transforming gene product. Cell 15:1363–1369.

Collett MS, Purchio AF, Erikson RL (1980): Avian sarcoma virus-transforming protein, $pp60^{src}$, shows protein kinase activity specific for tyrosine. Nature 285:167–169.

Decker S (1981): Phosphorylation of ribosomal protein S6 in avian sarcoma virus-transformed chicken embryo fibroblasts. Proc Natl Acad Sci USA 78:4112-4115.

Erikson E, Erikson RL (1980): Identification of a cellular protein substrate phosphorylated by the avian sarcoma virus-transforming gene product. Cell 21:829–836.

Erikson E, Shealy DJ, Erikson RL (1981): Evidence that viral transforming gene products and epidermal growth factor stimulate phosphorylation of the same cellular protein with similar specificity. J Biol Chem 256:11381–11384.

Feldman RA, Hanafusa T, Hanafusa H (1980): Characterization of protein kinase activity associated with the transforming gene product of Fujinami sarcoma virus. Cell 22:757–765.

Feldman RA, Wang L-H, Hanafusa H, Balduzzi PC (1982): Avian sarcoma virus UR2 encodes a transforming protein which is associated with a unique protein kinase activity. J Virol 42:228–236.

Ghysdael J, Neil JC, Vogt PK (1981a): A third class of avian sarcoma viruses, defined by related transformation-specific proteins of Yamaguchi 73 and Esh sarcoma viruses. Proc Natl Acad Sci USA 78:2611–2615.

Ghysdael J, Neil JC, Wallbank AM, Vogt PK (1981b): Esh sarcoma virus codes for a gag-linked transformation-specific protein with an associated protein kinase activity. Virology 111:386–400.

Gilmer TM, Erikson RL (1981): The Rous sarcoma virus transforming protein, pp60^{src}, expressed in Escherichia coli functions as a protein kinase. Nature 294:771–773.

Gilmer TM, Parsons JT, Erikson RL (1982): Construction of plasmids for the expression of the Rous sarcoma virus transforming protein, p60^{src}, in Escherichia coli. Proc Natl Acad Sci USA 79:2152–2156.

Hanafusa H (1977): Cell transformation by RNA tumor viruses. In Fraenkel-Conrat H, Wagner RP (eds): "Comprehensive Virology." New York: Plenum Publishing Corporation, pp 401–483.

Hunter T, Sefton BM (1980): The transforming gene product of Rous sarcoma virus phosphorylates tyrosine. Proc Natl Acad Sci USA 77:1311–1315.

Kawai S, Hanafusa H (1971): The effects of reciprocal changes in temperature on the transformed state of cells infected with a Rous sarcoma virus mutant. Virology 46:470–479.

Kawai S, Yoshida M, Segawa K, Sugiyama H, Ishizaki R, Toyoshima K (1980): Characterization of Y73, an avian sarcoma virus: A unique transforming gene and its product, a phosphopolyprotein with protein kinase activity. Proc Natl Acad Sci USA 77:6199–6203.

Krebs EG, Beavo JA (1979): Phosphorylation-dephosphorylation of enzymes. Annu Rev Biochem 48:923–959.

Lee W-H, Bister K, Pawson A, Robins T, Moscovici C, Duesberg PH (1980): Fujinami sarcoma virus: An avian RNA tumor virus with a unique transforming gene. Proc Natl Acad Sci USA 77:2018–2022.

Levinson AD, Oppermann H, Levintow L, Varmus HE, Bishop JM (1978): Evidence that the transforming gene of avian sarcoma virus encodes a protein kinase associated with a phosphoprotein. Cell 15:561–572.

Mathey-Prevot B, Hanafusa H, Kawai S (1982): A cellular protein is immunologically crossreactive with and functionally homologous to the Fujinami sarcoma virus transforming protein. Cell 28:897–906.

Neil JC, Ghysdael J, Vogt PK (1981): Tyrosine-specific protein kinase activity associated with p105 of avian sarcoma virus PRCII. Virology 109:223–228.

Pawson T, Guyden J, Kung T-H, Radke K, Gilmore T, Martin GS (1980): A strain of Fujinami sarcoma virus which is temperature-sensitive in protein phosphorylation and cellular transformation. Cell 22:767–775.

Purchio AF, Erikson E, Brugge JS, Erikson RL (1978): Identification of a polypeptide encoded by the avian sarcoma virus *src* gene. Proc Natl Acad Sci USA 75:1567–1571.

Radke K, Martin GS (1979): Transformation by Rous sarcoma virus: Effects of *src* gene expression on the synthesis and phosphorylation of cellular polypeptides. Proc Natl Acad Sci USA 76:5212–5216.

Gene Structure and Regulation in Development, pages 95–111
Published 1983 Alan R. Liss, Inc., 150 Fifth Avenue, New York, NY 10011

Steroid-Controlled Gene Expression in a *Drosophila* Cell Line

Peter Cherbas, Charalambos Savakis, Lucy Cherbas, and M. Macy D. Koehler

Department of Cellular and Developmental Biology, Harvard University, Cambridge, Massachusetts 02138

How gene activity is regulated has long been a central concern of developmental biology. Intense interest has been focused on two questions that can be approached by the molecular techniques now available: (a) How are genes coordinately regulated in eukaryotic cells, and (b) how is it that different genes are active in different cells and that these apparently epigenetic restrictions are passed from a parental cell to its progeny? We will concern ourselves here with the ways in which the actions of steroid hormones exemplify and may help to illuminate these processes.

To illustrate steroid action it is useful to consider the ecdysteroid-induced* development of the salivary glands in the fruitfly *Drosophila melanogaster* [for review, see Ashburner and Berendes, 1978; Berendes and Ashburner, 1978]. Throughout the final larval stage, the giant polytene gland cells are engaged in the synthesis of glue polypeptides, which are temporarily sequestered in secretory vesicles. When molting hormone (the steroid 20-hydroxyecdysone, 20-HE) appears in the hemolymph 4 to 8 hr prior to the onset of metamorphosis, it causes the gradual exocytosis of the secretory product (glue) into the gland lumen. These events can be reproduced in organ culture where, as in the animal, hormone addition provokes exocytosis after a 3- to 4-hr lag. Of course, what makes this system informative is that some of the

The major portions of this chapter were published previously in Molecular Genetic Neuroscience, edited by Francis O. Schmidt, Stephanie J. Boyd, and Floyd E. Bloom, © 1982 Raven Press, New York, and are reproduced here by permission of publisher.

*Ecdysteroid is the generic name for steroids related to 20-hydroxyecdysone, the insect molting hormone.

major transcriptional changes that occur can be monitored by observing the puffing pattern of the polytene chromosomes. Puffing at a site is correlated with increased transcriptional activity at that site [Ashburner, 1977]; therefore, a record of puffing during the ecdysteroid response is a sampling of transcriptional events. M. Ashburner, H.D. Berendes, and their colleagues have analyzed in detail the effects of ecdysteroids on puffing.

When salivary glands from a mature larva are cultured, the addition of 20-HE leads to a sequence of changes in the puffing pattern that is identical to that occurring *in vivo* during metamorphosis [Ashburner, 1972a]. Sites whose puffing is altered by 20-HE belong, in general, to one of three classes, examples of which are illustrated in Fig. 1. Prior to hormone addition, only about half a dozen "intermolt puffs" are present; these correspond to the structural genes for the glue polypeptides. Ecdysteroid treatment leads to the rapid regression of these puffs and, within minutes, to the appearance of a small number of "early puffs" at new locations. The early puffs reach maximal size within 4 hr; they then regress and are replaced by a very large number (of the order of 100) of "late puffs," which appear and regress on individual schedules during the next 10 to 20 hr [Ashburner, 1974].

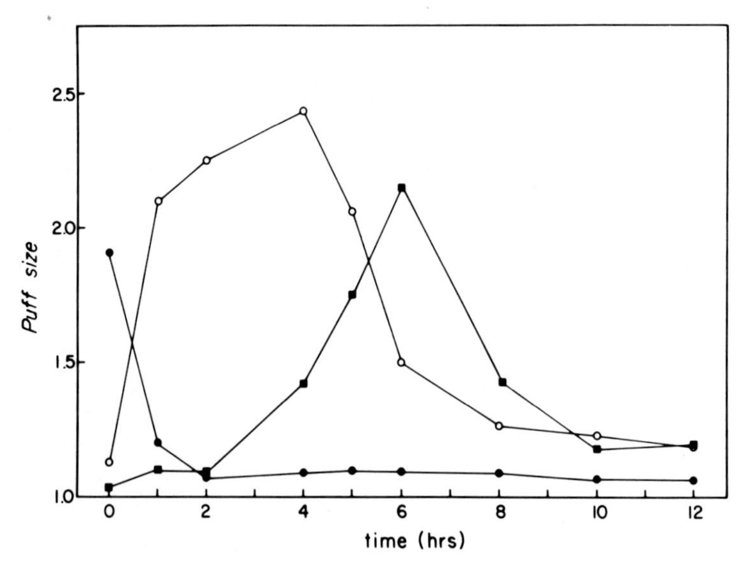

Fig. 1. Time courses of the intermolt puff 25B (closed circle), the early puff 74EF (open circle), and the late puff 78D (square). Puff size is expressed as the ratio of puff diameter to the diameter of a nearby constant region. (Data from Ashburner, 1972a; figure republished from Cherbas and Cherbas, 1981.)

There is persuasive evidence that at least some of the early puffs are primary responses, whereas most late puffs are secondary. Aside from timing, early puffs are distinguishable from late puffs by three criteria: (1) the sizes of early puffs show a graded response to increasing hormone dose; late puffs exhibit an all or nothing dependence on ecdysteroid concentration [Ashburner, 1973]; (2) early puffs regress rapidly after hormone withdrawal; after 5 hr late puffing becomes hormone-independent [Ashburner and Richards, 1976]; and (3) protein synthesis inhibitors do not prevent the induction of early puffs but block the eventual regression of the early puffs and the induction of the late puffs [Ashburner, 1973]. These observations, among others, have led to the conclusion that ecdysteroids induce transcription at a small group of primary-responsive loci, including the early puff sites, and inhibit transcription of a second set of primary loci, including the intermolt puff sites. Induction of transcription at secondary—late—puff sites and inhibition of transcription at the early puff sites are caused by products of the primary induced loci [Ashburner et al., 1973].

The outline of steroid hormone action implied by such systems as the *Drosophila* salivary gland suggests an experimental approach to the question about gene activation with which we began. These questions, restated in terms of steroid hormones, become: (1) how do steroid-receptor complexes recognize the primary responsive genes in a given cell, and (2) if the set of primary responsive genes is tissue-specific, what is the molecular basis of this specificity and of its maintenance through cell division?

Ecdysteroid Receptors in *Drosophila*

The lack, until recently, of a suitable radiolabeled ecdysteroid hormone has delayed progress in the biochemical study of ecdysteroid receptors. Recently, by use of the extremely active analog ponasterone A, radiolabeled at a high specific radioactivity, ecdysteroid-binding proteins were detected and characterized in cells of the *Drosophila* cell line Kc [Maroy et al., 1978] and in *Drosophila* imaginal disks [Yund et al., 1978]. The receptors from the two cell types have properties similar to those of vertebrate steroid receptors. The idea that ecdysteroid-receptor complexes participate directly in puff induction is supported by the experiments of Gronemeyer and Pongs [1980], which show that in salivary glands ecdysteroids become concentrated at distinct chromosomal sites, including the early puffs. Thus, one site of ecdysteroid action appears to be the chromosome. (This does not, of course, exclude the possibility that the same hormone also acts at other sites, such as the cell or nuclear membrane [for a discussion of models for ecdysteroid action on the membrane, see Kroeger and Lezzi, 1966; Ashburner and Cherbas, 1976].)

Ecdysteroid Effects on the *Drosophila* Cell Line Kc

Efforts have been made to identify and isolate primary responsive genes from a *Drosophila* cell line. Dozens of independent cell lines have been established from dissociated embryos of *D. melanogaster,* and many of these lines exhibit some sensitivity to ecdysteroid hormones [for recent reviews, see Cherbas and Cherbas, 1981; Sang, 1981]. In general, these lines are near-diploid, have stable phenotypic and karyotypic properties, and are relatively easy to grow. They are probably derived from one or more imaginal cell types; imaginal cells in *Drosophila* are undifferentiated, dividing diploid cells in the larva, and are differentiated during metamorphosis to form the adult.

The effects of ecdysteroids have been studied most extensively on the Kc line and its subline Kc-H. When these cells are exposed to ecdysteroids, they differentiate morphologically and enzymatically and cease to proliferate. The morphological response becomes visible within several hours after addition of ecdysteroid and consists of the extension of long processes that often contain multiple inclusions (Fig. 2) [Cherbas et al., 1980]. Acetylcholinesterase activity, which is not detectable in untreated Kc-H cells, appears in these cells after 24 hr of continuous exposure to ecdysteroids [Cherbas et al., 1977]. In flies, AChE activity is apparently confined to the nervous system [Hall and Kankel, 1976]; this, along with the morphology of the treated cells, suggests that the Kc line may be a neural derivative.

These effects of ecdysteroids on Kc-H cells are reproducible, synchronous, and specific to active ecdysteroids. Two lines of evidence suggest that these effects are mediated by the ecdysteroid receptor that has been detected in Kc cells. First, the relative affinities for the receptor of a number of ecdysteroids are indistinguishable from their relative activities in inducing the various responses [Beckers et al., 1980; Cherbas et al., 1980a; Cherbas et al., 1980b]. Second, two ecdysteroid-resistant clones that were selected for growth in the presence of 20-HE and failed to show any response to ecdysteroid are deficient in ecdysteroid-binding activity [Cherbas et al., 1980b; J.D. O'Connor and L. Cherbas, unpublished observations].

Ecdysteroid-Responsive Genes

The pattern of protein synthesis in a clone of Kc-H cells was examined by pulse-labeling with [^{3}H]leucine and separation of the labeled polypeptides on SDS-polyacrylamide gels or on two-dimensional gels [O'Farrell, 1975]. Within 1 hr of hormone addition, increased relative synthesis can be detected of polypeptides that fall into three molecular weight classes: 40, 29, and 28 kD [Savakis et al., 1980]. We call these the ecdysteroid-inducible polypep-

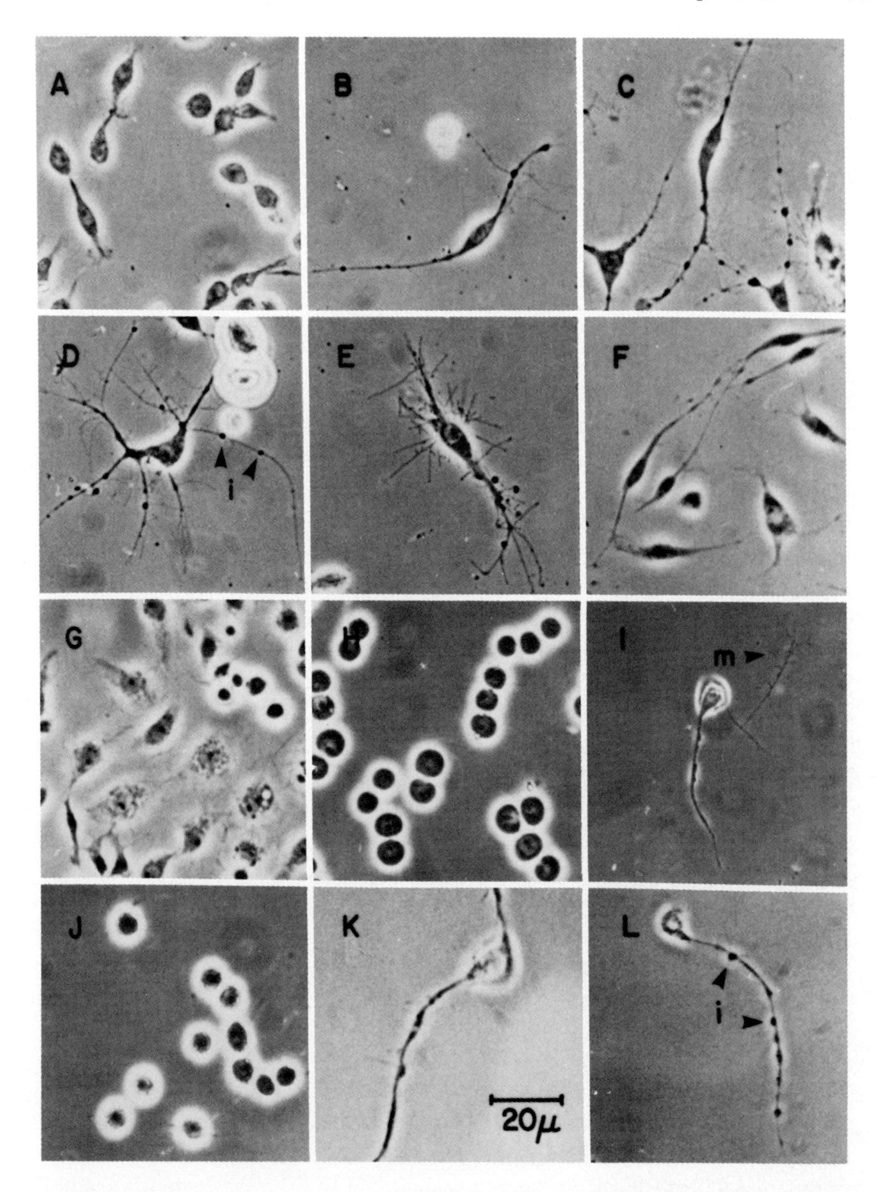

Fig. 2. Kc-H cells before and after treatment with 20-HE. A to G are cells growing on glass surfaces; H to L are cells growing in suspension. Cells in A, H, and J are untreated controls. Those in G were treated for 3 days with 10^{-8} M 20-HE. All others were exposed for at least 3 days to concentrations of hormone sufficient to elicit a maximal response (10^{-7}—10^{-5} M 20-HE). Examples of inclusions (i) and membranous extensions (m) are indicated by arrows. (Figure republished from Cherbas et al., 1980a.)

tides (EIPs) and distinguish EIP 40, EIP 29, and EIP 28. In fact, as shown in Fig. 3A, two-dimensional electrophoresis resolves each of the EIPs into more than one form in the isofocusing dimension, and the catalog of changes compromises eight major spots—EIP 40 (I,II,III), EIP 29 (I,II), and EIP 28 (I,II,III)—where the most acidic form of each molecular weight class is designated I [Cherbas et al, 1981; C. Savakis, M.M.D. Koehler, and P. Cherbas, unpublished data]. Two-dimensional electrophoresis resolves an additional major ecdysteroid-inducible polypeptide of approximately 35 kD (Fig. 3A). Except for EIP 28 III, all EIPs are stable during prolonged chases. Increased synthesis of these polypeptides becomes detectable within 1 hr, reaches a maximum within 4 to 8 hr, and continues for 2 to 3 days. During this time, few additional ecdysteroid-provoked changes, and none of comparable magnitude, are detectable using the two-dimensional separation [M. M. D. Koehler and P. Cherbas, unpublished data]. The induction of the EIPs is ecdysteroid-specific, increases with hormone concentration over at least a 10-fold range of concentrations, and does not occur in ecdysteroid-resistant cell clones [Savakis et al., 1980].

In vitro translations of Kc-H cell RNAs have demonstrated that the EIP mRNAs are predominantly polyadenylated, and that the levels of translatable EIP mRNAs increase essentially in parallel with the *in vivo* synthesis of these polypeptides (Fig. 3B) [Savakis, 1981]. All eight forms of the EIPs are represented in two dimensional separations of translation products in approximately the same relative ratios as in *in vivo* synthesized polypeptides (see Fig. 3). This, combined with the results of the *in vivo* pulse-chase experiments, suggested that the various EIP forms represent independent products of translation rather than products of posttranslational modifications. The 35-kD ecdysteroid-inducible polypeptide was not detectable among the translation products of polyadenylated RNAs. Therefore, for technical reasons, efforts were concentrated on EIPs 40, 29, and 28.

Since the translation results indicated that the EIP mRNAs are among the 10 most abundant polyadenylated mRNA species in hormone-treated cells and that the titers of these mRNAs are elevated in hormone-treated cells severalfold relative to untreated cells, molecular clones of the EIP mRNA sequences were isolated by the following method [for details, see Savakis, 1981]. Polyadenylated RNA from ecdysteroid-treated cells was used as template for the synthesis of double-stranded copy DNA (cDNA). This was used to prepare a cDNA library by inserting the cDNA at the Pst I site of plasmid pBR322 and transforming *E.coli* cells with the hybrid plasmid molecules. The resulting tetracycline-resistant clones were screened by comparing their patterns of hybridization to radioactive probes representing (a) control cell polyadenylated RNA and (b) 4-hr ecdysteroid-treated cell polyadenylated RNA. In a screen of around 5,000 bacterial colonies, 22 that showed reproducibly greater hybridization to probe (b) were recovered. The plasmids from these bacterial colonies were tested for the presence of EIP cDNA sequences by hybrid selection and translation: plasmid DNA immobilized on nitrocellulose filters was challenged with polyadenylated RNA from Kc-H cells and the hybridized RNA was eluted from the filter and translated *in vitro*.

Three EIP-related plasmids were recovered, and two-dimensional patterns of the translation products of the corresponding mRNAs were produced (Fig. 4). Plasmid pKc441 selected RNAs encoding all the EIPs 40. Plasmids pKc252 and pKc 191 are homologous to RNAs encoding all the EIPs 29 and 28. Plasmid pKc191 contained a cDNA insert homologous to that of pKc252, but considerably shorter. In addition to the five EIP 28/29 polypeptides, mRNA hybridizing to pKc252 and pKc191 directed the synthesis of two other 29-kD polypeptides with isoelectric points intermediate between the EIP 28/29 and EIP 40 clusters. These polypeptide spots are detectable among the *in vitro* translated, but not among the *in vivo* synthesized, polypeptides (see Fig. 3).

Properties of the prototype EIP-specific plasmids pKc252 and pKc441 are summarized in Table I. Each of these plasmids is homologous to mRNAs encoding a family of polypeptides. It was initially hypothesized that two gene families exist, one homologous to pKc252, encoding EIPs 28/29, and another homologous to pKc441, encoding EIPs 40. However, each haploid genome has been shown to contain only one gene homologous to pKc252 and one gene homologous to pKc441, mapping at chromosomal sites 71C3.4-D1.2 and 55B-D, respectively. The situation with the gene encoding EIPs 28/29 is summarized in Fig. 5A. Labeled plasmid pKc252 DNA was used as a probe to isolate clones containing the EIP 28/29 genes from a library of *D. melanogaster* genomic DNA fragments cloned in a bacteriophage λ vector [L. Cherbas, unpublished observations]. All clones recovered from this screen were overlapping, and the restriction map of the genomic region covered by these clones is shown at the top of Fig. 5A. When Kc-H cell DNA, digested with various restriction endonucleases, was subjected to electrophoresis on agarose gels, and then blotted onto nitrocellulose filters and challenged with labeled pKc252 DNA, hybridization patterns such as those shown in Fig. 5B were obtained. In this experiment, DNA from a genomic clone was restricted with the same enzyme and run in parallel with the genomic DNA. The amounts of genomic and phage DNA loaded in each lane were adjusted so that approximately equal levels of hybridization should be obtained if one copy of pKc252-specific sequences existed per haploid genome. The absence of heterogeneity in the restriction patterns of the cloned genomic sequences (Fig. 5A) and in the Kc cell DNA (Fig. 5B) imply that there is, per haploid genome, only one gene encoding the five EIP 28/29

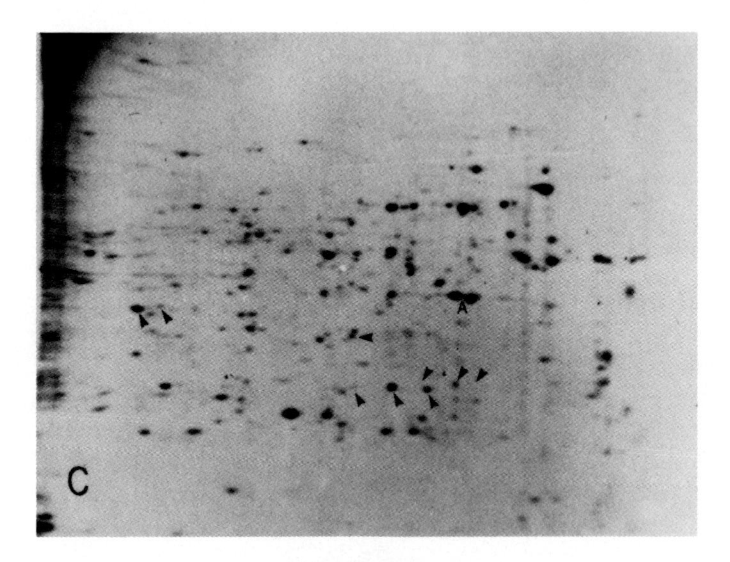

IN VIVO

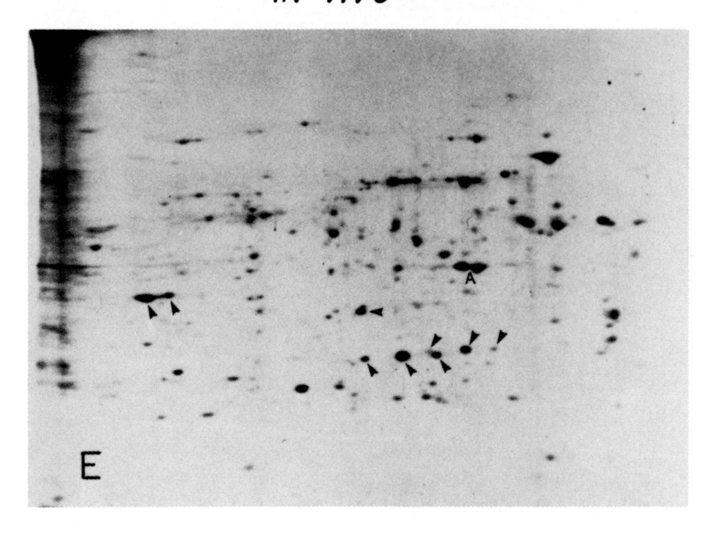

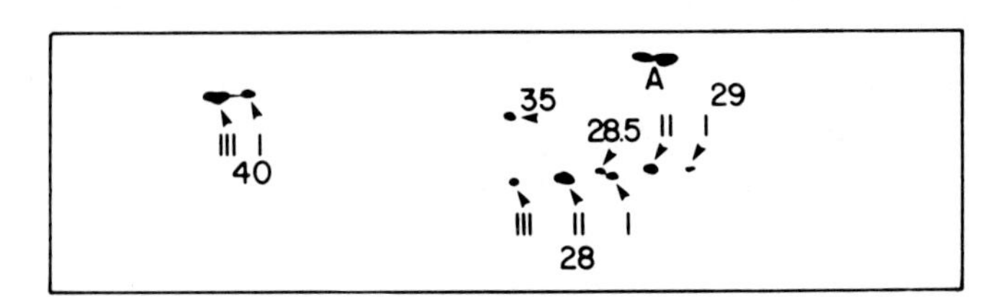

Figure 3.

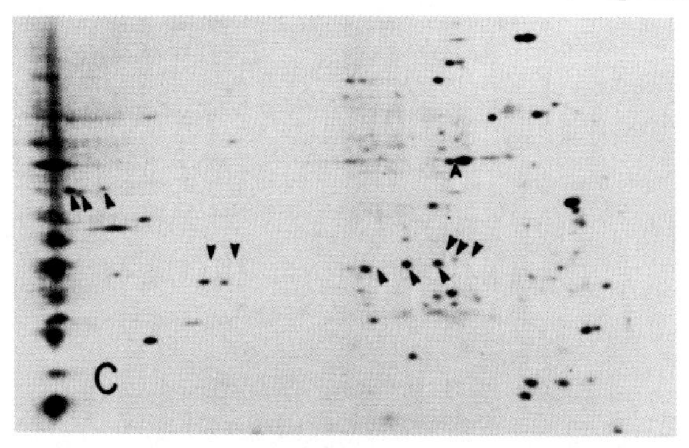

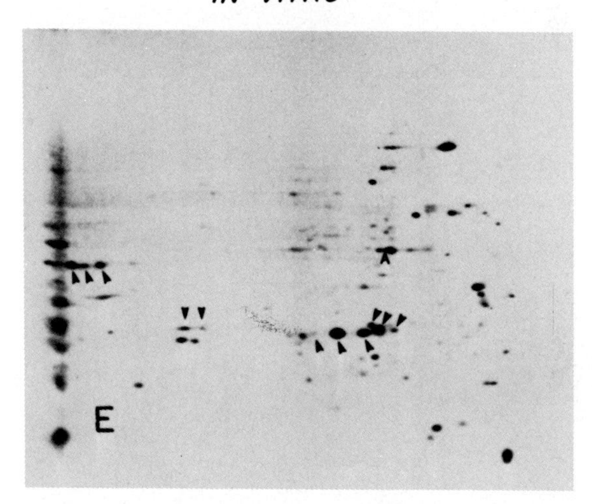

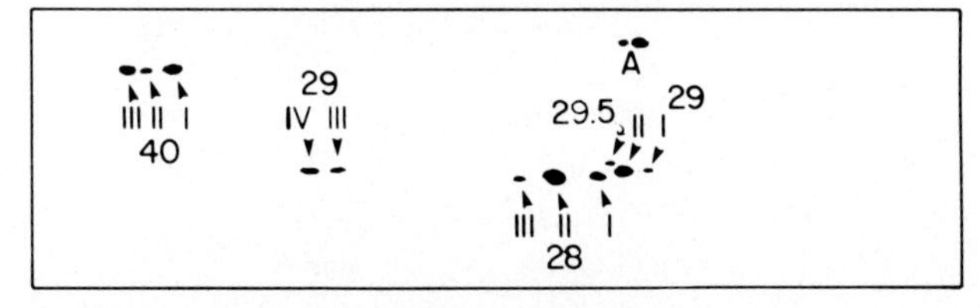

Fig. 3. Two dimensional separations of pulse-labeled Kc-H cells and of translation products of their RNAs. Left: Control cells (C) or cells treated 4 hr with 10^{-6} M 20-HE (E) were labeled 20 min with ^{3}H-leucine, immediately solublized in detergent-urea, and separated according to the procedures of O'Farrell (1975). The polypeptides were detected by autofluorography. Above: Polyadenylated RNAs from untreated cells (C) and from cells treated 4 hr with 10^{-6} M 20-HE (E) were translated in the RNA-dependent rabbit reticulocyte lysate (Pelham and Jackson, 1976). The [^{3}H]-labeled polypeptides were separated and detected as in the left panel. Isofocusing was in the horizontal dimension with the anode (acidic) at the right; the subsequent SDS-polyacrylamide separation ran from top to bottom. Arrows indicate spots dicussed in the text. The names of the individual spots are given in the keys below. "A" designates the two spots corresponding to actin. (Figure republished from Cherbas et al., 1981.)

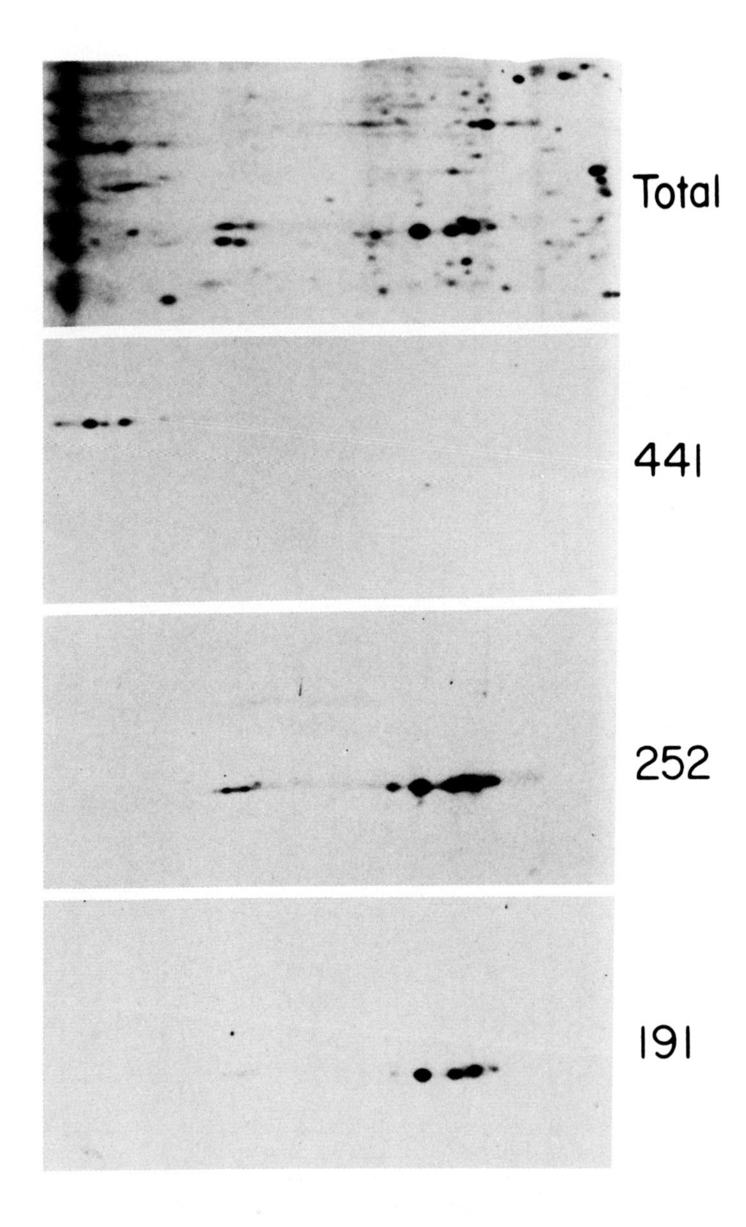

Fig. 4. Two-dimensional separations of hybrid-selected translation products. Polyadenylated RNA from 20-HE-treated Kc-H cells was translated without treatment (Total) or after selection by hybridization to pKc252, pKc441, or pKc191. The translation products were displayed as in Fig. 3.

TABLE I. Properties of two plasmids containing EIP cDNA inserts.

Plasmid	EIPs selected	Insert length (nucleotides)	Chromosome site
pKc252	28 (I,II,III) 29 (I,II)	850	71C3,4-D1,2
pKc441	40 (I,II)	450	55B-D

polypeptides. Similar results have been obtained for the EIP 40 gene [J. Rebers and L. Cherbas, unpublished observations].

The apparent paradox of one gene encoding a family of polypeptides is under investigation. We believe that the multiple EIP forms are not due to artifactual modifications such as those described by O'Farrell (1975), because the two-dimensional pattern is very reproducible and is seen in gels where other polypeptides do not exhibit heterogeneity in the isofocusing dimension (see Fig. 5). Possible explanations for the EIP heterogeneity include: (1) alternative mRNA splicing events; (2) alternative translation modes of the same message; and (3) some kind of posttranslational modification—such as phosphorylation or autophosphorylation—that occurs in the reticulocyte lysate as well as in intact Kc cells, and that gives rise very rapidly to a terminal equilibrium pattern.

Using plasmids pKc252 and pKc441 as probes, the EIP mRNAs were quantified during the early phase of the ecdysteroid response [Savakis, 1981]. The titers of mRNAs encoding the EIPs are approximately 50% higher than control levels 30 min after hormone addition. Preliminary experiments indicate that the presence of cycloheximide administration during this period, at concentrations sufficient to inhibit about 97% of the total protein synthetic activity, does not prevent the EIP mRNA induction. Cycloheximide alone appears to cause increases in the EIP mRNA titers, by an unknown mechanism; the effects of cycloheximide and ecdysteroid are additive [Savakis, 1981].

In general, the properties of EIP induction are very similar to those of early puff induction in salivary glands. The induction is rapid; it occurs in the presence of cycloheximide, and the extent of induction increases with ecdysteroid concentration over at least a 10-fold range of concentrations. Therefore, the EIP genes, like the salivary gland early puff sites, are probably primary ecdysteroid-responsive loci.

Tissue-Specificity of the Primary Responsive Genes

Early ecdysteroid-responsive genes have been identified in larval tissues (salivary glands, prothoracic glands, and fat body) by analysis of puffing

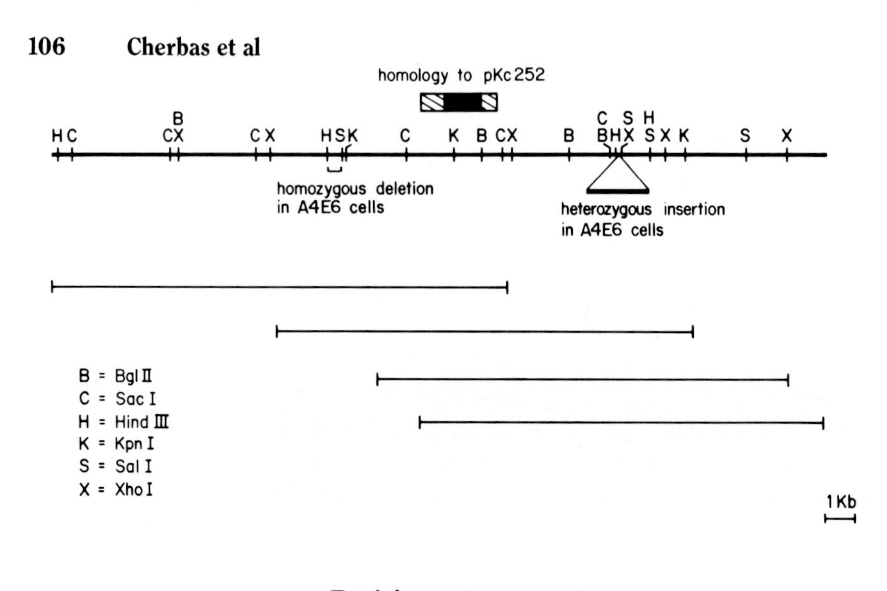

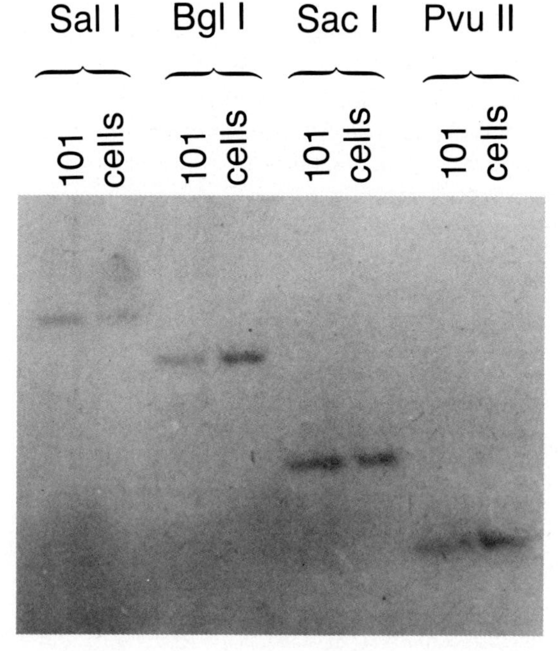

Fig. 5. Restriction patterns of genomic DNA homologous to pKc252. Top: Restriction maps of four overlapping cloned fragments. The limits of the individual cloned fragments are indicated by the horizontal lines below the map. Minimum and maximum limits of the region that hybridizes to pKc252 DNA are indicated by the solid and hatched bars, respectively. B, Bgl II; C, Sac I; H, Hind III; K Kpn I; S, Sal I; X, Xho I. Bottom: DNA extracted from Kc-H cells and DNA of the clone λ (Charon 4)252–101 (the second of the four clones above) were digested with the restriction endonucleases Sal I, Bgl I, Sac I, and Pvu II. The fragments were separated by electrophoresis on an agarose gel, blotted onto nitrocellulose, and hybridized to [^{32}P]-labeled pKc252.

patterns, and in imaginal disks by *in situ* hybridization of newly synthesized RNA [Ashburner, 1972a; Bonner and Pardue, 1976; Holden and Ashburner, 1978; Richards, 1980]. Puffs are induced at regions 74EF and 75B in the three larval tissues. Other regions puff in one tissue but not in another; 71CD, for example, puffs rapidly in salivary glands but not in the prothoracic gland. (Of course, one cannot tell from such studies whether apparently identical early puff sites are identical loci or are merely closely linked. Resolution of this problem will require the use of cloned sequences). Early induced RNAs in imaginal disks do not hybridize to any of the regions that give early puffs in salivary glands. The EIP 40 gene maps to 55B-D, a site that is not among the early induced regions in the three larval tissues or in imaginal disks. The EIP 28/29 gene maps to 71C3.4-D1.2, very close to a salivary gland early puff site, but none of the EIP genes appears to be expressed in ecdysteroid-treated salivary glands [W. Koerwer and C. Savakis, unpublished data].

The region 71C-E is of particular interest (see Fig. 6). In salivary glands, this region is the site of a large complex puff that has been analyzed by Ashburner and his colleagues [Ashburner, 1972b; Velissariou, 1980]. At 71CD there is a salivary gland early puff; immediately next to it is an intermolt puff corresponding to a glue polypeptide gene. The site of the largest late salivary gland puff is contained in 71E. The structural gene for EIPs 28/29 is located between the early and intermolt puffs, as demonstrated by *in situ* hybridization of pKc252. Thus, three ecdysteroid-controlled genes, two of which appear to be tissue-specific primary responsive genes, are

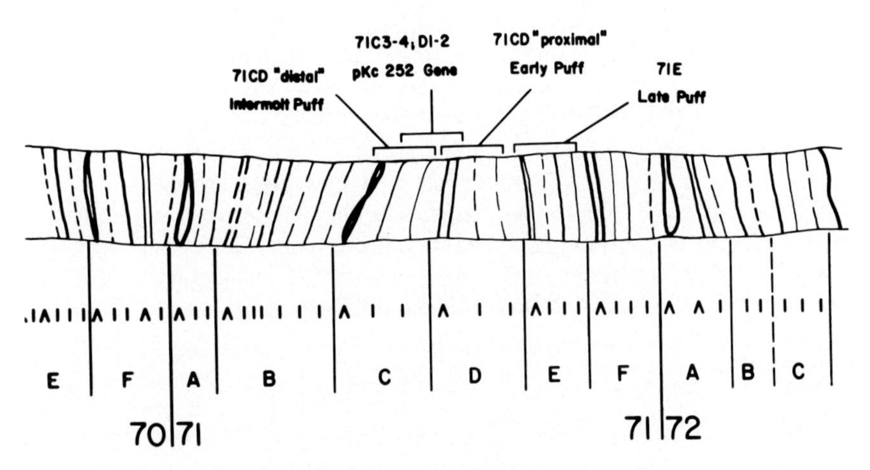

Fig. 6. Cytogenetic map of the region 71C-E. (After Bridges, 1941.) Loci that are ecdysteroid-responsive in salivary glands or in Kc cells are indicated.

located within a four to five band region. It is intriguing to speculate about possible evolutionary or functional significance for a clustering of tissue-specific hormone-responsive genes; in any case, such a clustering will be useful in the isolation of tissue-specific ecdysteroid-responsive genes.

Prospects

The availability of cloned EIP sequences will enable researchers to map the pattern of EIP mRNA in larval and adult tissues and to examine the regulation of EIP mRNA during development. This type of analysis may give some clues about the origin of Kc cells and may indicate whether their response to ecdysteroids has anything to do with neural development. More important, cloned EIP genes can be used in functional assays of gene control using *in vitro* reconstituted transcription systems and DNA-mediated transformation. Such experimental techniques should enable us to begin to answer basic questions about the control of primary responsive genes by a steroid-receptor complex and the basis for tissue specificity in the steroid-responsiveness of individual genes.

Discussion

W. Hahn asked whether, in Kc cells, there are any translatable mRNA species that lack poly(A) and are distinct from the polyadenylated species. P. Cherbas answered that there are many mRNAs in the so-called poly(A)$^-$ fraction (i.e., RNAs not retained by oligo(dT)-cellulose) that are translated into discrete products that are not detectable among the translation products of oligo(dT)-binding RNAs. On this basis, there seems to exist a discrete population of poly(A)$^-$ mRNAs in Kc-H cells. It is not known, however, whether these mRNAs lack poly(A) completely or have very short poly(A) tracts.

It was also asked whether juvenile hormone blocks any of the ecdysteroid effects in Kc cells. P. Cherbas answered that juvenile hormone analog ZR515, at 10^{-9} M, blocks many aspects of the ecdysteroid response including the ability of short pulses of ecdysteroids to reduce the cloning efficiency of Kc-H cells. It has no detectable effect on EIP induction [L. Cherbas, M.M.D. Koehler, and P. Cherbas, unpublished data].

Summary

The Kc-H cell line is one of numerous *Drosophila melanogaster* cell lines that are responsive to the steroid molting hormone 20-hydroxyecdysone. In Kc-H cells, the hormone provokes morphological and enzymatic differentiation and proliferative arrest. The earliest biochemical changes so far detected consist of increases in the synthesis of a few ecdysteroid-inducible polypeptides (EIPs). Molecular cloning has shown that 9 of the 10 EIPs are encoded by just two genes. Hormone treatment leads to rapid increases in the transcripts for these EIPs.

Addendum: An Update

Since the completion of the review reprinted above we have continued to study the EIP genes and their regulation in Kc-H cells with the following pertinent results:

(i) In Kc-H cells the major EIP 28/29 transcript is 976N long (exclusive of poly-(A); it derives from a transcription unit 2136 N in length in which 4 exons are interrupted by introns of 992, 110, and 59 N. These structures have been determined by sequencing the complete gene, by sequencing portions of several cDNA clones, by S1 nuclease analysis of RNA:DNA hybrids, and by primer extension experiments.

One important conclusion from these experiments is that the basal or uninduced transcripts of the 28/29 gene do not differ in structure from those synthesized during hormonal induction.

(ii) We believe that EIP 28III and EIP 29III (EIP 28.5 in the text above) are primary translational products and that the other EIPs 28/29 derive from them by post-translational or co-translational acetylation. EIP 28III and EIP 29III are synthesized in the ratio 3:1 and this ratio is not altered by ecdysteroids. Instead the hormone elevates the synthesis of all the EIP 28/29 species by about 8-fold (after 4 hr hormone treatment).

There exists a minor EIP 28/29 transcript which is generated by a variant splicing pathway at the boundary between the second and third exons. This minor species represents approximately one-quarter of the 28/29 transcript. Although its structure remains to be solved in detail, we believe that this species encodes EIP 29III.

(iii) Careful measurements of the abundance of the EIP 28/29 transcripts have reinforced the idea that induced transcription at this gene is indeed a primary response to ecdysteroids. By titrating transcripts with suitable probes in solution and measuring the amounts of the resulting hybrids after S1 digestion and electrophoresis we have obtained a suitably quantitative record of the induction. There is a lag of about 15 min followed by an essentially linear increase in the EIP 28/29 transcripts up to 4 hr when the induction is about 10-fold, ie, essentially equivalent to the measured increase in synthesis of the polypeptides. At 4 hr the amount of the increase is related to the ecdysteroid concentration over the broad range typical of early steroid effects.

With these results we believe that we can now turn our attention to the mechanism of the induction and, equally important, to the developmental specificity of the induction in the tissues of flies.

We gratefully acknowledge research support from the American Cancer Society, the NSF, and from a program project grant of the NIH.

Original Addendum by
P. Cherbas A. Bieber
L. Cherbas C. Savakis
R. Schulz D. Eickbush
M.M.D. Koehler

ACKNOWLEDGMENTS

Work unpublished prior to this article's appearance in Molecular Genetic Neuroscience was supported by grants from the National Science Foundation (PCM-80-03931 and PCM-78-07614) and the American Cancer Society (VC316 and CD107A).

REFERENCES

Ashburner M (1972a): Patterns of puffing activity in the salivary gland chromosomes of *Drosophila*. VI. Induction of ecdysone in salivary glands of *D melanogaster* cultured in vitro. *Chromosoma* 38:255–281.

Ashburner M (1972b): Puffing patterns in *Drosophila melanogaster*. In: *Developmental Studies on Giant Chromosomes,* edited by W. Beermann, pp. 101–151. Springer-Verlag, New York.

Ashburner M (1973): Sequential gene activation by ecdysone in polytene chromosomes of *Drosophila melanogaster*. I. Dependence upon ecdysone concentration. Dev Biol 35:47–61.

Ashburner M (1974): Sequential gene activation by ecdysone in polytene chromosomes of *Drosophila melanogaster*. II. The effects of inhibitors of protein synthesis. *Dev Biol* 39:141–157.

Ashburner M (1977): Happy birthday-puffs! In: *Chromosomes Today,* edited by A. de la Chapelle and M. Sorsa, pp 213–222. Elsevier, Amsterdam.

Ashburner M, Berendes HD (1978): Puffing of polytene chromosomes. In: *"The Genetics and Biology of Drosophila," Vol 2b,* edited by M Ashburner, TRF Wright, pp 315–395. Academic Press, New York.

Ashburner M, Cherbas P (1976): The control of puffing by ions—The Kroeger hypothesis: A critical review. *Mol Cell Endocrinol* 5:89–107.

Ashburner M, Chihara C, Meltzer P, Richards G (1973): Temporal control of puffing activity in polytene chromosomes. *Cold Spring Harbor Symp Quant Biol* 38:655–662.

Ashburner M, Richards G (1976): Sequential gene activation in polytene chromosomes of *Drosophila melanogaster*. III. Consequences of ecdysone withdrawal. *Dev Biol* 54:241–255.

Beckers C, Maroy P, Dennis R, O'Connor JD, Emmerich H (1980): Uptake and release of ^{3}H-ponasterone A by the Kc cell line of *Drosophila melanogaster*. In: *"Progress in Ecdysone Research,"* edited by JA Hoffman, pp 335–347. Elsevier, Amsterdam.

Berendes HD, Ashburner M (1978): The salivary glands. In: *"The Genetics and Biology of* Drosophila,*" Vol 2b,* edited by M Ashburner and TRF Wright, pp 453–498. Academic Press, New York.

Bonner JJ, Pardue ML (1976): Ecdysone-stimulated RNA synthesis in imaginal discs of *Drosophila melanogaster.* Assay by in situ hybridization. *Chromosoma* 58:87–99.

Bridges PN (1941): A revised map of the left limb of the third chromosome of *Drosophila melanogaster. J Hered* 32:64–65.

Cherbas L, Cherbas P (1981): The effects of ecdysteroid hormones on *Drosophila melanogaster* cell lines. *Adv Cell Culture* 1:91–124.

Cherbas L, Yonge CD, Cherbas P, Willliams CM (1980a): The morphological response of Kc-H cells to ecdysteroids: Hormonal specificity. *Wilhelm Roux Arch* 189:1–15.

Cherbas P, Cherbas L, Demetri G, Manteuffel-Cymborowska M, Savakis C, Yonge CD, Williams CM (1980b): Ecdysteroid hormone effects on a *Drosophila* cell line. In: *"Gene Regulation by Steroid Hormones,"* edited by Roy AK and Clark JH pp 278–305. Springer-Verlag, New York.

Cherbas P, Cherbas L, Savakis C, Koehler MMD (1981): Ecdysteroid-responsive genes in a *Drosophila* cell line. *Am Zool* 21:743–750.

Cherbas P, Cherbas L, Williams C (1977): Induction of acetylcholinesterase activity by β-ecdysone in a *Drosophila* cell line. *Science* 197:275–277.

Gronemeyer H, Pongs O (1980): Localization of ecdysterone on polytene chromosomes of *Drosophila melanogaster. Proc Natl Acad Sci USA* 77:2108–2112.

Hall JC, Kankel DR (1976): Genetics of acetylcholinesterase in *Drosophila melanogaster. Genetics* 83:517–535.

Holden JJ, Ashburner M (1978): Patterns of puffing activity in the salivary gland chromosomes of *Drosophila.* IX. The salivary and prothoracic gland chromosomes of a dominant temperature sensitive lethal of *D. melanogaster. Chromosoma* 68:205–227.

Kroeger H, Lezzi M (1966): Regulation of gene action in insect development. *Ann Rev Entomol* 11:1–22.

Maroy P, Dennis R, Beckers C, Sage BA, O'Connor JD (1978): Demonstration of an ecdysteroid receptor in a cultured cell line of *Drosophila melanogaster. Proc Natl Acad Sci USA* 75:6035–6038.

Pelham HRB, Jackson RJ (1976): An efficient mRNA-dependent translation system from reticulocyte lysates. *Eur J Biochem* 67:247–257.

Richards G (1980): Ecdysteroids and puffing in *Drosophila melanogaster.* In *"Progress in Ecdysone Research,"* edited by J. Hoffmann, pp 363–378. Elsevier, Amsterdam.

Sang JH (1981): *Drosophila* cells and cell lines. *Adv Cell Culture* 1:125–177.

Savakis C (1981): Studies on Ecdysteroid-Inducible Polypeptides and their mRNAs in a *Drosophila melanogaster* Cell Line. PhD Thesis, Harvard University.

Savakis C, Demetri G, Cherbas P (1980): Ecdysteroid-inducible polypeptides in a *Drosophila* cell line. *Cell* 22:665–674.

Velissariou V (1980): The Cytogenetics of the Salivary Gland Glue Proteins of *Drosophila melanogaster.* PhD Thesis, University of Cambridge.

Yund MA, King DS Fristrom JW (1978): Ecdysteroid receptors in imaginal discs of *Drosophila melanogaster. Proc Natl Acad Sci USA* 75:6039–6043.

Gene Structure and Regulation in Development, pages 113–133

Manipulating the Genotype of Developing Mice

Beatrice Mintz
Institute for Cancer Research, Fox Chase Cancer Center, Philadelphia, Pennsylvania 19111

INTRODUCTION

Manipulation of the genotype in vivo encompasses a collection of experimental procedures—all recently shown to be practicable—that hold great promise for an understanding of the developmental regulation of gene expression in mammals. The genotypic change might involve either actual introduction of exogenous gene sequences, or selection for a specific new endogenous mutation or genetic rearrangement, or establishment of new associations of cells of different genotypes within mosaic individuals.

According to analyses of cell lineages in allophenic mice with lifelong cellular genetic markers, each specialized kind of cell arises relatively early from a small, fixed number of specific precursor cells; these become committed to a particular transcriptional pattern and are able to transmit that commitment, as a stable cellular heredity, to their mitotic descendants [Mintz, 1971, 1974]. The fact that the precursor cells have not been directly identified in most tissues is due to the delay (for unknown reasons) in the full expression of their specialized phenotypes. In any case, the early inception of "determination" of cell type requires that any intervention aimed at modifying the genotypes or genotypic associations of cells be carried out so that the change is already present when the relevant gene-function decisions are being made in the establishment of the lineages of interest. This might be accomplished either by changing the genetic composition in early cells that are still developmentally *totipotent* or in the *pluripotent* stem cells that are the precursors of the lineages in which that gene will be expressed.

We have defined four portals of entry whereby such potentially revealing experimental intervention is possible in the development of a mouse [Mintz, 1982]. The first possibility—production of mice from aggregated blastomeres of different genotypes—allows cells of two (or more) genotypes to participate and interact in the differentiation of all tissues. Results of studies on the allophenic animals have been summarized elsewhere [Mintz, 1971, 1974] and will not be further discussed here. Of the remaining three routes (Fig.

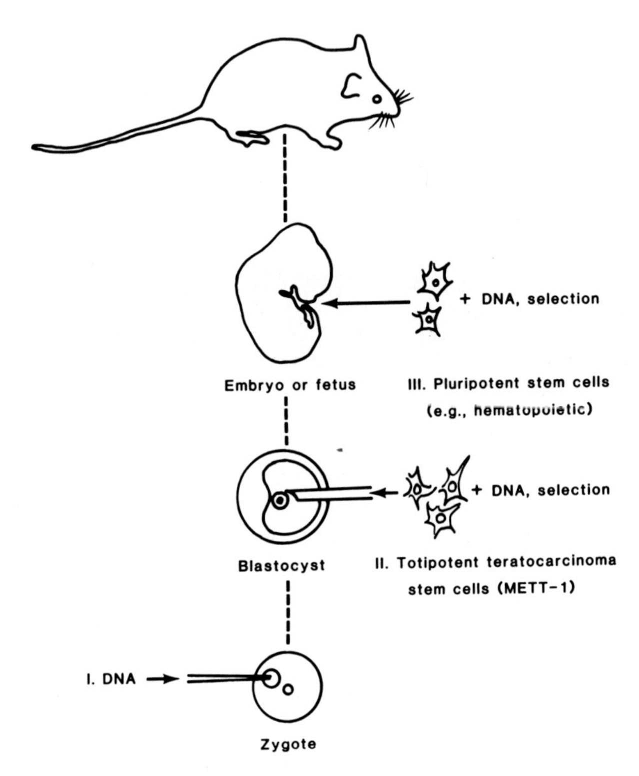

Fig. 1. Three routes for experimental introduction of exogenous genetic material. The paths labeled I and II allow the donor gene to be present throughout differentiation and therefore may involve all cell types. Only specific cell lineages will receive the gene via route III (from Mintz [1982]).

1)—the subject of the present discussion—two offer the prospect that the gene change may ultimately be carried into germ cells; thus, a new strain of mice of predetermined genotype could result. The earliest practical time for such intervention is in the fertilized egg. However, cellular totipotency continues for some time during cleavage [Mintz, 1974] and even persists in some embryonic ("ectoderm") cells following implantation, substantially after the embryonic and extraembryonic lineages have diverged [Diwan and Stevens, 1976]. Therefore, a gene change in such a totipotent embryo cell—eg, in the blastocyst [Jaenisch and Mintz, 1974]—can still be transmitted to the full spectrum of somatic cells and to germ-line progeny. The totipotent stem cells of mouse teratocarcinomas [Stevens, 1967; Pierce, 1967] have provided a unique substitute for totipotent embryo cells: The stem cells of these tumors apparently arise by neoplastic conversion of initially normal early embryo cells [Mintz et al, 1978]. Some (not all) transplant or in vitro lines retain and express their totipotentiality if injected into early embryos, where they are "normalized" and brought into the mainstream of development, as first observed with a transplanted line of stem cells [Mintz and Illmensee, 1975]. Thus, a genetic change might be effected in the stem cells of an appropriate cell culture line and then become a part of a mouse and its progeny [Mintz, 1978a].

Another portal of entry, applicable to only a few cell lineages, is exemplified by the hematopoietic system but might become feasible for some other tissues. The definitive myeloid and lymphoid blood-cell lineages are thought to be derived from primitive hematopoietic stem cells [Metcalf and Moore, 1971]. If one wished to introduce, say, a recombinant globin gene, the pluripotent stem cells would be the recipients of choice because expression of this gene would presumably be regulated later, when the committed erythroid lineage becomes established. Introduction of the gene into an already committed erythroid cell, eg, an erythroblast, would not be expected to be an optimal entry point for analysis of the regulatory mechanisms in globin gene expression, although the donor gene might in fact become successfully activated.

In the following sections, introduction of genetic changes via the three portals of entry shown in Figure 1 will be successively discussed.

INJECTION OF CLONED DNA SEQUENCES

The first successful attempt to introduce foreign DNA into a mammal was carried out by microinjection of purified simian virus 40 (SV40) DNA into the fluid-filled cavity of early mouse embryos at the blastocyst stage [Jaenisch and Mintz, 1974]. The basic micromanipulation techniques had been introduced earlier (for other purposes) by Lin [1966]. The objectives of our study [Jaenisch and Mintz, 1974] were to investigate the mechanisms of incorporation of tumor viral genetic material, its vertical transmission, and, ultimately, control of its expression. The DNA from some tissues in approximately 40% of the resultant mice contained SV40-specific DNA sequences, as judged from reannealing kinetics in molecular hybridization tests with a radiolabeled SV40 DNA probe.

Those experiments were carried out before the availability of recombinant DNA. As a result of recombinant DNA technology, it is now possible to introduce any cloned gene in intact or specifically modified form. Injection into the uncleaved fertilized egg, rather than the blastocyst, would be expected to increase the number of tissues in which the donor genetic material might be found. Synthesis of endogenous DNA is known to occur in the male and female pronuclei before they fuse to form the zygote nucleus [Mintz, 1965a], and this pronucleate stage therefore provides a convenient and promising stage for injection of DNA before cleavage.

The DNA construct obtained from T. Maniatis and used in our first series of injections [E. Wagner et al, 1981] contained the human β-globin gene [Fritsch et al, 1980], which would be expected to have tissue-specific expression in erythroid cells; and the herpes simplex viral (HSV) thymidine kinase *(tk)* gene, which would be constitutively expressed. Both genes were ligated in a shared pBR322 plasmid vector, designated PtkHβ1. The vector included the 7.6 kilobase (kb) *Hind*III fragment of the human adult genomic β-globin gene (including the entire gene plus approximately 6 kb of flanking sequences) and the 3.6-kb *Bam*HI fragment of the HSV *tk* gene.

Approximately 2,500–3,000 copies of the DNA were injected into the male (larger) pronucleus, and the eggs, after brief reincubation in vitro, were then surgically transferred to the oviducts of pseudopregnant females. In the first series, 33 animals were examined in late fetal stages. The high molecular weight DNA of five individuals (15%), and of their corresponding placentas, was positive for the foreign globin and *tk* genes and also for pBR322 sequences. This is seen in Figures 2 and 3, after the DNAs were digested with the designated restriction enzyme, electrophoresed in agarose gel, transferred to nitrocellulose filters [Southern, 1975], tested for hybridization with the appropriate ^{32}P-labeled restriction fragment of the respective genes, and autoradiographed. The five positive fetal-placental pairs contained 3–50 copies per cell of the human β-globin gene, of which the 4.4-kb *Pst* I diagnostic fragment spans the entire coding region; and 3–20 copies per cell of the 3.6-kb *Bam*HI fragment of the HSV *tk* gene [E. Wagner et al, 1981].

Both the globin and *tk* genes were present as intact copies and, in some individuals, also in higher molecular weight DNA fragments. These larger fragments (eg, in animal #25 in Fig. 2) probably arise as a result of changes such as deletions and duplications. Such rearrangements could involve breakage and loss of at least one specific restriction site and utilization of a corresponding cleavage site in adjacent host DNA. The evidence is therefore consistent with occurrence of integration of the foreign sequences into mouse chromosomal DNA. Moreover, undigested DNA did not yield any hybridizable bands that would correspond to the free DNA of the plasmid.

A later series of postnatal mice from eggs injected with DNA from the same plasmid revealed animals in which donor gene sequences had persisted into adult life in stable form in all somatic tissues tested. In one animal, the intact human and viral genes were found in its tissue DNA; in another, at least some of the 3′ sequences flanking the coding region of the human β-globin gene were present. Transmission of these sequences occurred through the germ line of the latter individual, to approximately half of its progeny.

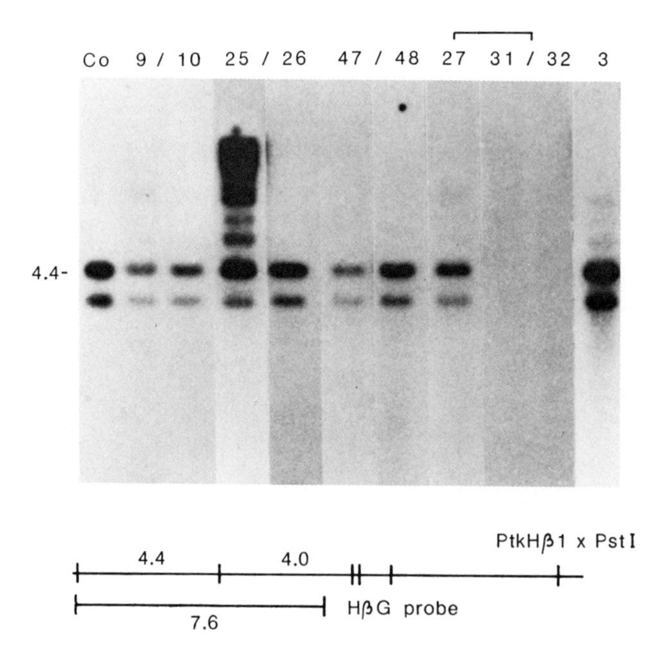

Fig. 2. Intact human β-globin (HβG) gene sequences found in DNA of near-term mouse fetuses (odd numbers) and their placentas (even numbers) after injecting the fertilized eggs with DNA from the PtkHβ1 plasmid containing the HβG and herpes simplex viral (HSV) thymidine kinase *(tk)* genes as well as pBR322 sequences. After digestion with the restriction endonuclease *Pst* I and hybridization with the ^{32}P-labeled 7.6-kb *Hind*III fragment of the plasmid DNA, the diagnostic 4.4-kb fragment spanning the human β-globin coding region was found in five animals and their placentas. Co, control *Pst* I digest of the plasmid DNA (from Wagner et al [1981]).

This frequency is indicative of a Mendelian pattern of inheritance for a single gene at a heterozygous locus and is strong evidence for integration in host chromosomal DNA [Stewart et al, 1982].

Among the five prenatal animals that were positive for the HSV *tk* gene, one clearly positive case of function of that gene was identified [E. Wagner et al, 1981]. An initial test was based on differences in substrate specificity of HSV and mouse *tk* enyzmes; the result was confirmed in an independent test involving enzyme neutralization with antiserum specifically directed against the HSV-type of enzyme. Thus, the donor gene was mediating production of its specific functional protein in vivo, even in the absence of any selective pressure or experimental induction.

Successful introduction and apparent integration and germ-line transmission of recombinant DNA from other species into mice, after microinjection

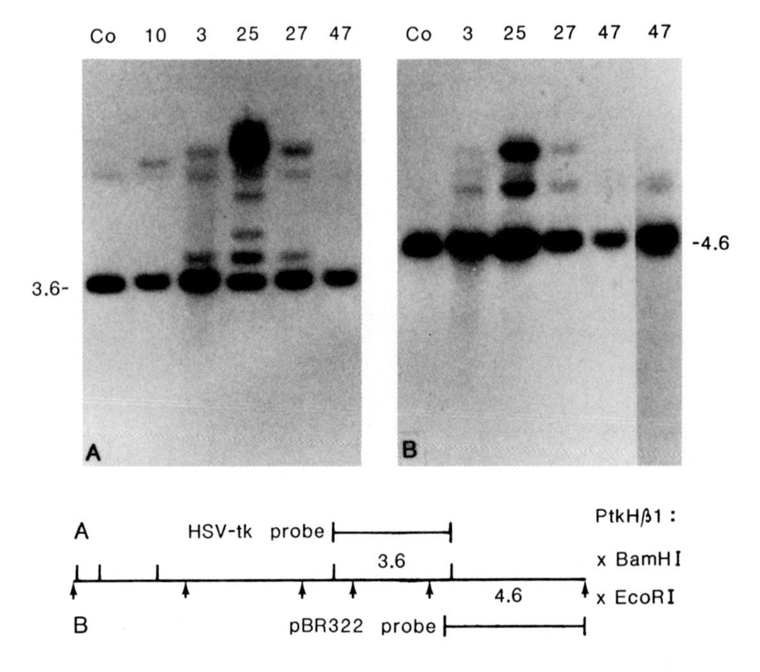

Fig. 3. The foreign intact thymidine kinase gene (A) and pBR322 plasmid sequences (B) were found in the same animals whose DNA was positive for human β-globin sequences (see Fig. 2). The labeled HSV *tk* probe revealed the presence of the diagnostic 3.6-kb *Bam*HI fragment; the pBR322 probe disclosed the 4.6-kb fragment. Co, control digests of the plasmid DNA (from Wagner et al [1981]).

into eggs, has also recently been reported for the rabbit β-globin gene [Costantini and Lacy, 1981], the HSV *tk* gene fused to the promoter of the mouse metallothionein gene [Brinster et al, 1981], and some SV40, HSV *tk*, and human interferon sequences [Gordon and Ruddle, 1981]. The fusion gene comprising the foreign *tk* also gave evidence of function [Brinster et al, 1981]. In another report on injection of the rabbit β-globin gene [T. Wagner et al, 1981], preliminary data were interpreted by the authors as evidence for expression of the rabbit gene. However, a Southern blot analysis of the DNA was carried out for only one experimental mouse and the hybridization revealed only internal sequences of the probe; there were no changes in the flanking regions of the sort that might indicate the presence of integrated (rather than free) DNA of the donor type.

Experiments with introduction of recombinant DNA into mouse eggs have thus had a very promising beginning. They have led to retention and transmission of foreign genes, as well as some expression of a constitutive gene.

It remains to be seen whether integration into a favorable chromosomal site will play a major role in the usefulness of this approach for analyzing control of tissue-specific expression of genes.

DEVELOPMENTALLY TOTIPOTENT TERATOCARCINOMA STEM CELLS AS VEHICLES FOR GENETIC ENGINEERING OF MICE

The second point of entry (Fig. 1) that we have devised for manipulation of the mouse genome is through mouse teratocarcinoma stem cells. These cells are apparently the malignant counterparts of normal early somatic embryo cells [Mintz, 1978a; Mintz et al, 1978] whose proliferation has been sustained although their differentiation has become restricted and chaotic (Fig. 4). Various tissues are found in the solid tumors formed when the cells are grown subcutaneously in a syngeneic host; however, these tissues are

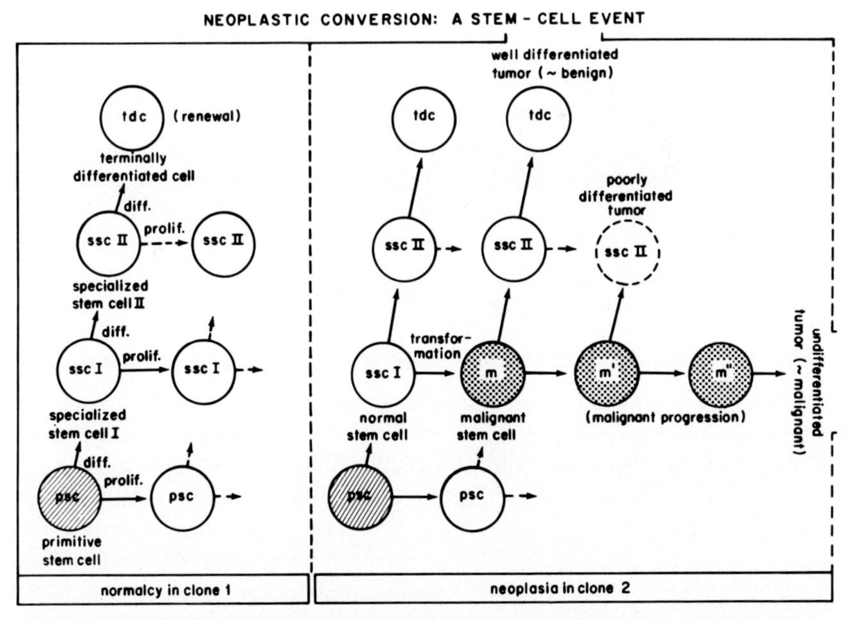

Fig. 4. Diagrammatic representation of neoplasia as a lessening or loss of differentiation in the mitotic progeny of an initially normal target stem cell, without loss of its proliferative capacity. Teratocarcinoma formation is attributable to neoplastic conversion of a primitive stem cell that was developmentally totipotent (from Mintz [1978], with permission of Academic Press, Inc).

usually immature and there are many types that are never present [Stevens, 1967]. The fact that the differentiated cells are derived from the primitive stem cells was demonstrated by the range of types obtained after single cells were transplanted to adult recipients [Kleinsmith and Pierce, 1964]. The stem cells are capable of being propagated as transplant lines in vivo (either in the solid or modified-ascites form) and also as cell culture lines.

If the stem cells were in fact the neoplastic derivatives of developmentally totipotent embryo cells, it seemed possible that their differentiation might become more normal if they were in the company of the corresponding normal stem cells, ie, if they were placed in early embryos. Such an experiment would be comparable to the production of allophenic mice—animals with cells of different genotypes, produced by bringing together blastomeres from two different embryos [Mintz, 1965b, 1974]. In this case, however, the teratocarcinoma stem cells would be used instead of one of the normal embryonic contributors. The experiment was carried out by injecting tumor stem cells from a karyotypically normal in vivo *transplant* line into blastocysts of another strain identifiable by numerous genetic markers. The tumor cells were indeed stably "normalized" by the accompanying embryo cells (Fig. 5), after almost a decade of serial transplantation in the malignant state [Mintz and Illmensee, 1975; Cronmiller and Mintz, 1978]. The neoplastic defect was evidently one of aberrant control of gene expression. Tumor-lineage cells were able to contribute, in the best cases, not only to all somatic tissues (along with embryo-lineage cells) but also to functional germ cells and, therefore, to progeny in which the tumor-cell genome was perpetuated. Successful but more limited differentiation was obtained in related experiments in which the tumor stem cells contributed to formation of one [Brinster, 1974] or of several [Papaioannou et al, 1975] somatic tissues, but not of all somatic tissues and never to germ cells.

Teratocarcinoma Cells as Precursors of New Mutant Mouse Strains

The capacity for normal and complete somatic and germinal differentiation in our experiments formed the basis for a novel experimental scheme: Inasmuch as the stem cells could be grown in culture, they might there serve as "surrogate eggs" in which selected genetic changes could be made at will; after injection of the cells into blastocysts, the change could ultimately be represented in the germ line and lead to new strains of mice of predetermined genotype [Mintz, 1978a,b; 1979]. Large numbers of stem cells could be subjected to specific selection or screening procedures in vitro and only cells with the gene change of interest would be injected into blastocysts. In the case of DNA transformation, the teratocarcinoma route would have certain advantages over the egg-injection route: Many tumor cells could be treated, and selected or screened; moreover, selected cell clones could be precharac-

Fig. 5. Orderly differentiation versus malignant proliferation of mouse teratocarcinoma cells as a function of the tissue environment in vivo. The normal animal on the right was derived from a blastocyst (C57BL/6 black strain) injected with teratocarcinoma stem cells (129 agouti strain). In all of its tissues, including the largely agouti-colored coat, tumor-lineage cells predominated and developed normally. However, when tumor stem cells were injected subcutaneously into a syngeneic adult, the cells formed a large tumor (left) (from Mintz [1978], with permission of Academic Press, Inc).

terized, eg, to ascertain the chromosomal site of DNA integration, by means of in situ hybridization. However, direct injection of DNA into the egg might still offer the advantage of immediate transition into all cells of an individual.

Among the range of options that the teratocarcinoma experimental system offers is production of animals with murine mutations that would yield mouse models of a corresponding human genetic disease. Production of animals with mutations in mitochondrial genes would also be possible [Mintz, 1979].

An example of a relevant human genetic disease is Lesch-Nyhan disease, due to a severe deficiency of the enzyme hypoxanthine phosphoribosyltransferase (HPRT). The human disorder results from an X-linked recessive defect that is usually fatal in affected males [Lesch and Nyhan, 1964; Seegmiller et al, 1967]. By first selecting mutagenized mouse teratocarcinoma stem cells of an in vitro culture line for resistance to the purine base analogue 6-thioguanine, an HPRT-deficient cell clone was isolated. When these cells were injected into blastocysts of another strain, mosaic animals were obtained [Dewey et al, 1977]. Autoradiographs of cells from explanted connective tissue showed that some colonies failed to take up ^{3}H-hypoxanthine, and were therefore HPRT-deficient, whereas others were normal (Fig. 6). Confirmation of the HPRT enzyme deficiency in cells of the tumor lineage was

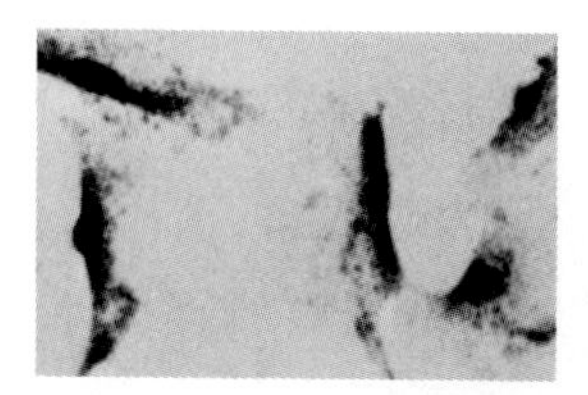
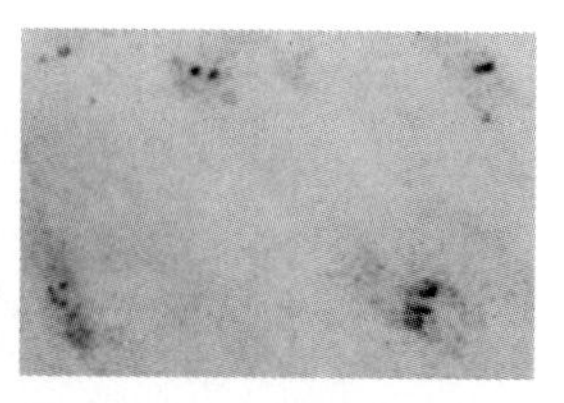

Fig. 6. In vitro cultures of cells from explanted connective tissue of a mosaic mouse. After incubation in [^{3}H]hypoxanthine, two kinds of colonies are seen in an autoradiograph. (Left) Part of a normal colony with heavily labeled cells. (Center) Part of an unlabeled colony that is markedly deficient in HPRT. These originate, respectively, from normal blastocyst-lineage (C57BL/6 strain) cells and mutant HPRT$^-$ teratocarcinoma cells (129 strain), after injection of the tumor stem cells into a blastocyst. The lineage of the two types of colonies in the culture is confirmed by starch gel electrophoretic separation (right) of extracts showing the strain-specific variants of glucosephosphate isomerase (from Dewey et al [1977]).

obtained by biochemical measurements. The tumor-strain cellular population (distinguishable by an independent strain-specific isozyme marker) had clearly retained the HPRT defect throughout differentiation. Strong selective pressure against the defective cells in blood of the mosaic animals provided an interesting parallel to selection against cells of the defective phenotype in the blood of human heterozygous female carriers.

The cell line used in this HPRT-mutant experiment, and in all other teratocarcinoma mutagenesis experiments until recently, was not karyotypically normal. The abnormality in the HPRT-deficient case was a trisomy that would be incompatible with production of viable germ-line progeny [Gropp, 1975], although it obviously permitted formation of all somatic tissues [Dewey et al, 1977]. We shall return presently to this important question of the requirement for a euploid culture line of teratocarcinoma stem cells in such experiments.

Another example in which cell differentiation has been obtained in vivo from teratocarcinoma cells with a specifically selected genetic change concerns chloramphenicol resistance, due to a mitochondrial gene mutation. After teratocarcinoma-cell fusion with cytoplasts from another cell line containing the mitochondrial mutation, the gene was stably incorporated into in vivo development by injection of the cybrid cells into blastocysts [Watanabe et al, 1978]. (This method of cybrid formation was previously employed by others [Bunn et al, 1974] for fibroblast cell lines.) We have also isolated mutant teratocarcinoma cells deficient in other enzymes, such as adenine phosphoribosyltransferase [Reuser and Mintz, 1979] or thymidine kinase

[Pellicer et al, 1980]; and have laid some of the groundwork for seeking teratocarcinoma mutants with deficiencies in specific receptor systems, such as the low-density lipoprotein receptor [Goldstein et al, 1979] and the transferrin receptor [Karin and Mintz, 1981].

Recombinant Gene Transfer Into Mice Via Teratocarcinoma Cells

Transfer of recombinant DNA into teratocarcinoma cells has been accomplished for selectable as well as unselectable genes. In our earlier experiments, mutant teratocarcinoma cells that were stably deficient in thymidine kinase activity (tk^-) were first isolated. They were then used as recipients, after addition of DNA in a calcium phosphate precipitate, and selection in HAT (hypoxanthine/aminopterin/thymidine) medium, to obtain cell clones that had taken up the foreign HSV *tk* gene [Pellicer et al, 1980], as in previous experiments [Wigler et al, 1977, 1978] carried out with fibroblast cell lines. Cotransfer of the unselectable human β-globin gene, by inclusion in the precipitate along with the HSV *tk* gene, was successful in some tk^+ transformants [Pellicer et al, 1980].

More recently, we have treated tk^- teratocarcinoma stem cells with DNA from the same plasmid vector, PtkHβ1, as was used for DNA injection into the pronucleus of fertilized eggs. In this case, the unselectable human β-globin gene was linked, in the same plasmid, to the selectable HSV *tk* gene. A high transformation efficiency was obtained after selection in HAT [Wagner and Mintz, 1982]. Hybridization tests disclosed the presence of intact copies (3–6 copies per cell) of the human gene in the majority of transformants. That these donor sequences were associated with cellular DNA was inferred from the presence of some new high molecular weight fragments, seen in Southern blots, and from stability of the donor sequences in tumors resulting from subcutaneous injection of the cells (Fig. 7).

In order to test for production of mRNA transcripts of the human gene, total polyadenylate-containing RNA was examined in four of the transformed cultured cell clones. Two of these showed hybridization to the human-gene probe. The RNA species from one of them resembled mature transcripts of the human β-globin gene; the others were larger in size [Wagner and Mintz, 1982]. Erythroid development was seen in some of the tumors, although it is not yet known whether any of the hemoglobin in them was attributable to the foreign gene.

Dominant-selection vectors enable recombinant genetic material to be transferred to wild-type cells, thereby circumventing the initial mutagenesis and selection steps (eg, to obtain tk^- mutant recipient cells). This could help to avoid occurrence of extraneous and undesirable genetic changes during the initial phases of gene transfer into teratocarcinoma cells in culture. The plasmid DNA vectors (pSV-*gpt*, from P. Berg), carrying the Escherichia coli

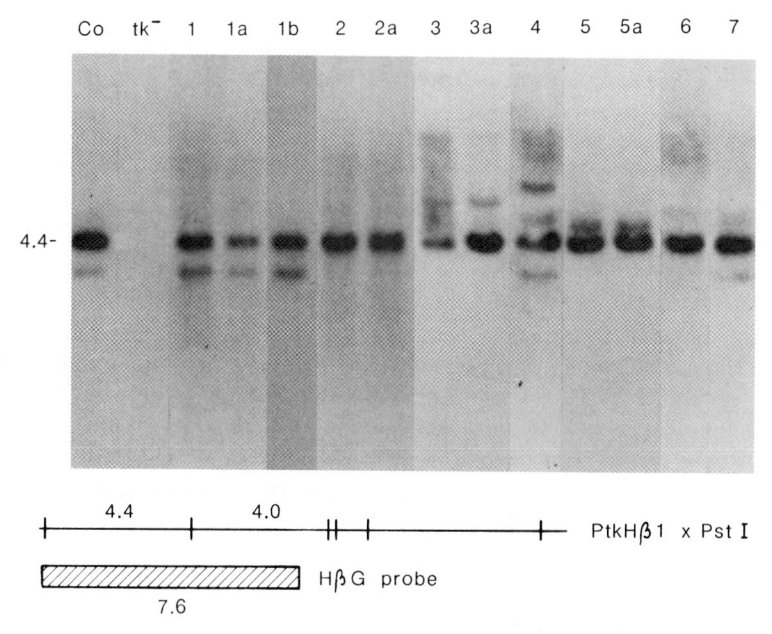

Fig. 7. By Southern blot hybridization, the diagnostic 4.4-kb fragment spanning the human β-globin coding region is present in cultured teratocarcinoma stem cells of seven separately isolated cell clones (# 1–7) and from tumors (1a, 2a, 3a, 5a) derived from four of them, after subcutaneous inoculation; one of the tumors was retransplanted, ultimately to a third host (1b). The PtkHβ1 plasmid was used for transfection; the restriction enzyme and labeled probe were the same as in Figure 2 (from Wagner and Mintz [1982], with permission of the American Society for Microbiology).

bacterial gene for xanthine-guanine phosphoribosyltransferase and some regulatory sequences from SV40, in addition to pBR322 [Mulligan and Berg, 1980, 1981], have provided such an opportunity. We isolated numerous transformants after addition of pSV2-*gpt* DNA in calcium phosphate to the cell culture medium and growth in mycophenolic acid, xanthine, hypoxanthine, aminopterin, and thymidine [Wagner and Mintz, 1982]. The donor DNA sequences were apparently stably integrated into the stem cells and their differentiated tumor derivatives. We may conclude that any cloned gene can be introduced into teratocarcinoma stem cells, even without mutagenesis of the cells, by cotransfer with, or linked transfer in, such vectors. New hybridization methods allowing in situ chromosomal visualization of genes [eg, Robins et al, 1981] should make it possible to characterize the transformants and to choose those of interest for experiments involving further differentiation in the soma and in the germ line in vivo.

A Culture Line of Developmentally Totipotent Teratocarcinoma Cells

A limitation in the teratocarcinoma "portal of entry" scheme has heretofore been the lack of a cell culture line characterized by karyotypic normalcy and also by developmental totipotency. Until recently, the only teratocarcinoma stem-cell sources that have proved capable of yielding all somatic tissues as well as germ cells, after injection into blastocysts, were two in vivo transplant lines—one chromosomally male (X/Y) and one female (X/X); these generated germ-line progeny in some mosaic phenotypic males and females, respectively [Mintz and Illmensee, 1975; Cronmiller and Mintz, 1978]. We have now been able to surmount this difficulty by establishing a cell culture line with the desired properties [Mintz and Cronmiller, 1981]. The line originated from a tumor produced by the method [Stevens, 1970] of transplanting an embryo to an ectopic site under the testis capsule, where embryogenesis becomes disorganized and early stem cells can persist. The grafted embryo from which the cell line arose was fortuitously X/X. Tumor cells were explanted in a culture medium enriched with various supplements [Rizzino and Sato, 1978; Oshima, 1978] which may have contributed to retention of karyotypic normalcy (Fig. 8) and enabled omission of a feeder-cell layer. The cell line has been designated METT-1 (the first Mouse Euploid Totipotent Teratocarcinoma cell line) [Mintz and Cronmiller, 1981].

The developmental potential of these stem cells (of the 129 strain) was assayed by microinjecting them into blastocysts of the C57BL/6 strain [Stewart and Mintz, 1981]. Among the postnatal survivors, 13% exhibited the coat colors of the tumor strain as well as those of the embryo strain. This frequency is much higher than in all previous experiments, whether with cell culture lines or transplant lines of teratocarcinoma stem cells. Biochemical markers documented the capacity of tumor-lineage cells to contribute to formation of all somatic tissues and to remain normal. Of ten mosaic-coat females that were test-mated to males of the blastocyst (recessive-color) strain, two have produced some offspring of the diagnostic tumor-strain agouti color; F_1 heterozygotes, in turn, transmitted their tumor-strain genes to F_2 homozygous segregants (Fig. 9) [Stewart and Mintz, 1981; and unpublished data].

The METT-1 line of teratocarcinoma stem cells therefore bridges the gap between soma and germ line, and between propagation and genetic manipulation of cells in vitro and embryogenesis in vivo (Fig. 10).

REPLACEMENT OF TOTIPOTENT HEMATOPOIETIC STEM CELLS IN EARLY FETAL LIFE

The hematopoietic lineage is of special interest with respect to its differentiation, pathology, and normal and defective genes. It is not yet possible to

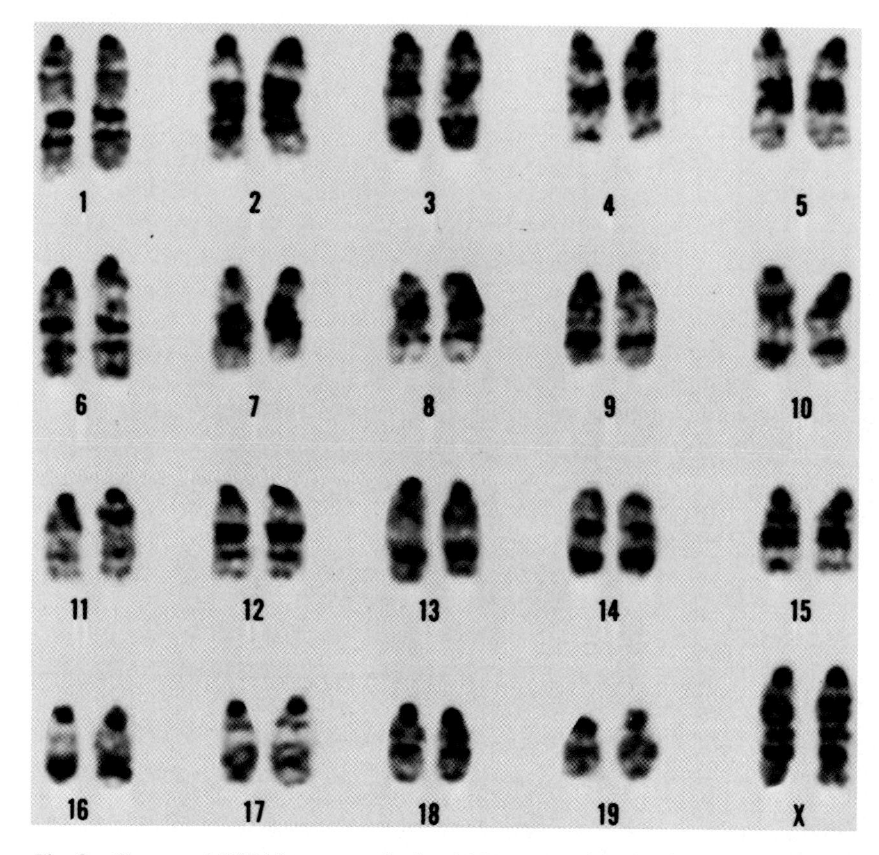

Fig. 8. The normal X/X (chromosomally female) karyotype of the developmentally totipotent METT-1 teratocarcinoma stem cell culture line is seen in a G-banded preparation (from Mintz and Cronmiller [1981], with permission of Plenum Publishing Corp).

introduce recombinant genes into fetal liver hematopoietic stem cells, from which the definitive bone marrow and spleen cells are derived. However, we have been able to replace the cells themselves in utero with genetically marked donor cells that permit many questions concerning gene control of blood development and disease to be analyzed in vivo.

Transfer of Normal Totipotent Hematopoietic Stem Cells Into Genetically Defective Fetuses

Microinjection into the efferent blood vessels of the fetal placenta results in rapid entry of the injected material into the circulation of the early fetus,

Fig. 9. Three generations of mice derived from the karyotypically normal METT-1 teratocarcinoma stem cell line. The animal at the left is a female that was produced from a blastocyst (C57BL/6 strain) injected with teratocarcinoma cells of this line; those cells developed normally, along with embryo-lineage cells, and contributed the agouti hair clones to its black-and-agouti coat. Some of the mosaic animal's germ cells were also of tumor-strain lineage, as seen in the all-agouti color of an offspring (center) from a mating to the recessive-color C57BL/6 strain. The agouti animal at the right, from the following generation, also displays the tumor-strain color, transmitted from the F_1 (from Stewart and Mintz [1981]).

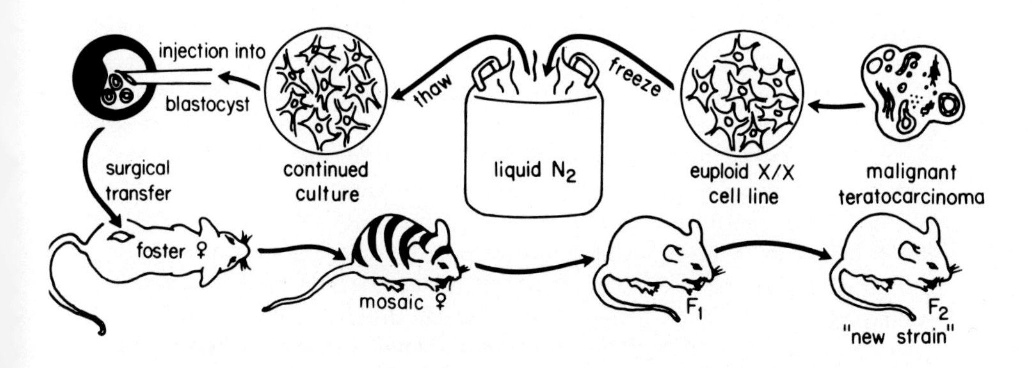

Fig. 10. Diagrammatic scheme of the experiment with teratocarcinoma stem cells of the METT-1 line (see Fig. 9). The occurrence of germ-line progeny of the tumor strain serves as a model for production of new strains of mice with perdetermined genetic changes. The changes would be introduced while the teratocarcinoma cells are still in culture, eg, by gene transfer, or selection for cells with a specific murine nuclear or mitochondrial gene change of interest (from Stewart and Mintz [1981]).

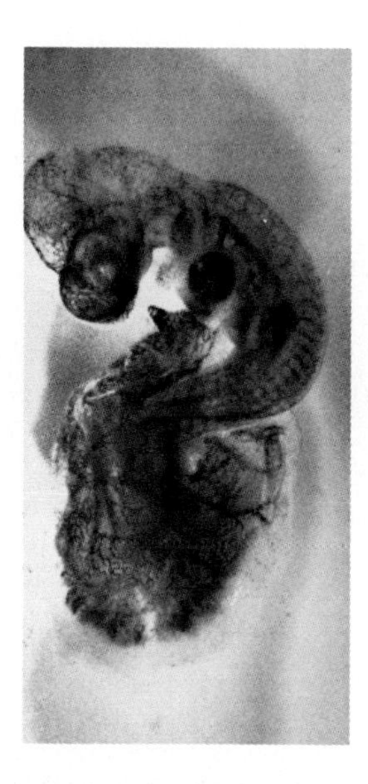

Fig. 11. The feasibility of introducing material into the circulatory system of a ten-day mouse embryo is shown by the presence of India ink in the heart and blood vessels of the embryo (and of associated extraembryonic tissue) after microinjection of ink into a placental blood vessel (from Fleischman and Mintz [1979]).

as seen after injection with India ink (Fig. 11). Allogeneic normal fetal liver cells have been successfully introduced by this route [Fleischman and Mintz, 1979]. In order to provide the relatively small number of donor cells with a selective advantage over the endogenous proliferating hematopoietic cells, a genetically disadvantaged recipient fetus was used, ie, one whose own hematopoietic stem cells were defective. The *W*-series anemic mutants [Russell and Bernstein, 1966] have such a stem-cell defect [McCulloch et al, 1964]. When recipients at 11 days' gestation were injected with normal fetal liver cells from slightly older donors, strain-specific hemoglobin markers demonstrated that the erythroid lineage had been successfully replaced—in some cases completely so, even before birth [Fleischman and Mintz, 1979]. The genetically defective recipients were then no longer anemic (Fig. 12) and were able to live a full lifespan. The rate of replacement varied directly with the severity of the genetic defect in the *W*-allelic series (Fig. 13); segregants of nonanemic genotypes did not undergo detectable replacement [Fleischman et al, 1982]. Further analyses with concurrent markers in granulocytic cells (with *beige*-genotype giant lysosomes) and lymphoid cells (by means of serum allotypes) have shown that cell replacement in fact occurred at the

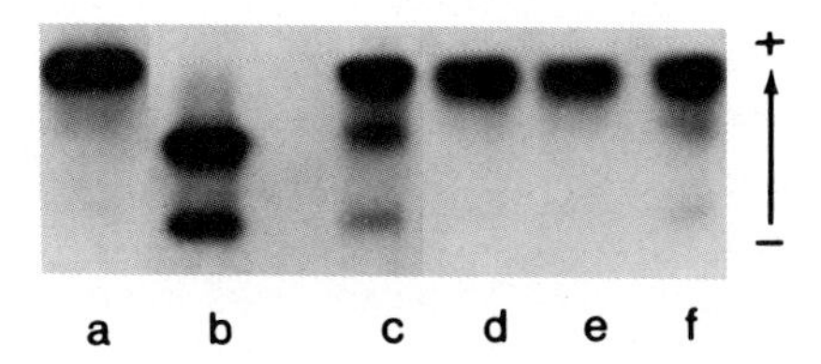

Fig. 12. Analyses of strain-specific electrophoretic patterns of hemoglobin (shown in controls in lanes a and b) demonstrate that four recipient animals (of the *W/W* genotype, with a hematopoietic stem-cell defect) that were analyzed five days after birth (lanes c-f) already have hemoglobin, and therefore red blood cells, largely or entirely of the normal fetal-liver-donor strain. Cell inoculation via the placenta was carried out at 11 days of fetal life (from Fleischman and Mintz [1979]).

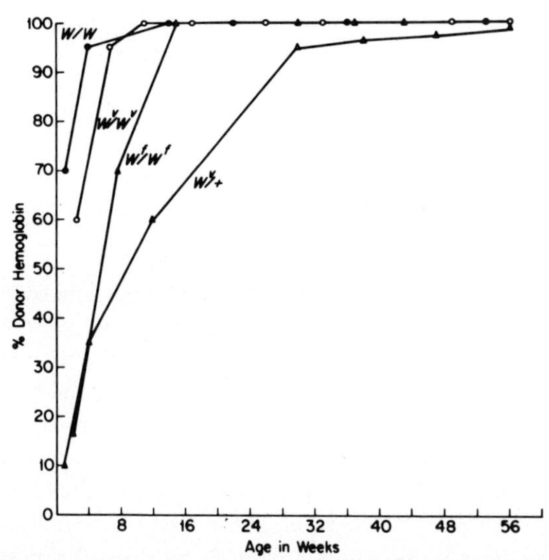

Fig. 13. Postnatal kinetics showing complete substitution of donor- for host-strain red blood cells (judged from hemoglobin variants). Rate of replacement is directly correlated with decreasing severity of the defect in recipients of the *W/W*, W^v/W^v, W^f/W^f, and $W^v/+$ genotypes (from Fleischman et al [1982]).

totipotent hematopoietic stem-cell level [Fleischman et al, 1982]. These stem cells were capable of sustained long-term self-renewal as well as differentiation. Thus, this experimental system uniquely provides a means of placing early-stage, genetically marked, totipotent hematopoietic stem cells in the normal developmental sequence of tissues and microenvironments, progressing from fetal liver to spleen, thymus, and bone marrow. There is no graft-versus-host reaction or immunological rejection of these young cells and no irradiation of the host is needed. Cell and tissue interactions critical for normal differentiation and gene expression can therefore be examined in the various hematopoietic derivatives, and the etiology of many blood and immune diseases or malignancies can be analyzed in vivo.

Adult Hematopoietic Cells in Fetuses

One of the questions thus far investigated is whether hematopoietic cells from adult bone marrow can resume fetal differentiation if they mature in a fetal environment. An antigen, Ft, present on fetal erythrocytes and absent on adult erythrocytes was used as the discriminant. Adult marrow cells were introduced into fetuses via the placenta and, after birth, erythrocytes were fractionated by fluorescence-activated cell sorting based on antibody binding to Ft. Lysates of Ft-positive (ie, fetal) cells did not detectably contain donor-strain hemoglobin. The adult cells therefore did not resume fetal-specific Ft expression [Blanchet et al, 1982]. The longer-term fate of allogeneic bone marrow cells in fetal recipients is currently under investigation.

SUMMARY

We have demonstrated that pure genes can be introduced into mouse eggs, remain into adult life, and be transmitted to progeny. Moreover, at least one foreign gene (HSV-*tk*) can be fully functional in the animals that develop, even in the absence of induction or selection. A second route for predetermined genetic change is defined by mouse teratocarcinoma stem cells, of which a new cell culture line (METT-1) is developmentally totipotent. Changes to mutant forms of murine genes, or introduction of recombinant cloned genes, can be effected while these cells are in culture; after transfer of the cells to early embryos, it should be possible to produce, through the germ-line, strains of mice whose genotype has been established in advance. An entry route specific for early totipotent hematopoietic stem cells has also been shown to be feasible, by microinjection of these cells into the early fetus via the placental circulation in utero. Allogeneic combinations are accepted with impunity.

Thus, even in an organism as complex as a mammal, there are now promising new options for analyzing in vivo the developmental regulation of gene expression and for realizing the experimental replacement of defective genes.

ACKNOWLEDGMENTS

This program has been supported by US Public Health Service grants HD-01646, CA-06927, and RR-05539, and by an appropriation from the Commonwealth of Pennsylvania. The participation of my many colleagues and collaborators is gratefully acknowledged.

REFERENCES

Blanchet JP, Fleischman RA, Mintz B (1982): Murine adult hematopoietic cells produce adult erythrocytes in fetal recipients. Dev Genet 3:197–205.

Brinster RL (1974): The effect of cells transferred into the mouse blastocyst on subsequent development. J Exp Med 140:1049–1056.

Brinster RL, Chen HY, Trumbauer M, Senear AW, Warren R, Palmiter RD (1981): Somatic expression of herpes thymidine kinase in mice following injection of a fusion gene into eggs. Cell 27:223–231.

Bunn CL, Wallace DC, Eisenstadt JM (1974): Cytoplasmic inheritance of chloramphenicol resistance in mouse tissue culture cells. Proc Natl Acad Sci USA 71:1681–1685.

Costantini F, Lacy E (1981): Introduction of a rabbit beta-globin gene into the mouse germ line. Nature (London) 294:92–94.

Cronmiller C, Mintz B (1978): Karyotypic normalcy and quasinormalcy of developmentally totipotent mouse teratocarcinoma cells. Dev Biol 67:465–477.

Dewey MJ, Martin DW Jr, Martin GR, Mintz B (1977): Mosaic mice with teratocarcinoma-derived mutant cells deficient in hypoxanthine phosphoribosyltransferase. Proc Natl Acad Sci USA 74:5564–5568.

Diwan SB, Stevens LC (1976): Development of teratomas from the ectoderm of mouse egg cylinders. JNCI 57:937–942.

Fleischman RA, Mintz B (1979): Prevention of genetic anemias in mice by microinjection of normal hematopoietic stem cells into the fetal placenta. Proc Natl Acad Sci USA 76:5736–5740.

Fleischman RA, Custer RP, Mintz B (1982): Totipotent hematopoietic stem cells: Normal self-renewal and differentiation after transplantation between mouse fetuses. Cell 30:351–359.

Fritsch EF, Lawn RM, Maniatis T (1980): Molecular cloning and characterization of the human β-like globin gene cluster. Cell 19:959–972.

Goldstein JL, Brown MS, Krieger M, Anderson RGW, Mintz B (1979): Demonstration of the low density lipoprotein receptors in mouse teratocarcinoma stem cells and description of a method for producing receptor-deficient mutant mice. Proc Natl Acad Sci USA 76:2843–2847.

Gordon JW, Ruddle FH (1981): Integration and stable germ line transmission of genes injected into mouse pronuclei. Science 214:1244–1246.

Gropp A (1975): Morphogenesis in trisomic mouse embryos. Clin Genet 8:839.

Jaenisch R, Mintz B (1974): Simian virus 40 DNA sequences in DNA of healthy adult mice derived from preimplantation blastocysts injected with viral DNA. Proc Natl Acad Sci USA 71:1250–1254.

Karin M, Mintz B (1981): Receptor-mediated endocytosis of transferrin in developmentally totipotent mouse teratocarcinoma stem cells. J Biol Chem 256:3245–3252.

Kleinsmith LJ, Pierce GB Jr (1964): Multipotentiality of single embryonal carcinoma cells. Cancer Res 24:1544–1551.

Lesch M, Nyhan WL (1964): A familial disorder of uric acid metabolism and central nervous system function. Am J Med 36:561–570.

Lin TP (1966): Microinjection of mouse eggs. Science 151:333–337.

McCulloch EA, Siminovitch L, Till JE (1964): Spleen-colony formation in anemic mice of genotype WW^v. Science 144:844–846.

Metcalf D, Moore MAS (1971): "Haemopoietic Cells." Amsterdam: Elsevier/North Holland.

Mintz B (1965a): Nucleic acid and protein synthesis in the developing mouse embryo. In Wolstenholme GEW, O'Connor M (eds): "Ciba Foundation Symposium on Preimplantation Stages of Pregnancy." London: J & A Churchill, pp 145–155.

Mintz B (1965b): Genetic mosaicism in adult mice of quadriparental lineage. Science 148:1232–1233.

Mintz B (1971): Clonal basis of mammalian differentiation. In Davies DD, Balls M (eds): "Control Mechanisms of Growth and Differentiation." 25th Symposium of the Society for Experimental Biology. England: Cambridge University Press, pp 345–370.

Mintz B (1974): Gene control of mammalian differentiation. Annu Rev Genet 8:411–470.

Mintz B (1978a): Gene expression in neoplasia and differentiation. Harvey Lect 71:193–246.

Mintz B (1978b): Genetic mosaicism and in vivo analyses of neoplasia and differentiation. In Saunders GF (ed): "Cell Differentiation and Neoplasia." 30th Annual Symposium on Fundamental Cancer Research, M.D. Anderson Hospital and Tumor Institute. New York: Raven Press, pp 27–53.

Mintz B (1979): Teratocarcinoma cells as vehicles for introducing mutant genes into mice. Differentiation 13:25–27.

Mintz B (1982): Putting genes into mice. In Robberson DL, Saunders GF (eds): "Perspectives on Genes and the Molecular Biology of Cancer." 35th Annual Symposium on Fundamental Cancer Research, M.D. Anderson Hospital and Tumor Institute. New York: Raven Press, pp. 207–224.

Mintz B, Cronmiller C (1981): METT-1: A karyotypically normal in vitro line of developmentally totipotent mouse teratocarcinoma cells. Somatic Cell Genet 7:489–505.

Mintz B, Cronmiller C, Custer RP (1978): Somatic cell origin of teratocarcinomas. Proc Natl Acad Sci USA 75:2834–2838.

Mintz B, Illmensee K (1975): Normal genetically mosaic mice produced from malignant teratocarcinoma cells. Proc Natl Acad Sci USA 72:3585–3589.

Mulligan RC, Berg P (1980): Expression of a bacterial gene in mammalian cells. Science 209:1422–1427.

Mulligan RC, Berg P (1981): Selection for animal cells that express the *Escherichia coli* gene coding for xanthine-guanine phosphoribosyltransferase. Proc Natl Acad Sci USA 78:2072–2076.

Oshima R (1978): Stimulation of the clonal growth and differentiation of feeder layer dependent mouse embryonal carcinoma cells by beta-mercaptoethanol. Differentiation 11:149–155.

Papaioannou VE, McBurney MW, Gardner RL, Evans MJ (1975): Fate of teratocarcinoma cells injected into early mouse embryos. Nature (London) 258:70–73.

Pellicer A, Wagner EF, El Kareh A, Dewey MJ, Reuser AJ, Silverstein S, Axel R, Mintz B (1980): Introduction of a viral thymidine kinase gene and the human β-globin gene into developmentally multipotential mouse teratocarcinoma cells. Proc Natl Acad Sci USA 77:2098–2102.

Pierce GB (1967): Teratocarcinoma: Model for a developmental concept of cancer. Curr Top Dev Biol 2:223–246.

Reuser AJJ, Mintz B (1979): Mouse teratocarcinoma mutant clones deficient in adenine phosphoribosyltransferase and developmentally pluripotent. Somatic Cell Genet 5:781–792.

Rizzino A, Sato G (1978): Growth of embryonal carcinoma cells in serum-free medium. Proc Natl Acad Sci USA 75:1844–1848.

Robins DM, Ripley S, Henderson AS, Axel R (1981): Transforming DNA integrates into the host chromosome. Cell 23:29–39.

Russell ES, Bernstein SE (1966): Blood and blood formation. In Green EL (ed): "Biology of the Laboratory Mouse." New York: McGraw-Hill, pp 351–372.

Seegmiller JE, Rosenbloom FM, Kelly WN (1967): Enzyme defect associated with a sex-linked human neurological disorder and excessive purine synthesis. Science 155:1682–1684.

Southern EM (1975): Detection of specific sequences among DNA fragments separated by gel electrophoresis. J Mol Biol 98:503–517.

Stevens LC (1967): The biology of teratomas. Adv Morphog 6:1–31.

Stevens LC (1970): The development of transplantable teratocarcinomas from intratesticular grafts of pre- and postimplantation mouse embryos. Dev Biol 21:364–382.

Stewart TA, Mintz B (1981): Successive generations of mice produced from an established culture line of euploid teratocarcinoma cells. Proc Natl Acad Sci USA 78:6314–6318.

Stewart TA, Wagner EF, Mintz B (1982): Human β-globin gene sequences injected into mouse eggs, retained in adults, and transmitted to progeny. Science 217:1046–1048

Wagner EF, Mintz B (1982): Transfer of nonselectable genes into mouse teratocarcinoma cells and transcription of the transferred human β-globin gene. Mol Cell Biol 2:190–198.

Wagner EF, Stewart TA, Mintz B (1981): The human β-globin gene and a functional viral thymidine kinase gene in developing mice. Proc Natl Acad Sci USA 78:5016–5020.

Wagner TE, Hoppe PC, Jollick JD, Scholl DR, Hodinka RL, Gault JB (1981): Microinjection of a rabbit β-globin gene into zygotes and its subsequent expression in adult mice and their offspring. Proc Natl Acad Sci USA 78:6376–6380.

Watanabe T, Dewey MJ, Mintz B (1978): Teratocarcinoma cells as vehicles for introducing specific mutant mitochondrial genes into mice. Proc Natl Acad Sci USA 75:5113–5117.

Wigler M, Pellicer A, Silverstein S, Axel R (1978): Biochemical transfer of single-copy eucaryotic genes using total cellular DNA as donor. Cell 14:725–731.

Wigler M, Silverstein S, Lee L-S, Pellicer A, Cheng Y-C, Axel R (1977): Transfer of purified herpes virus thymidine kinase gene to cultured mouse cells. Cell 11:223–232.

III. Oogenesis and Early Development

Gene Structure and Regulation in Development, pages 137–146

The Transcription Unit of Lampbrush Chromosomes

Joseph G. Gall, Manuel O. Diaz, Edwin C. Stephenson, and Kathleen A. Mahon
Department of Biology, Yale University, New Haven, Connecticut 06511

Few other chromosomes so insistently demand a functional interpretation as the lampbrush chromosomes of oocytes. Their chief defining feature, the lateral loops, represent regions of intense RNA synthesis, and since hundreds of loop pairs occur on the longer chromosomes of some animals, such as salamanders, the variety of transcription products must be very large indeed. Questions about lampbrush chromosome function can be posed at several levels, corresponding roughly to levels of morphological complexity: (1) What is the nature of a single transcription unit (TU)? (2) What is the relationship between a loop and its constituent TUs? (3) What is the biological significance of so many active TUs in a single nucleus?

This review will largely bypass the second and third questions, as important as they are, and concentrate on the first, the nature of the transcription unit. Not only is the TU an obvious unit of synthetic activity, but it corresponds to a morphological unit definable at the light microscope level. This happy circumstance greatly facilitates a variety of cytochemical tests, particularly in situ nucleic acid hybridization [Callan, 1982; Varley et al, 1980a,b; Diaz et al, 1981] and immunofluorescence detection of proteins [Sommerville, 1981].

During the 1950s and early 1960s several important features of lampbrush chromosome transcription were established [reviewed by Callan, 1982]. RNA was shown to be present in the loops of salamander chromosomes by staining with basic dyes and by UV absorption and to be synthesized in these loops by its rapid labeling with radioactive precursors. In addition, detailed morphological studies demonstrated that the ribonucleoprotein matrix was not uniformly distributed along a loop, but occurred in one or more gradients (thin-thick regions).

These observations were nicely tied together in the electron microscope studies of Miller and his co-workers [Miller and Hamkalo, 1972; Hamkalo and Miller, 1973], who first clearly recognized the relationship between RNA synthesis and loop morphology. Using the same hypotonic spreading technique that was so successful in elucidating the structure of rDNA TUs, Miller found that lampbrush chromosome loops consisted of a very delicate axis, presumably containing DNA, from which innumerable fibrils extended laterally. Like the "Christmas trees" of the rDNA, but on a much grander scale, these fibrils were arranged in gradients of increasing length. Miller postulated that the fibrils represented nascent RNA transcripts (associated with protein) and that one thin-thick gradient corresponded to a TU. The basic correctness of this view has been accepted by all later workers.

An important feature of Miller's observations, confirmed by several subsequent studies [Angelier and Lacroix, 1975; Scheer et al, 1976, 1979; Sommerville, 1977; Hill, 1979], is the enormous length of the nascent transcripts. Lateral fibers of 5–10 μm are easily found in spread preparations of newt chromosomes, and longer ones certainly exist although they are difficult to measure. Assuming that the nascent RNA molecules are some 2–5 times longer than the fibers seen with the electron microscope [Miller and Hamkalo, 1972; Hill, 1979], we can calculate that some transcripts must be 100 kilobases (kb) or more in length (1μm of RNA = 3 kb). A second striking feature of lampbrush chromosome transcripts is their close spacing along the DNA axis. Whereas transcripts on somatic TUs are often widely separated from one another, those on lampbrush TUs are usually maximally packed. In this respect they resemble the TUs of the ribosomal RNA genes. The picture of a lampbrush chromosome TU that emerges is thus of an unusually long segment of DNA with an unusually high density of nascent transcripts. The TU is, in fact, so massive that its dimensions are easy to estimate from light microscopy; that is, from the lengths of the thin-thick regions (Fig. 1).

Earlier accounts of lampbrush chromosomes either assumed or stated that most loops contained only one thin-thick region of matrix [Callan, 1955; Callan and Lloyd, 1960; Gall, 1956]. A few cases were known where multiple thin-thick regions occurred on one loop, but these were thought to be exceptions or to have resulted from the bases of loops pulling away from the main chromosome axis. Recent studies are unanimous in finding many loops with multiple thin-thick regions [Makarov and Safronov, 1976; Scheer et al, 1976, 1979; Diaz et al, 1981]. The change in viewpoint is due to the fact that lampbrush chromosomes are now usually studied after centrifugation onto a supporting microscope slide, whereas previously they were examined free in the isolation medium. In the latter case it is almost impossible to follow thinner loops along their entire length, because they extend well

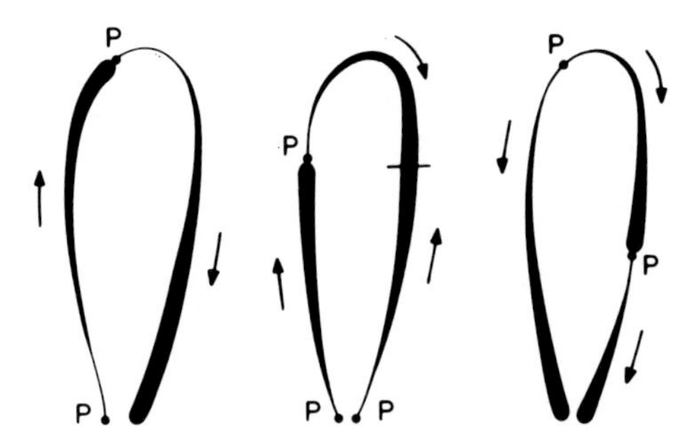

Fig. 1. Diagram of transcription units within lampbrush chromosome loops. In the first case, two promoters (P) are active and transcription proceeds in the same direction in both units. In the second case, two units of opposite polarity "run into" each other, whereas in the third case two units of opposite polarity diverge from closely adjacent origins.

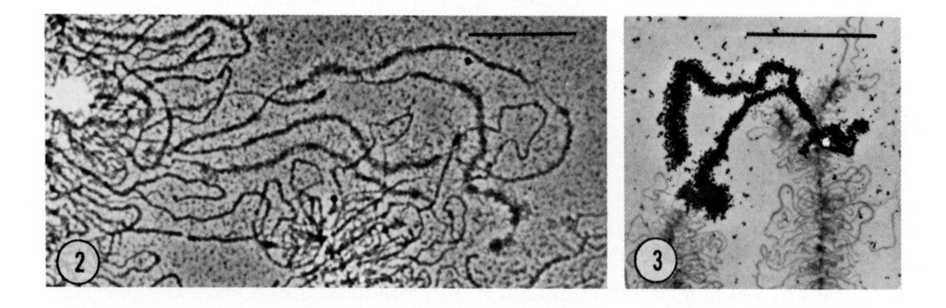

Fig. 2. Lampbrush chromosome loops at the histone locus of the newt, Notophthalmus. Several very long transcription units occur on the same loop. Bar = 20 μm.

Fig. 3. A pair of very long loops at the histone locus of the newt, Notophthalmus. ^{3}H-labeled cRNA, transcribed in vitro from satellite 1 DNA, was hybridized in situ to the nascent RNA transcripts of the chromosome. In this case each loop contains one very long transcription unit. Bar = 50 μm.

beyond one focal plane and they are in constant Brownian motion. Thicker loops with massive matrix were the easiest to study, and these, almost by definition, are ones in which a single TU extends all the way around the loop. There was thus a strong observational bias against seeing multiple TUs until the centrifugation technique came into general use and thinner loops were closely observed (Fig. 2).

What has just been said about multiple TUs does not materially alter the conclusion that lampbrush chromosome TUs are unusually long. In Xenopus, Hill [1979] measured the lengths of TUs at various times during oocyte development and arrived at an average length of 6.3 μm for Stage 3 oocytes and similar values for earlier stages. The average is certainly greater for tailed amphibians such as Triturus, Pleurodeles, and Notophthalmus, whose loops are considerably longer than those of Xenopus. Individual TUs of up to 50 μm are easy to find in Notophthalmus (Figs. 2, 3). In what follows we will assume that a loop contains some integral number of TUs. Usually that number is small, but instances of 8–10 TUs on one loop certainly occur. Our studies of the histone locus in the newt Notophthalmus have led to a general model of lampbrush chromosome TUs in which failure of transcription termination is a key element. Specifically, we propose the following model for lampbrush chromosomes (Fig. 1):

1) Transcription begins at a promoter sequence 5′ to a structural gene.
2) Transcription continues past the normal (somatic?) termination signals at the 3′ end of the structural gene.
3) Transcription terminates at one of three places:
   a) where two transcription units of opposite polarity "run into" each other;
   b) where the next downstream TU has already initiated transcription; or
   c) where the loop enters the chromomere.

The key element in development of this model was our ability to correlate the detailed molecular structure of the histone genes of Notophthalmus with their lampbrush chromosome transcripts as studied by in situ hybridization [Diaz et al, 1981; Gall et al, 1981; Stephenson et al, 1981a,b]. Molecular studies showed that most histone genes in Notophthalmus are organized as 9-kb clusters containing five coding regions (Fig. 4), four on one strand (H1, H2A, H3, and H4) and one on the other (H2B). These clusters are not closely adjacent as in Drosophila and several sea urchins, but are separated by long tracts of a simple sequence DNA having a 222-bp repeat (satellite 1). The tracts of satellite 1 range in length from less than 10 kb to at least 200 kb. By in situ hybridization of cloned DNA probes to the nascent RNA transcripts on the chromosomes, we can identify the sequences that are being transcribed on a given set of loops. In the case of the histone genes, we showed that transcripts of the 9-kb histone repeat and transcripts of satellite 1 are found on the same cluster of loops. This was the basic cytological observation that suggested some type of failure of transcription termination.

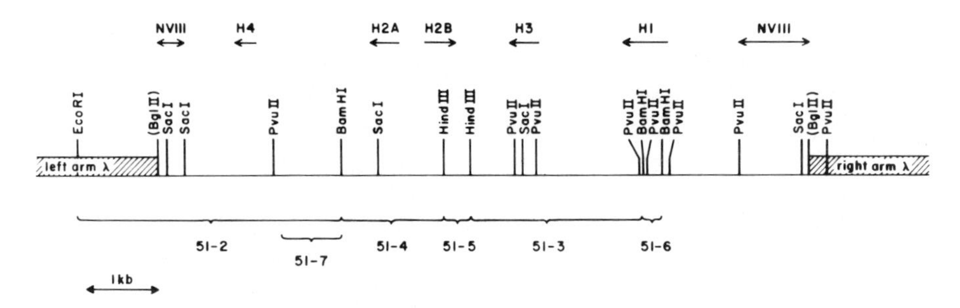

Fig. 4. Map of ES6 Nv51, a clone containing the major histone gene repeat of the newt, Notophthalmus viridescens. The positions and polarities of the five histone genes are indicated by arrows. 51–2 through 51–7 represent segments that have been subcloned in pBR322 or M13 (in the text 51–7A and 51–7B refer to the two strands of 51–7 cloned in M13).

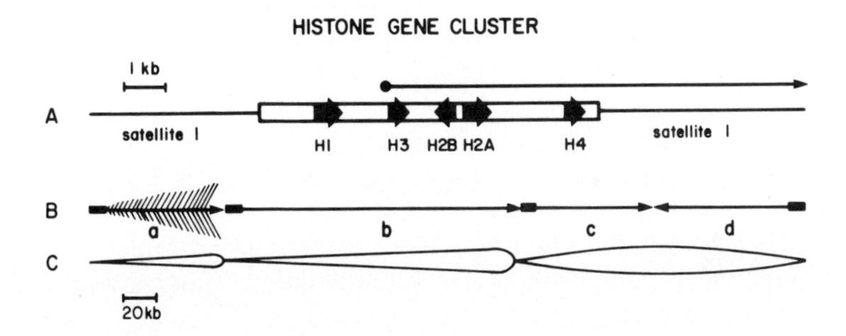

Fig. 5. Interpretation of transcription at the histone gene locus of the newt, Notophthalmus. Transcription initiates at a histone gene promoter but fails to terminate at the end of the gene. Long transcripts are formed, which contain both histone and satellite sequences. A) Transcription beginning at the H3 promoter and continuing through downstream sequences in the histone gene cluster into the satellite 1 spacer region. B) Four successive histone gene clusters (boxes) separated by satellite 1 sequences. Arrows represent possible directions of transcription, leading to four transcription units (a, b, c, d). Transcription units c and d "run into" each other between the histone clusters. Unit d must originate at the histone H2B promoter, the only promoter leading to leftward transcription in this diagram, whereas a, b, and c could originate at any of the other four promoters. Nascent transcripts are indicated on unit a. C) Light microscope appearance of the transcription units diagrammed in B.

From a detailed in situ analysis using various single-stranded probes in the vector M13, we have concluded that transcription initiates within a histone cluster at one of the five histone promoters, but then continues past the normal termination signals at the 3′ end of the gene into the rest of the histone cluster and eventually into the satellite region (Fig. 5). Several lines of evidence support this view [Diaz et al, 1981, and in preparation]:

1) Transcripts from both strands of satellite 1 occur at the histone locus. Where multiple TUs occur on a loop, TUs of one morphological polarity all contain transcripts of the same satellite strand, whereas those of the opposite polarity contain transcripts of the other strand. This observation is consistent with the view that transcription can initiate on either strand.
2) Transcripts of the satellite strand attached to the H2B coding region are rarer than transcripts of the satellite strand attached to the H1, H2A, H3, and H4 coding regions. This finding suggests that the frequency of transcription initiation sites on a given strand is related to the number of histone gene promoters.
3) Transcripts of sequences between the coding regions of the histone genes occur on all TUs. In fact, one probe that contains only such a spacer region (51–7A, Fig. 4) hybridizes as strongly to lampbrush chromosome loops as a probe containing coding regions.
4) One of the strongest lines of evidence that transcription begins within the histone cluster is the simple observation that entire TUs are labeled when one hybridizes with histone coding regions. The argument is as follows: Total labeling of a TU with a given probe implies that all transcripts in the unit contain the sequence. This will be true if the sequence in question is just downstream from the promoter. If transcription began randomly, for instance, if it began in the middle of the satellite region, then many TUs would be labeled along only part of their length when hybridized with probes from within the histone cluster.
5) The transcription of a region depends strongly on its position in the histone cluster, whereas random initiation would make all regions essentially equally represented in the transcripts. Clones 51–7A and 51–7B are particularly instructive. Since 51–7A is downstream of three promoters (H1, H3, and H2A), one would expect its transcripts to be common; conversely, since 51–7B is upstream of the only promoter on its strand, one would not expect it to be transcribed at all. In situ hybridization confirms these predictions with the exception that very rare transcripts of 51–7B occur. The latter presumably result from readthrough of an entire histone cluster.

In summary, in situ hybridization experiments demonstrate that all of the following sequences are represented in nascent transcripts at the histone loci: (1) the coding strand of all the histone genes; (2) the noncoding strand of some of the histone genes; (3) spacers between the histone genes (within the clusters); and (4) both strands of satellite 1. Not all of these sequences occur in every TU; instead the patterns found by in situ hybridization are consistent with random initiation at one of the five promoters within each cluster followed by readthrough into adjacent satellite regions.

It is tempting to suppose that the features revealed by the histone loci are common to lampbrush TUs, ie, initiation at structural gene promoters, failure of termination, and readthrough into adjacent noncoding sequences. Histone TUs are morphologically similar in all obvious respects to other lampbrush TUs; indeed, without in situ hybridization it is impossible to recognize them on the basis of structure alone. A common pattern of transcription would explain, at least in a descriptive sense, the generally large size of the lampbrush TUs. However, this interpretation must be further tested, because the transcription and processing of histone mRNA in *other* tissues differ from the usual pattern for polymerase II genes. Specifically, histone mRNA in somatic tissues is not polyadenylated and does not contain the 3′ AAUAAA sequence that is thought to be a polyadenylation signal for other genes [Proudfoot and Brownlee, 1976]. Moreover, transcription termination appears to be at or near the 3′ end of the histone mRNA and to depend on a 23-bp sequence near that end [Birchmeier et al, 1982]; other genes lack this sequence and probably terminate well past the eventual 3′ end of the mRNA [Hofer and Darnell, 1981]. For these reasons it is not clear what unusual feature of oocyte nuclear transcription might lead to the simultaneous production of very long transcripts from histone genes and from other polymerase II genes.

The very large size of lampbrush TUs suggests that they will generally include transcripts of interspersed repetitive elements. Several studies have shown that newt lampbrush chromosome loops hybridize with a variety of repeated sequences. These include simple synthetic polymers such as poly(dCT):poly(dAG) and poly(dC):poly(dG) [Callan and Old, 1980; Callan et al, 1980], intermediate repetitive DNA isolated on the basis of rapid reassociation (low Cot DNA) [Varley et al, 1980b], and tandemly repeated satellite sequences [Varley et al, 1980a; Diaz et al, 1981]. Varley et al [1980b] pointed out that failure of transcription termination would lead to the production of transcripts from various types of repetitive sequences, particularly in a genome like that of the newt, which contains such a high proportion of repetitive DNA. This would be true whether or not transcription initiation was specific, so long as the TUs were of great length. Our work with the histone loci suggests that initiation in the lampbrush chromosome occurs at promoters 5′ to structural genes, but that termination does not occur at the 3′ ends. In order to test the hypothesis further it will be necessary to study additional structural genes and their 3′ flanking sequences. Unfortunately at present there are no other defined genomic clones of newt genes with which appropriate in situ studies could be performed. The enormous size of the genome prevents one from obtaining such clones readily and it may be necessary to attempt in situ studies on lampbrush chromosomes of Xenopus. Xenopus has a much smaller genome than the newt and for this reason appropriate genomic clones are available. Despite the much smaller size of Xenopus lampbrush loops, Jamrich and Warrior [1982] have recently obtained good in situ hybridizations with this organism, and it is probable that detailed studies on the TUs will be feasible.

Our model of the lampbrush TU may help to explain another well-known feature of lampbrush chromosomes — the general correlation between TU length, loop length, and C-value. It has been known for many years that lampbrush chromosomes are largest and have the longest loops in tailed amphibians like the newt and axolotl, which have extraordinarily high C-values (C = 20–50 pg), whereas the chromosomes and loops of anurans like Xenopus (C = 3.2 pg) are much smaller. Even smaller loops are found in the alga Acetabularia (C = 0.92 pg), where they have been studied by both light and electron microscopy [Scheer et al, 1979]. So long as TU length was regarded as a measure of gene length, this correlation between loop size and C-value was puzzling, because there was no reason to suppose that genes in high-C-value organisms were especially long. The discovery of introns provided a source of variability in gene length, and some surprisingly long genes are now known. However, there does not seem to be a simple relationship between C-value and number of introns or average gene length. It is more reasonable to suppose that structural genes are simply more widely spaced in organisms with high C-values than in those with low values; in other words, most of the variability in DNA content arises from regions between the genes. If this is the case and if the lengths of lampbrush TUs reflect the distance between structural genes, as they almost certainly do for the histone genes of Notophthalmus, then organisms with high C-values should generally have the longest lampbrush TUs and the longest loops.

The realization that loops often contain two or more TUs raises new questions about loop organization. In the case of the histone loci of Notophthalmus, multiple TUs in the same loop are clearly related: regardless of their polarity, they all contain histone and satellite 1 transcripts. However, it is not possible to predict whether loops with multiple TUs always contain clusters of related genes. The alternative interpretation is that a loop is a structural unit not necessarily related to the particular genes that it contains. The available in situ hybridization data are not definitive on this question. Many partially labeled loops were seen when repetitive sequence probes were used for in situ hybridization. Some of these must represent loops in which only one of several TUs is labeled (eg, Fig. 11, in Morgan et al [1980]), and they demonstrate that the sequence used as a probe is not transcribed on all TUs of a given loop. Nevertheless, these cases do not prove that the multiple TUs on these loops are unrelated. According to the model we have discussed, two TUs on the same loop might begin at two copies of the same structural gene, but read into different downstream sequences. Such sequences might contain a motley array of repetitive elements, pseudogenes, transposable elements, etc. For this reason partial labeling of a loop in a hybridization experiment might be misleading with respect to the structural genes present. It will be necessary to study various cloned genes and their downstream sequences in detail before an adequate picture of lampbrush loop organization emerges.

In summary, we suggest that a lampbrush chromosome TU consists of a structural gene and a segment of downstream DNA that may be as long as the distance to the next structural gene (less in the case of TUs of opposite polarity). The length of the downstream segment will obviously vary from gene to gene, but its average length will be correlated with the C-value of the organism under consideration. We suggest that transcription initiates correctly at the 5′ end of the structural gene but fails to terminate at its 3′ end. Our model at present contains no functional explanation for the transcriptional readthrough, nor do we know what happens to the majority of the transcription products.

ACKNOWLEDGMENTS

Supported by research grant GM 12427 from the National Institute of General Medical Sciences. The technical assistance of Lois Nichols is gratefully acknowledged. Giuseppina Barsacchi-Pilone and Harry Erba contributed important data to this study.

REFERENCES

Angelier N, Lacroix JC (1975): Complexes de transcription d'origines nucléolaire et chromosomique d'ovocytes de Pleurodeles waltlii et P. poireti (Amphibiens, Urodèles). Chromosoma 51:323–335.

Birchmeier C, Grosschedl R, Birnstiel ML (1982): Generation of authentic 3′ termini of an H2A mRNA in vivo is dependent on a short inverted DNA repeat and on spacer sequences. Cell 28:739–745.

Callan HG (1955): Recent work on the structure of cell nuclei. In "Fine Structure of Cells," Symposium of the VIIIth Congress in Cell Biology, Leiden 1954. Gronigen, Noordhof, pp 89–109.

Callan HG (1982): Lampbrush chromosomes. Proc R Soc Lond [Biol] 214:417–448.

Callan HG, Lloyd L (1960): Lampbrush chromosomes of crested newts Triturus cristatus (Laurenti). Philos Trans R Soc Lond [Biol] 243:135–219.

Callan HG, Old RW (1980): In situ hybridization to lampbrush chromosomes: A potential source of error exposed. J Cell Sci 41:115–123.

Callan HG, Old RW, Gross KW (1980): Problems exposed by the results of in situ hybridization to lampbrush chromosome. Eur J Cell Biol 22:21.

Diaz MO, Barsacchi-Pilone G, Mahon KA, Gall JG (1981): Transcripts from both strands of a satellite DNA occur on lampbrush chromosome loops of the newt Notophthalmus. Cell 24:649–659.

Gall JG (1956): On the submicroscopic structure of chromosomes. Brookhaven Symp Biol 8:17–32.

Gall JG, Stephenson EC, Erba HP, Diaz MO, Barsacchi-Pilone G (1981): Histone genes are located at the sphere loci of newt lampbrush chromosomes. Chromosoma 84:159–171.

Hamkalo BA, Miller OL (1973): Electronmicroscopy of genetic activity. Annu Rev Biochem 42:379–396.

Hill RS (1979): A quantitative electron-microscope analysis of chromatin from Xenopus laevis lampbrush chromosomes. J Cell Sci 40:145–169.

Hofer E, Darnell JE (1981): The primary transcription unit of the mouse β-major globin gene. Cell 23:585–593.

Jamrich M, Warrior R (1982): Study of transcription and protein distribution on Xenopus lampbrush chromosomes. Abstracts of 11th Annual UCLA Symposium. "Gene Regulation." J Cell Biochem (Suppl) 6:317.

Makarov VB, Safronov VV (1976): Functional organization of chromomere. III. Analysis of transcriptional units in chromomeres of Triturus cristatus cristatus. Tsitologiya 18:290–295.

Miller OL, Hamkalo BA (1972): Visualization of RNA synthesis on chromosomes. Int Rev Cytol 33:1–25.

Morgan GT, Macgregor HC, Colman A (1980): Multiple ribosomal gene sites revealed by in situ hybridization of Xenopus rDNA to Triturus lampbrush chromosomes. Chromosoma 80:309–330.

Proudfoot NJ, Brownlee GG (1976): 3′ Non-coding region sequences in eukaryotic messenger RNA. Nature 263:211–214.

Scheer U, Spring H, Trendelenburg MF (1979): Organization of transcriptionally active chromatin in lampbrush chromosome loops. In Busch H (ed): "The Cell Nucleus," Vol VII. New York: Academic Press, pp 3–47.

Scheer U, Franke WW, Trendelenburg MF, Spring H (1976): Classification of loops of lampbrush chromosomes according to the arrangement of transcriptional complexes. J Cell Sci 22:503–520.

Sommerville J (1977): Gene activity in the lampbrush chromosomes of amphibian oocytes. In Paul J (ed): "Biochemistry of Cell Differentiation II." Intern Rev Biochem. Baltimore: University Park Press, 15:79–156.

Sommerville J (1981): Immunolocalization and structural organization of nascent RNP. In Busch H (ed): "The Cell Nucleus," Vol VIII. New York: Academic Press, pp 1–57.

Stephenson EC, Erba HP, Gall JG (1981a): Histone gene clusters of the newt Notophthalmus are separated by long tracts of satellite DNA. Cell 24:639–647.

Stephenson EC, Erba HP, Gall JG (1981b): Characterization of a cloned histone gene cluster of the newt Notophthalmus viridescens. Nucleic Acids Res 9:2281–2295.

Varley JM, Macgregor HC, Erba HP (1980a): Satellite DNA is transcribed on lampbrush chromosomes. Nature 283:686–688.

Varley JM, Macgregor HC, Nardi I, Andrews C, Erba HP (1980b): Cytological evidence of transcription of highly repeated DNA sequences during the lampbrush stage in Triturus cristatus carnifex. Chromosoma 80:289–307.

Gene Structure and Regulation in Development, pages 147–170
Published 1983 Alan R. Liss, Inc., 150 Fifth Avenue, New York, NY 10011

Molecular Biology of the Sea Urchin Embryo

Eric H. Davidson, Barbara R. Hough-Evans, and Roy J. Britten
Division of Biology, California Institute of Technology, Pasadena, California 91125

The early development of the sea urchin has been more intensively studied at the molecular level than has any comparable embryonic system. The reasons for this are partly historical in that a deep background of useful developmental knowledge has accumulated since sea urchin embryos became a favorite subject of experimentation about a century ago. Modern investigators capitalize on a number of practical advantages. The eggs of certain sea urchin species can be obtained in the laboratory the year round [1], and other conveniently accessible species are gravid at alternate seasons. In the United States the main, although not the only, sea urchins used for research are *Strongylocentrotus purpuratus*, *Lytechinus pictus*, *Arbacia punctulata*, and *Tripneustes gratilla*, referred to in this article respectively as *Sp*, *Lp*, *Ap*, and *Tg*. Eggs can be fertilized in vitro, and rapid, highly synchronous development ensues. Cultures in excess of 10^8 embryos (*Sp*) are routinely grown in our laboratories. The ready availability of such large amounts of material facilitates quantitative studies of the macromolecules of the embryo, as do the ease of demembranation, homogenization, and cell fractionation, the relative lack of yolk, the permeability of the embryos to radioisotopes, and the relatively small genome size, which is about 8×10^8 nucleotide pairs or approximately one-fourth the size of the mammalian genome. The embryo is rather transparent and (in *Sp*) has only about 600 cells at the gastrula stage, compared, for example, to more than 10^4 in amphibian gastrulas. Not only can all the cell types be observed in living embryos throughout early development, but the embryo can be disaggregated in ways that permit preparative recovery of some of these cell types and the structures they form. There are also some disadvantages, the principal one perhaps being the lack of genetic data, although sea urchin embryos can easily be carried through advanced larval stages including metamorphosis in the laboratory [2], and adults can

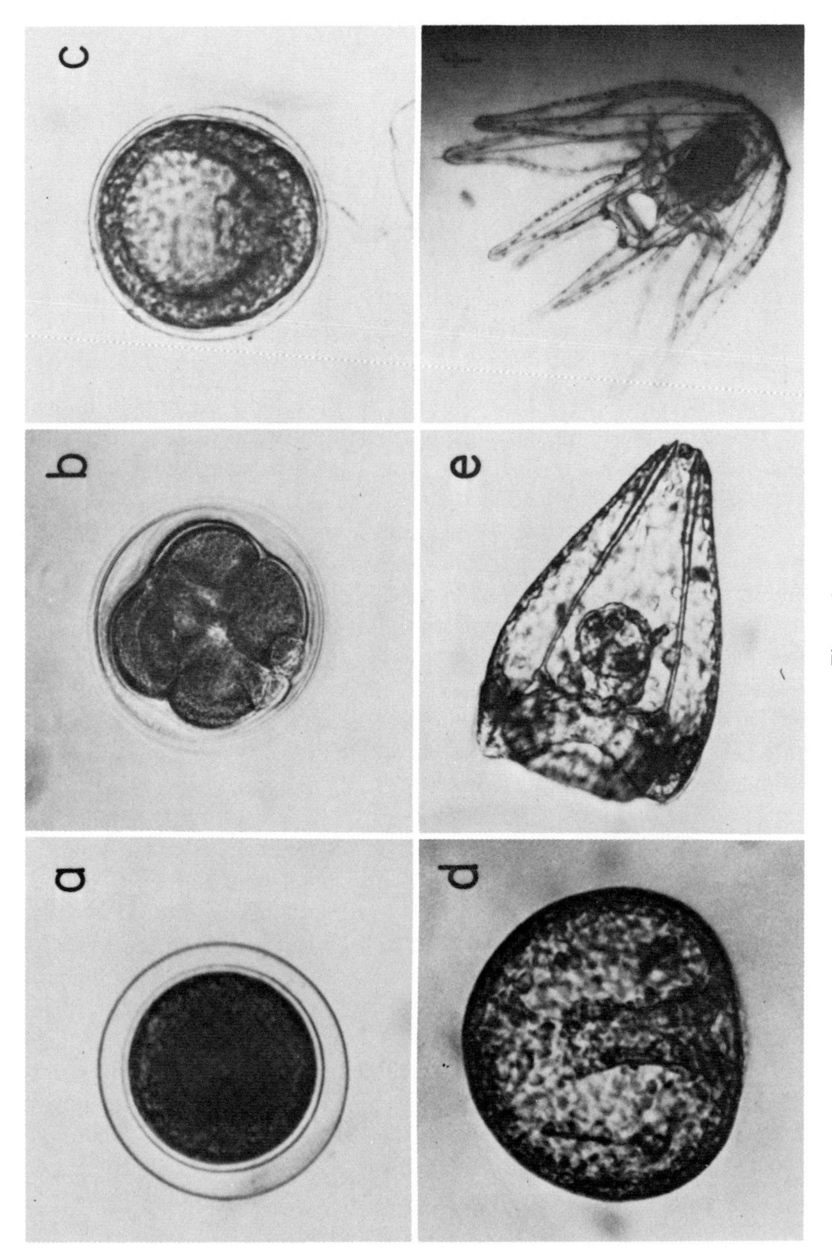

Figure 1.

be maintained routinely in suitable culture systems for years [1]. Some inbred strains have indeed been produced, including several displaying developmental mutations, but inbred strains have so far been used only for a few specific investigations [3].

Embryonic and Larval Development

Most research on sea urchin embryos concerns the period between fertilization and the stage when feeding can begin. The rate of development is temperature-dependent, and the following abbreviated description refers in particular to embryos of the purple California sea urchin (*Sp*), cultured at 15°C. Immediately after fertilization a thick membrane rises off the surface of the egg (Fig. 1a), and within a short time sclerotizes, thus providing physical protection during cleavage. Cleavage begins about 1½ hours after fertilization. The first cleavage furrow divides the egg into equal halves along what is called the "animal-vegetal" axis. The top of the egg (animal pole) later becomes the site of a plate of apical cells bearing a tuft of especially long cilia, and the opposite pole (vegetal pole) later becomes the site of gastrular invagination. At fourth cleavage (5 hours after fertilization) four very small cells, the micromeres, are formed at the vegetal pole (Fig. 1b). Together these include only about 8 percent of the total volume of the embryo, and they are of particular interest because they are the specific progenitors of a lineage of cells that ultimately gives rise to the larval skeleton.

Fig. 1. Normal development of the *Strongylocentrotus purpuratus* embryo at 15°C. (a) Fertilized egg. The vitelline layer has lifted off the egg surface to form the fertilization envelope. The egg diameter is 80 μm. (a to c) All are at about the same magnification (×400). (b) The 16-cell embryo (5.5 hours after fertilization). The four small transparent cells at the lower left are the micromeres. [From [78]; courtesy of *Developmental Biology*] (c) Mesenchyme blastula (18 hours after fertilization). Primary mesenchyme cells can be seen within the vegetal pole region of the blastocoel. Hatching occurs at about this time. (d) Late gastrula (46 hours after fertilization). The primitive gut extends most of the way across the blastocoel. On the left, one of the two early spicules is visible, with several primary mesenchyme cells associated with it. The embryos are covered with cilia and swim about vigorously. This photograph and the succeeding ones are at progressively lower magnifications. (e) Early pluteus (72 hours after fertilization), ventral view. The embryo has elongated, the gut is differentiated, and the mouth has formed between the four nascent arms toward which extend the prominent skeletal rods. The long axis of the pluteus larva is about twice the diameter of the gastrula. (f) Fed pluteus larva 3 weeks after fertilization (about 600-μm diameter). The eight arms, skeletal rods, and stomach, which is filled with ingested algae, can be observed. Dark spots along the arms are pigment granules. On the left of the stomach can be observed a small pouch that is the beginning of an imaginal structure which at metamorphosis will give rise to part of the adult.

A logarithmic rate of cell division is maintained until there are about 200 cells (12 hours after fertilization). Shortly thereafter the embryo secretes a protease that dissolves the fertilization membrane and hatching occurs (about 18 hours after fertilization). The swimming blastula stage embryo contains about 400 cells, and is organized as a hollow ciliated ball bearing the prominent ciliary tuft at the animal pole. During the blastula stage approximately 30 primary mesenchyme cells, descendants of the original micromeres, invade the blastocoelic cavity (Fig. 1c). The development of a thickened, flat plate of cells at the vegetal pole marks the site of invagination and the subsequent onset of gastrulation. This process leads to the formation of the archenteron (embryonic gut), which can be seen in Figure 1d (46 hours after fertilization). Morphogenesis of the differentiated structures of the larva ensues. The gut tube bends forward across the blastocoel, preceded by strands of secondary mesenchyme cells, and where it makes contact with the embryo wall the mouth forms inductively. The initial site of invagination becomes the anus. The internal skeleton forms as a pair of tripartite spicules, secreted initially as a calcium carbonate–protein complex within the vacuoles of primary mesenchyme cells that are situated in linear arrays along the blastocoel walls. During the pluteus stage (Fig. 1e) the embryo elongates, a stomach region differentiates, and the four skeletally supported arms grow anteriorly, bearing rows of cilia that beat in a coordinated fashion. At about 70 hours after fertilization the pluteus contains approximately 1500 cells, and is sufficiently developed to feed and exist as a free living, pelagic larva. Until feeding begins the mass of the protein and of the RNA in the embryo remains essentially constant. Most of the ribosomes used throughout embryogenesis were originally present in the unfertilized egg. Prior to feeding, embryogenesis is a process of cell division, reorganization of maternal components, and the appearance of new transcripts and their products, all occurring within a closed system that is not increasing in mass.

In the laboratory, *Sp* larvae require about 6 weeks of feeding to attain the stage at which metamorphosis may occur [2,4]. The larva at this point contains about 5×10^4 cells [4]. Figure 1f shows an *Sp* larva about halfway through its growth period, in which the ventral portions of the imaginal structures that on metamorphosis will give rise to the juvenile sea urchin can be seen [5]. Metamorphosis is dramatic and rapid, occurring within a few hours after the larva settles down on an appropriate surface. Most of the external larval structures, such as the eight arms, disappear, although in addition to dorsal and ventral imaginal derivatives, it is probable that some coelomic elements and portions of the larval gut are also included in the emergent juvenile sea urchin. Further morphogenesis takes place after metamorphosis, including completion of the adult digestive tract and mouth, and gonadal differentiation [6].

The general form of embryogenesis in echinoderms places this group in the great branch of the animal kingdom, the deuterostomes, that also includes chordates. In deuterostomes the site of gastrular invagination becomes the anus, and the mouth forms secondarily. Other basic similarities between echinoderm and some chordate embryos include the manner in which gastrulation occurs, and the origin and disposition of the larval mesoderm. These classically observed morphological relations have been supported by a particular homology in the organization of the genes coding for actin. The sea urchin actin genes contain intervening sequences that are present at precisely the same locations as those in the coding regions of mammalian and avian actin genes, while intervening sequences appear at totally different positions in the actin genes of *Drosophila* (that is, a protostomial invertebrate), and in several lower invertebrates [7]. It is difficult to estimate the extent to which the molecular mechanisms of sea urchin embryogenesis will prove relevant to mammals because there is as yet insufficient knowledge of mammalian development to permit general comparisons.

Maternal Messenger RNA and Its Utilization After Fertilization

The sea urchin egg contains a mass of stored inactive maternal message sufficient to engage all of its ribosomes, although before fertilization < 1 percent of these ribosomes are actually assembled into polysomes [8]. The amount of RNA that might serve as maternal message has been calculated for *Sp* as 50 to 100 picograms per egg [9], included in a total of about 3 nanograms of RNA, of which 85 percent is ribosomal [10]. Maternal messenger RNA (mRNA) is identified as embryo polysomal message that derives from the unfertilized egg, or as unfertilized egg RNA that supports protein synthesis in an in vitro translation system. Much, although clearly not all, of the translatable mRNA in the unfertilized egg contains 3′-polyadenylic acid [poly(A)] tracts about 50 to 120 nucleotides long [11,12]. Although some RNA species are preferentially polyadenylated and others occur mainly as 3′-poly(A)–deficient egg RNA's, the same set of sequences are included in both RNA fractions [13,14].

The most direct demonstration of the protein-coding capacity of unfertilized egg mRNA comes from cell-free translation experiments. Infante and Heilman visualized several hundred protein spots in two-dimensional gel analyses of the translation products of poly(A) RNA extracted from unfertilized eggs of *Sp*, and showed that these are indistinguishable from the translation products coded by the polysomal message of early embryos [15]. There are several specific proteins for which maternal mRNA's are known to be stored. Tubulins are synthesized in the growing oocyte, and maternal tubulin message continues to be translated in the embryo even though a large quantity of tubulin protein is also inherited via the egg cytoplasm [9,16]. The unfertilized egg also contains maternal actin message [17,18]. The most detailed studies have been carried out on maternal mRNA's coding for histones. There are about 10^6 molecules of message for each of the four core histones in the egg of *Sp*, together constituting some 5 to 10 percent of the total quantity of maternal mRNA [9,10,19]. Maternal mRNA for histones is synthesized and stored in growing oocytes and is also made in mature unfertilized eggs [20,21]. However, in the latter all the newly synthesized histone mRNA appears to be translated immediately and is then turned over [21]. Thus the stored maternal histone message is made earlier. The maternal messages code for special early embryonic variants of the core histones and of histone H1 [22]. Some of these histone variants are found only in the nuclei of early cleavage stage embryos while others are synthesized well into the blastula stage.

The histones and tubulins are not exceptional in that many other embryonic proteins that are coded by maternal mRNA's are also synthesized prior to fertilization and inherited by the embryo. Brandhorst [23] showed for both *Lp* and *Sp* that of about 400 species of proteins resolved in two-dimensional gel analyses and synthesized within an hour of fertilization (that is, on maternal messages), almost all are also being translated at a low rate in unfertilized eggs. In the sea urchin fertilization triggers a dramatic increase in protein synthesis but does not significantly alter the set of sequences being translated. It was this striking quantitative feature, which is essentially a peculiarity of echinoderm biology, that led to the initial discovery of maternal mRNA [9].

The mature sea urchin egg has already completed its meiotic reduction divisions and is stored in the lumen of the ovary in a relatively quiescent state. RNA synthesis nonetheless continues in the pronucleus at about the same rate per haploid genome as is observed in a cleavage-stage embryo nucleus [21]. When fertilization occurs it sets off a train of cytological, physiological, and molecular changes in the state of the egg that begins within seconds of first contact between the sperm acrosome and the egg surface. Most of these events have been reviewed by Epel and by Vacquier [24]. Several hours after fertilization, the rate of protein synthesis in the sea urchin has increased by a factor of about 100, compared to its rate in the unfertilized eggs, as a result of two separate effects at the translational level. The major effect is on the assembly of polyribosomes from maternal components, a process that begins only minutes after contact with sperm. By the 16-cell stage about 30 percent of the egg ribosomes are involved in polysomal structures, and at most about 10 percent of the mRNA in these polysomes is of embryonic rather than maternal origin [8,9,21]. After this the polysome content continues to increase, probably due to the appearance and utilization of new embryonic mRNA, until about 60 percent of the egg ribosomes are included. Protein synthesis rate increases proportionately [25], and rises from about 120 pg per hour-embryo at first cleavage to $\geqslant$ 500 pg per hour-embryo at the blastula stage [9,25,26]. Fertilization also increases the translational elongation rate by a factor of about 2.5, compared to this rate in the polysomes of the unfertilized egg [27]. Neither the mechanism of this effect nor that responsible for mobilization of the stored maternal message are well understood. An immediate trigger might be the sharp increase in intracellular pH, from 6.84 to 7.27 (in *Lp* and similar changes take place in *Sp*) occurring about 60 seconds after fertilization [28]. Winkler, Steinhardt, and colleagues [29] directly demonstrated the increase in elongation rate in vitro by measuring ribosome transit times at pH 6.9 and pH 7.4 in a cell-free system derived from unfertilized *Lp* eggs. The pH change and other ionic alterations that follow fertilization may also affect the messenger ribonucleoprotein (mRNP) structures that include the maternal mRNA of unfertilized eggs. These particles are fairly stable, and differ in protein content and physical properties from those containing the newly synthesized mRNA of late embryos [30]. It has been proposed that these particles are responsible for inhibiting translation of their maternal mRNA in the unfertilized egg [31], but what actually happens to them in vivo remains to be discovered. There is some evidence from studies of specific actin [15] and histone [10,32] maternal mRNA's that the rate at which mobilization into polysomes occurs after fertilization may be quite different for certain particular sequences than for the majority.

Unexplained Features: Complexity and Sequence Organization of Maternal RNA

Were the heterogeneous RNA's stored in the egg simply a collection of messages to be translated after fertilization, its sequence complexity would be that of early embryo polysomal mRNA, and its size and sequence organization would be no different from that of the mRNA's found at later stages. Observations from our laboratory on the properties of sea urchin egg RNA's suggest a more complicated—and interesting—situation, although as the foregoing discussion shows, translatable maternal mRNA's are certainly present.

The complexity or total sequence length of unfertilized egg RNA has been measured by RNA excess hybridization reactions with single-copy DNA for three species (*Sp*, *Ap*, and *Tg*) and by the complementary DNA (cDNA) kinetic method for a fourth (*Lp*) [33]. The values reported all fall in the range 3.0×10^7 to 3.7×10^7 nucleotides. RNA complexities of this magnitude are characteristic of the eggs of other groups of animals as well, such as amphibians and the house fly [34]. The small pronucleus of the mature sea urchin egg cannot contribute sufficiently to the total mass of RNA to affect the

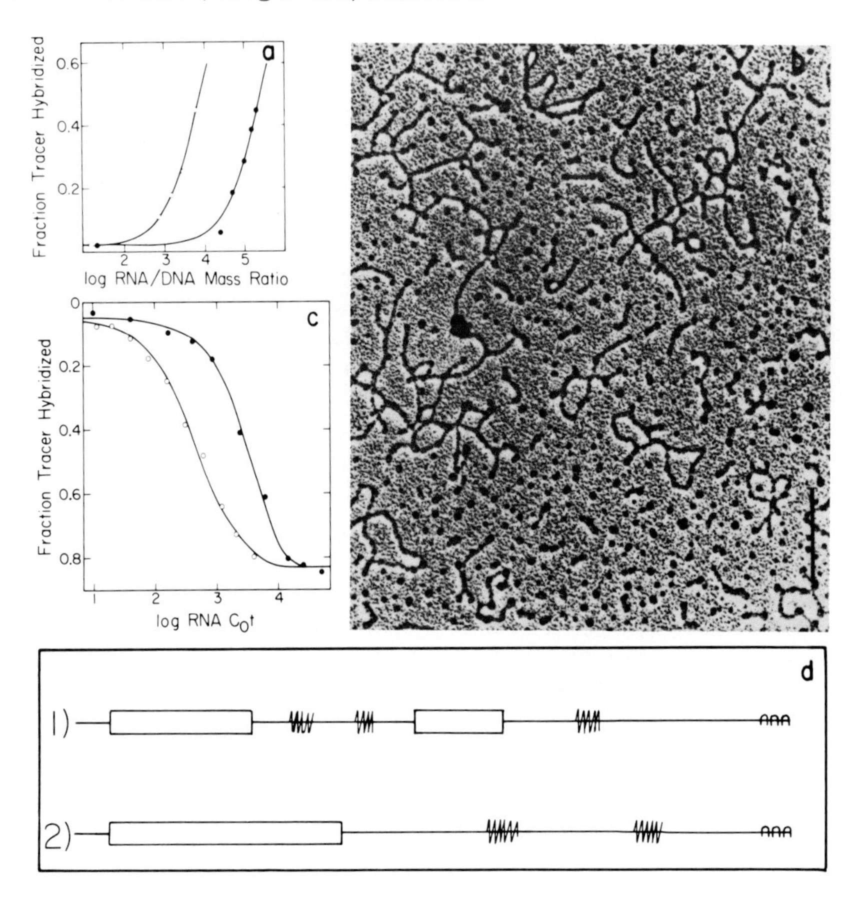

Fig. 2. Characteristics of maternal poly(A) displaying an interspersed sequence organization. (a) Titration experiment demonstrating that repetitive sequence transcripts are polyadenylated. A cloned repetitive sequence called CS2109B was labeled in vitro and the strands were then separated. (Open circles) One strand was hybridized to termination with increasing amounts of egg poly(A)$^+$ RNA or (closed circles) poly(A)$^-$ egg RNA, and the reactions were assayed by hydroxyapatite chromatography. The specific fraction of poly(A)$^+$ RNA-containing transcripts of the CS2109B repeat was calculated as 1.3×10^{-4}, compared to 4.1×10^{-6} for the poly(A)$^-$ RNA. [From [13]; courtesy of *Nature (London)*]. (b) Electron microscopy of renatured interspersed poly(A) RNA molecules of sea urchin egg RNA. The RNA was renatured to C_0t 600 (concentration × time; mole/liter-second), and duplex-containing molecules were eluted from an ethanol-cellulose column, precipitated with ethanol, and spread for electron microscopy. A representative field is shown. The circular molecules are ϕX174 DNA added as a length standard. The bar equals a single-stranded RNA length of 1000 nucleotides. [From [42]; courtesy of *Journal of Molecular Biology*] (c) Complexity of single-copy sequence transcripts included in a selected, interspersed RNA fraction. A ^{32}P-labeled single-copy DNA tracer was prepared by hybridizing total single-copy ^{32}P-DNA with total egg RNA and purifying the hybridized fraction. The tracer was enriched about 30-fold for egg RNA

overall observed complexity, which is in any case only about one-fifth of that measured for the nuclear RNA of immature sea urchin oocytes or of embryos [35]. However, Hough-Evans et al. found that in *Sp* the complexity of the maternal mRNA, defined as that population of sequences loaded on early embryo polysomes, is only about 70 percent of that of the total egg RNA [36]. This observation was the first indication that egg RNA might include heterogeneous sequences not directly explicable as maternal message.

Further evidence derives from a series of measurements carried out with repetitive DNA sequence probes. Costantini et al. [13,37] showed that the RNA of unfertilized *Sp* eggs includes transcripts of at least several hundred different repetitive sequence families. Reactions with strand-separated repeat tracers demonstrated that both strands of each given repetitive sequence family are represented in the egg RNA, although in general in different molecules. Furthermore, it was found that at least 70 percent of the mass of the egg poly(A) RNA contains interspersed repeat sequence transcripts. In Figure 2a is reproduced an experiment demonstrating that a specific repeat sequence is represented in the poly(A) RNA fraction. On the average about 10 percent of the length of a poly(A) RNA molecule consists of repetitive sequence transcript, and the remainder of single-copy sequence. If allowed to renature, such poly(A) RNA molecules will react at the complementary repeat sequences to form partially duplexed, multimolecular structures, such as are shown in Figure 2b. Evidently more than one repeat sequence element exists in many of the egg poly(A) RNA molecules, and the repeats are frequently located in internal positions. The experiment reproduced in Figure 2c demonstrates that the interspersed egg RNA molecules include virtually all the sequence complexity of total egg RNA. It follows that there are sets of diverse transcripts in the egg that share homologous repeat sequences, and several examples belonging to such sets have been cloned and their structure verified directly [38]. While the number average length of *Sp* egg poly(A)

---

sequences. It was reacted with total egg RNA (closed circle) and with RNA fractions containing interspersed repetitive sequence transcripts that had been selected by prior binding to an ethanol-cellulose column (open circle). All of the tracer that reacted with total egg RNA (about 80 percent) also reacted with the interspersed RNA. The latter reaction occurs at a faster rate because the reactive single-copy sequences are also concentrated when RNA's containing interspersed repeats are purified away from the bulk (ribosomal) RNA. [From [13]; courtesy of *Nature (London)*] (d) Two of the possible forms of sequence organization in interspersed maternal poly(A) RNA. Diagram 1 depicts an RNA molecule in which the interspersed repetitive sequence elements are included in an intervening sequence, while diagram 2 shows an RNA molecule in which the repeat sequences are all located in a long 3′ untranslated tail. Open rectangles, coding sequence; jagged line, repetitive sequence element; *AAA*, poly(A).

RNA is about 3 kilobases (kb) [13,37], RNA gel blots show that the interspersed molecules are usually much longer, often from 5 to 15 kb [38]. The presence of interspersed repeats in such long transcripts is directly reminiscent of the structure of nuclear RNA, although as already pointed out the egg RNA complexity differentiates it sharply from total nuclear RNA. Figure 2d displays two possible kinds of sequence organization that are consistent with these observations. In one case, the interspersed repeats are located in long 3′-terminal tails, and in the other they are embedded in intervening sequences that have not yet been processed out. The repetitive sequence elements are figured outside the coding regions, because primary sequence determinations carried out on eight of nine different cloned repeats of genomic families represented in egg RNA revealed translation stop codons in all possible reading frames [38].

An observation that is probably related has been reported by Duncan and Humphreys in studies of *Tg* egg RNA [11]. The fraction of egg RNA that lacks long 3′ poly(A) tracts is approximately equal in mass to the adenylated component. An interesting feature of RNA lacking 3′ poly(A) tracts is that it contains internal A_{10} sequences (polyadenylic acid containing ten residues), just as does embryo nuclear RNA [39]. These sequences are absent from cleavage stage polysomal RNA [11], that is, from the maternal message by the time it is assembled into polysomes. Whether either maternal or newly synthesized repeat-containing poly(A) RNA's are loaded directly on polysomes after fertilization remains moot, because interspersed maternal RNA species persist in the cytoplasm far into embryogenesis [38,40], probably complexed with proteins, and these particles are difficult to separate completely from bona fide polysomal structures. However, embryo polysomal mRNA lacks the prevalent repetitive sequence component observed in egg poly(A) RNA [37], although by mass perhaps 15 percent of embryo message at gastrula stage continues to display an interspersed sequence organization, a somewhat larger figure than reported earlier [41].

Both the presence of A_{10} sequences not found in polysomes and the structure of the interspersed poly(A) RNA's suggest the rather startling conclusion that a major fraction of the stored maternal RNA's are not fully processed. Whatever their function, this phenomenon is not confined to sea urchins. We have found that a major fraction of the cytoplasmic poly(A) RNA of *Xenopus* oocytes displays the same interspersed sequence organization [42]. Possibly the only maternal messages (in the strictest sense) are those that lack both A_{10} and in most cases repeat sequences, and are available for immediate translation [43,44]. On the other hand, if these sequence features reside in 3′ terminal regions of the transcripts (Fig. 2d) they might not interfere with translation. In either case the fate of these maternal RNA's, and the biological significance of the "excess" (that is, nonunderstood) sequence they contain are likely to be of basic importance in understanding the role of maternal components in development.

There are a number of interesting possibilities worth considering. For example, cytoplasmic processing in the embryo could represent a control point in the utilization of maternal information, and might be accomplished differently at various stages or in the various regions of the embryo. The maternal transcripts might include sequences (perhaps the repeat elements themselves) recognized by other macromolecules, such as proteins, small nuclear RNA's or other RNA's [45], and such complexes might be involved in sequestration of maternal mRNA's; or in subsequent processing; or in determining their turnover rates; or in localization of the transcripts within the embryo; or even in regulation of structure or function within the blastomere nuclei. Conceivably, some fundamental change in the regulation of transcriptional termination occurs between oogenesis, when the usual signals are not recognized, and later embryonic development, when they are [46]. Another possibility is that these sequences are used by the embryo to monitor the quantity of maternal transcripts, and thus to set the timing of events that require the attainment of a certain nucleus-to-cytoplasm ratio as cell division proceeds within the constant mass of the maternal cytoplasm.

Patterns of Gene Expression in the Embryo

Species hybridization and actinomycin experiments initially indicated that for many hours the pattern of biosynthesis in the embryo remains essentially that determined by the maternal mRNA [9]. This view has been substantiated in elegant detail at the molecular level. The spectrum of proteins synthesized in the embryo and resolvable on two-dimensional gels ($\leqslant 10^3$ species) changes very little from fertilization until the blastula stage [15,23], and even thereafter the rate of synthesis of only about 15 to 20 percent of these proteins alters more than tenfold [47]. These observations were, of course, made on extracts of whole embryos, and it is quite possible that among specific cell lineages larger differences exist in the patterns of protein synthesis. The changes that have been reported occur mainly during the transition from blastula to early gastrula, when various forms of differentiation become apparent in the embryo. Direct measurements of the population of polysomal and cytoplasmic RNA's carried out in our laboratory have shown that no very extensive switching on of new genes occurs during embryonic development. Thus Galau et al. [48,49] discovered to their surprise that all or most gastrula mRNA sequences are already represented in the stored maternal RNA of the egg, and in later development the overall complexity of the polysomal RNA actually declines to a minor extent. About 60 percent of that set of sequences found in the maternal mRNA is still present in the newly

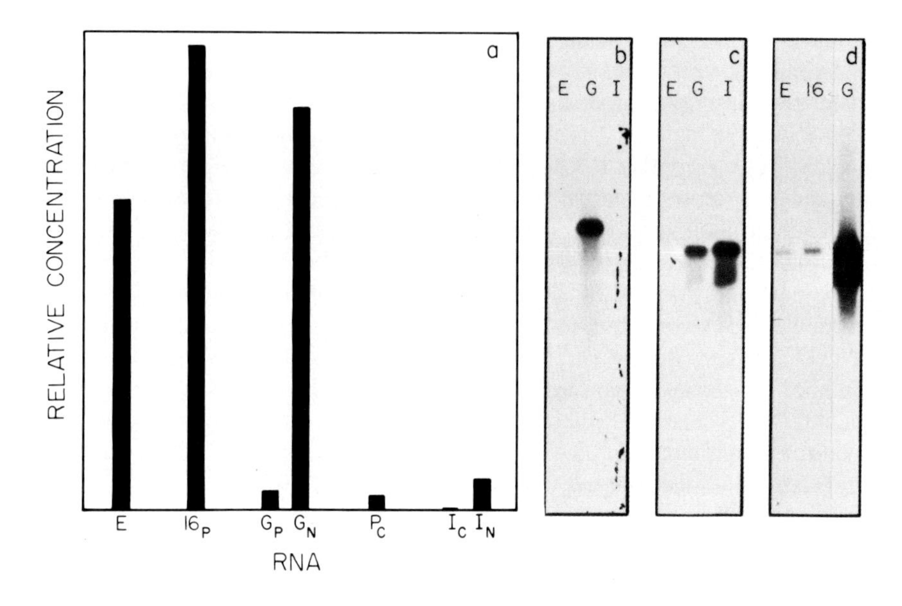

Fig. 3. Examples of patterns of gene activity observed with cloned sequence probes. (a) Representation of a rare maternal transcript in various RNA's. A cloned sequence called Sp88 that is complementary to about 1400 transcripts per egg was strand-separated and labeled in vitro. The expressed strand tracers were used to determine the concentration of complementary transcripts by titration. Data are expressed as relative concentration per unit mass RNA of Sp88 transcripts in *E*, total egg RNA; 16_P, 16-cell polysomal RNA; G_P, gastrula polysomal RNA; G_N, gastrula nuclear RNA; P_C, pluteus cytoplasmic RNA; I_C, intestine cytoplasmic RNA; I_N, intestine nuclear RNA. No Sp88 transcripts could be detected in cytoplasmic RNA of intestine cells. [From [50]; courtesy of *Developmental Biology*] (b to d) RNA gel blot hybridizations with cloned probes. (b) cDNA clone SpG4-B9 reacted with (from left to right) egg poly(A) RNA, gastrula polysomal poly(A) RNA, and intestine poly(A) RNA. This sequence is barely represented in the maternal RNA but is expressed strongly in gastrula. It is absent in intestine. The length of the transcript is 1.45 kb. (c) cDNA clone SpG2-D12, same RNA's. This sequence is not detectably represented in egg RNA but is prevalent in the gastrula and intestine RNA's. The length of the transcript is 1.35 kb. [Parts (b) and (c) are from [52]; courtesy of *Developmental Biology*] (d) Cloned actin gene probe, reacted with egg poly(A) RNA, 16-cell embryo poly(A) RNA, and gastrula poly(A) RNA [85]. The two main actin transcripts are 1.8 and 2.2 kb in length [17].

synthesized transcripts of the late embryo [36]. A specific rare maternal sequence that appears to be expressed only at the earliest stages was described by Lev et al. [50]. This sequence is loaded on polysomes in 16-cell embryos, but thereafter it disappears from the embryo cytoplasm (Fig. 3a).

The comparisons of polysomal RNA sequence sets referred to here concern mainly the low abundance transcripts that account for most of the mRNA complexity. Such sequences are present in only a few copies per gastrula cell on the average, that is, assuming that they are equally present in all the cells at this stage, which for any given example may not be the case. Low abundance sequences amount to about 40 percent of the total mass of cytoplasmic poly(A) RNA [48,49,51]. Higher prevalence RNA sequences of egg and embryo have also been compared, in measurements carried out with cDNA clone libraries [51,52]. Almost all of the genomic transcripts in the range 10^4 to 10^6 copies per embryo (that is, in the late embryo about 10 to 10^3 copies per cell on the average) are found to be represented to an approximately equal extent in the maternal RNA and in pluteus stage embryo cytoplasmic poly(A) RNA. Typically such transcripts decline in prevalence several-fold on a per embryo basis between egg and gastrula stages, whereupon their sequence concentration builds back up as a result of new synthesis in the late embryo [52]. However, there are also a limited number of moderate to low prevalence RNA species that appear for the first time during blastulation, in agreement with studies of protein synthesis patterns [47]. The appearance of poly(A) RNA sequences apparently representing newly activated genes has been verified by the RNA gel blot method with the use of cloned complementary DNA (cDNA) probes [52,53], as shown for example in Figure 3 b to d, and is also supported by a study of cDNA hybridization kinetics in blastula to gastrula stage embryos [54].

The high complexity of the maternal mRNA, and the persistence of much of the same complex set of polysomal RNA's throughout embryogenesis poses a very interesting developmental problem. Measurements indicate that the number average length of the polysomal poly(A) mRNA of early sea urchin embryos is about 2000 nucleotides [9] (that is, somewhat shorter than egg or embryo cytoplasmic poly(A) RNA, the number average length of which is $\geqslant$ 3000 nucleotides [13,14]). If each polysomal poly(A) RNA molecule contains a message sequence for one protein, the codogenic capacity of the protein synthesis apparatus is probably about 12,000 diverse polypeptides at blastula stage, when the complexity is 24×10^6 nucleotides, and is more than 8000 polypeptides at gastrula stage when the complexity is 17×10^6 nucleotides [48,49]. One hypothesis is that morphogenesis is extremely expensive in terms of protein diversity, in that a very large number of different proteins must cooperate in the construction of three-dimensional biological structures. Relevant examples include flagella, which contain more than 120 diverse protein species; the moth egg chorion, which contains more than 180 protein species; and the T4 bacteriophage head and tail, which include more than 50 different structural gene products [55]. It seems paradoxical that, while the structural complexity of the embryo progressively increases, the molecular complexity of the embryo mRNA is highest at the beginning, and that during development most maternal mRNA species are merely replaced by new transcripts of the same set of genes as were active in oogenesis. Perhaps a large fraction of the complex sets of proteins required for embryonic morphogenesis assembles into biological structures only long after they are synthesized. The mRNA's of the embryo might include programs for morphogenesis that begin to be transcribed and translated during oogenesis. Some of these mRNA's might continue to be expressed in the embryo genomes; perhaps different sets in different cell lineages. An example of a protein made in the oocyte for assembly in an embryonic structure might be provided by the tubulins. These maternal proteins are known to be utilized after fertilization for ciliary morphogenesis, and they continue to be synthesized during embryogenesis [9,16,56]. The hypothesis that the complexity of embryonic mRNA is required by morphogenetic function might also explain why a large fraction of embryo cytoplasmic transcripts (both rare and prevalent) are not found in those adult sea urchin cell types that have so far been examined, where the total polysomal message complexity is much lower [48,49,57].

Transcript Synthesis and Turnover Rates

It is interesting to consider more closely the role of structural gene transcription in the embyro nuclei. Measurements of the synthesis and turnover rates of embryo polysomal mRNA shows that most of the mass of this message is newly synthesized by the blastula stage [58,59]. However, Cabrera et al. [40] found that (in *Sp*) the actual point in development at which the maternal transcripts are replaced by new ones varies, depending on the particular sequence examined. In this work, using cloned cDNA probes we measured the cytoplasmic entry and turnover rates of a set of poly(A) RNA's of different prevalences. A general conclusion is that cytoplasmic transcript stability is an important determinant of the level of gene expression in sea urchin embryos in that most newly synthesized rare poly(A) RNA's turn over and are replaced with a half-life of several hours [9,40], while prevalent ones accumulate because they are stable. Therefore, low abundance sequences can be regulated either up or down by the embryo cells, while once transcribed, most abundant sequences are there to stay, and in general their prevalence can only rise, that is, as a result of new synthesis. Since the prevalence of most sequences in the embryo is similar to that in the maternal RNA of the egg, for each sequence much the same relation between turnover rate and cytoplasmic entry rate must have existed in oogenesis as in embryogenesis. The entry rates for specific sequences are sharply different, and are so low for some prevalent transcripts that it is clear that the majority of the cytoplasmic poly(A) RNA's of such species are still of maternal origin even in pluteus stage embryos [40].

By far the majority of genes that are active produce low abundance sequences, and in *Sp* entry rates for such transcripts are typically about one molecule per cell every one to several hours [40]. To maintain the steady-state concentrations of such cytoplasmic RNA's would require only occasional transcriptional initiations. Electron microscopic examinations of the transcription complexes of *Sp* embryo nuclei were carried out by Busby and Bakken [60]. They observed that more than 80 percent of the visualized complexes contain only single nascent transcripts. The average polymerase spacing even in the more intensely expressed regions displaying multiple transcripts was about 3.7 kb, suggesting an initiation rate of only about one molecule every 7 to 10 minutes [61]. The cytoplasmic entry rates measured by Cabrera et al. [40] even for prevalent sequences were of the same magnitude or lower. Thus the very high rates of initiation typical of genes transcribing superprevalent mRNA's in terminally differentiated mammalian or avian cells are not generally required for the maintenance or operation of a system that is undergoing embryonic morphogenesis.

The relation between the nuclear RNA synthesized in the embryo genomes and the population of mRNA's entering the cytoplasm is not well understood. There is evidence for conservation of a major fraction of 5′ terminal RNA "cap" structures, and their transfer to the cytoplasmic mRNA compartment [62]. However, both the complexity of the nuclear RNA and its rate of synthesis are at least tenfold higher than the corresponding parameters for the cytoplasmic mRNA of the same embryo cells [58,59,63]. In addition, Wold et al. [64] found that in *Sp* many low abundance, blastula polysomal RNA sequences which are absent from the cytoplasm of adult cells are nonetheless represented in the nuclear RNA of these cells. This was also observed in *Sp* for a set of blastula-specific sequences by Shepherd and Nemer [54]. In Figure 3a is illustrated an example of this phenomenon, in which a specific transcript found only in the cytoplasm of early embryos remains detectable in both gastrula and adult cell nuclear RNA's. The initiation and termination sites of this gastrula nuclear transcript are approximately the same as for the predominant homologous maternal transcript of unfertilized egg poly(A) RNA [65]. In general the nuclear RNA's of sea urchin embryo and adult cells display a very high degree of sequence overlap, although they are certainly not identical [64,66]. Since transcription rates (and steady-state levels) for regulated, cytoplasmically expressed polysomal sequences are so low, differential patterns of transcription that are physiologically meaningful could easily be obscured by random transcriptional "leakage" if such a phenomenon were to occur in some or all of these nuclei.

However, it cannot now be excluded that regulation in sea urchin embryo nuclei occurs at the posttranscriptional as well as at the transcriptional level.

Developmental Expression of Ribosomal, Actin, and Histone Genes

There are several interesting proteins known that are likely to be involved directly in formation of biological structures during development. Examples include the proteins associated with skeleton formation, proteins that might be involved in gastrulation and morphogenesis, and the vitelline layer proteins [67,68]. While these are the focus of ongoing research in several laboratories, little is as yet known of the respective genes. At present the best analyzed gene systems are the ribosomal RNA (rRNA) genes, the actin genes, and the histone genes. The quantitative details of expression in each of these systems provide an instructive paradigm for gene activity in the embryo.

The ribosomal genes are present in 50 to 200 copies per genome, depending on the sea urchin species. During oogenesis the ribosomal genes function at what is probably their maximum transcriptional rate. Griffith and Humphreys showed that in *Tg* the average rate of synthesis is about 1800 rRNA molecules of each subunit per minute-oocyte, or about 450 molecules per minute in each haploid genome set [69]. Ribosomal gene amplification apparently does not occur in sea urchin oogenesis, and the high rate of synthesis suffices to produce the maternal complement of ribosomes over the several months of oogenesis (in the egg of *Tg* there are about 4×10^8 ribosomes). In embryos the rate of rRNA synthesis per nucleus is reduced to only a few percent of that in the growing oocyte. In *Tg* this rate is about 60 molecules per minute in each (diploid) nucleus [69]; in *Lp* about 20 molecules per minute-nucleus; in *Sp* 18 to 40 molecules per minute-nucleus [70]. These rates are maintained all through embryological development without significant change. Electron microscope observations [60] suggest that in embryos a few of the ribosomal genes are highly active, while the vast majority remain repressed. Humphreys showed in *Lp* that on feeding as net growth resumes in the larva the rate of rRNA synthesis increases severalfold [70].

Actin is among the many maternal proteins inherited by the embryo [71], along with its maternal message. The egg of *Sp* contains relatively low quantities of a 2.2-kb transcript and an even smaller amount of a 1.8-kb species of maternal actin mRNA [17]. Cell free translation shows that messages for several actin isoforms are stored in the egg [15]. Crain et al. reported that between 8 and 18 hours after fertilization the polysomal concentration of the 1.8-kb mRNA form increases about 40-fold and that of the 2.2-kb form increases about 20-fold [18]. This sharp developmental increase can be seen in Fig. 3d. Adult tissues so far examined in our laboratory display only a 2.2-kb actin mRNA, and the 1.8-kb form could represent an actin gene (or genes) activated specifically in the embryo. Clustered sets of actin genes have been isolated from the genome of *Sp*, and some of these clusters include actin genes of diverse types [17]. The actin genes are distinguished from each other by nonhomologous transcribed sequences extending for several hundred nucleotides beyond the translation termination signal. Since they are expressed differentially in embryonic development and in adult tissues, these genes offer an excellent opportunity to determine the sequence features required for their developmental regulation.

The organization and expression during embryogenesis of the histone gene family have been intensively investigated, and this subject has been reviewed in detail [22,72]. After the 16-cell stage, the maternal mRNA's for the early histone variants are replaced by new transcripts deriving from the early histone genes active in the embryo nuclei, while the genes for histones found only in early cleavage nuclei are apparently functional exclusively in oogenesis and for a short time after fertilization, and are then shut off [22]. Recent results from the laboratories of Wilt and Kedes have added interesting quantitative details. Translation of histones rises sharply after fourth cleavage, and at its peak histone synthesis accounts for more than 30 percent of total protein synthesis [9,10]. At the 128-cell stage mRNA's for each of the core histone species are emerging from the average embryo nucleus at a rate of at least 800 molecules per minute [73] (there are about 400 genes for each of the early histone variants per haploid genome in *Sp*). Thus the number of histone mRNA molecules of each core species rises from about 10^6 per embryo at fourth cleavage, most of which are maternal in origin, to about 10^7 only 4 hours later, when 90 percent are the result of new synthesis [19,73,74]. The rate of histone synthesis is keyed to the rate of cell division [9], and when this falls, after about the 200-cell stage, the level of histone mRNA and of histone translation also decline, due to message turnover and a sharp decrease in the flow of new messages to about 15 percent of the peak rate [73,74]. At 300 cells there are again $\leqslant 10^6$ molecules of each core histone mRNA species present. Transcription of the early histone genes is repressed, and during blastulation there appear new "late" variants of each core histone and also histone H1 [22]. The late histones are transcribed from a different set of genes that are present in the genome in only 1 to 2 percent of the multiplicity of the early histone genes, and at least in some cases these genes are located distantly from the major early histone gene clusters [75]. Among the many fascinating problems of gene regulation posed by these data are the mechanisms by which first the cleavage stage histone genes and later the early histone genes are repressed, the relation between the patterns of expression and the genomic locations of these genes, the means by which early histone gene expression is coordinated with cell division rate, and the functional significance of the assembly into chromatin of the late as opposed to early histone variants. Early and late gene expression are also regulated differently during oogenesis, since the maternal message apparently does not include late histone gene transcripts. In contrast to the histones, many of the nonhistone nuclear proteins found in embryo chromatin as late as the gastrula stage are maternal in origin, and are not synthesized at all after fertilization [76].

The embryo applies diverse solutions to the logistic and informational demands of development. As additional genes and gene systems are investigated it becomes apparent that the basic components of these solutions, maternal mRNA's and proteins, and embryonic mRNA's and their translation products, are utilized to various extents and according to different developmental schedules.

Localization of Morphogenetic Fate in the Sea Urchin Embryo

Improvement of methods for obtaining mass preparations of micromeres and for recovering and culturing mesenchyme cells have led to a series of new experiments on the molecular basis of determination in the micromere cell lineage. An obvious proposal is that micromeres inherit a particular subset of maternal mRNA's, and thereby are endowed with special developmental potentialities. Rodgers and Gross [77] (in *Lp*) and Ernst et al. [78] (in *Sp*) found that only 70 to 80 percent of the maternal sequence complexity is represented in total micromere RNA, while the remainder of the embryo retains all the maternal sequences, at about the same concentrations as in the unfertilized egg. On the other hand, the sequence content of micromere polysomal RNA is just the same as that of the whole 16-cell embryo [78]. The similarity between the translated mRNA in micromeres and in the remainder of the embryo is observed as well at the protein synthesis level. Almost all the newly synthesized proteins resolved on two-dimensional gels are synthesized alike by micromeres, macromeres, and mesomeres [79]. While a small but potentially important subset of low abundance mRNA's would have been missed in any of these experiments, at least the simplest forms of the maternal mRNA segregation model for micromere commitment seem excluded. However, there are other differences between micromeres and the remainder of the fourth cleavage embryo, and these point in a different direction. At this stage micromere nuclei contain no high complexity nuclear RNA while nuclei of the other blastomeres synthesize such RNA actively [78]. In addition, Senger and Gross showed that the ratio of histone to nonhistone protein synthesis is higher in micromeres, and if transcription is blocked by actinomycin this difference is abolished [80].

One interpretation suggested by these observations is that the basic distinction between the micromere and other cell lineages derives from precocious differences in nuclear gene expression, due initially to a cytoplasmic environment in the micromeres very different from that to which the other blastomere nuclei are exposed. The micromeres are essentially budded off from the adjacent macromeres with a very small amount of cytoplasm. The absence of a class of maternal RNA sequences found in the rest of the embryo suggests that a special cytoplasmic domain has been created in the micromeres. Furthermore, nuclear gene expression in the micromeres might be expected to dominate the pattern of protein synthesis more rapidly than in cells that contain severalfold more cytoplasmic volume and maternal mRNA [80]. The micromere lineage is clearly committed at an early stage, and when cultured in isolation micromeres differentiate in vitro into skeleton-producing primary mesenchyme cells on the same temporal schedule as in situ [81]. By late in embryogenesis these and other differentiated cell types no doubt contain distinct populations of cytoplasmic poly(A) RNA's and synthesize distinct proteins. As one example, Bruskin et al. [53] have described a family of ten protein species translated specifically in late embryo ectoderm cells. Transcripts coding for these proteins increase in prevalence more than 100-fold after early blastula stage, and are already concentrated in ectoderm cells by the early gastrula stage.

The molecular basis of commitment on the part of early blastomere lineages is extremely subtle. When very early sea urchin embryos are treated with lithium chloride they undergo an abnormal form of development in which gastrula invagination and skeletal formation are suppressed. Yet no differences in the patterns of protein synthesis can be detected between lithium chloride-treated and normal embryos until after the catastrophic effects of this treatment have already occurred [82]. An approach that is promising in regard to commitment and morphogenesis is isolation of surface components from embryo cells that specify correct intercellular interactions. Giudice and his colleagues [83] showed that disaggregated sea urchin embryos will reassort to form blastula-like structures that subsequently complete embryological development. Recently, Noll et al. found that surface proteins can be extracted and then added back to promote such reaggregation [68], and the role and origins of these proteins can now be investigated.

The embryonic architecture of the sea urchin begins with the initial polarity of the egg. This subject was classically much investigated, and has been reexamined experimentally and reviewed by Schroeder [84]. The egg is endowed with an animal-vegetal polarity before fertilization, and this probably originates far back in oogenesis. At fertilization (or parthenogenic stimulation) a "clock" is activated, and on a timetable that is independent of the cleavage mitoses per se, cortical cytoplasmic elements of the vegetal pole are moved upward toward the equator of the egg. The residual polar cortex is the site where the micromeres will form. A fascinating challenge that can now be considered is definition of the molecular nature of this "clock," and of the primitive axial localization resident in the unfertilized egg.

Interpretation and Future Directions

We have focused on the major concepts and puzzles that have arisen with the growth of molecular level knowledge of sea urchin embryogenesis. Clearly we do not yet know how embryogenesis really "works" in this organism. Some major features are becoming evident, and other can perhaps be inferred. One simplified and hypothetical interpretation is as follows. During oogenesis a complex set of macromolecules that will be needed for morphogenesis and cell division begins to be transcribed and translated. They are accumulated in the growing oocyte in the quantities required for the mass of cytoplasm contained in the whole embryo, and following fertilization they are utilized until the embryo has acquired sufficient nuclei to replace them with its own transcription and translation products. The oocyte has only one (tetraploid) nucleus, and this process of accumulation requires a relatively long period of synthesis for those products needed in great abundance. Fertilization acts as a metabolic trigger that activates essentially the same pattern of biosynthesis as preceded it in the oocyte, but it also sets off a crucial series of spatial reorganizations of the cytoplasm. The result during cleavage is establishment of diverse domains in which, by some unknown process, the nuclei of different embryonic cell lineages are induced to function differentially. Most of the productive transcription (and processing) in the embryo nuclei is required for replacement of maternal RNA's and proteins, a task that is accomplished at various rates and by various means, depending on the sequence and the rate at which it is demanded. However, in these genomes there also occurs the activation of sets of cell lineage specific genes. The products of such genes might be needed for the cellular differentiations that result, by intercellular interaction, in the assembly of the three-dimensional multicellular structures of the embryo, though the actual construction of these may require extensive use of preformed components. Embryogenesis is terminated (pluteus stage) when a constant ratio of nucleus to cytoplasm is attained, when the transcripts on the polysomes are all of embryonic rather than maternal origin, when specific patterns of gene expression characteristic of the major cell lineages are established, and when the morphogenetic program that began to be read out in the oocyte nucleus has been carried to completion.

Although no other early embryonic system is nearly as well understood at the molecular level, it seems likely that additional approaches will be required if we are to be able to assay directly the functional meaning in development of the particular proteins, transcripts, and active genes of the embryo. It can be anticipated that there will be developed subembryonic and cell free systems in which the assembly of morphological structures can be studied, and other systems in which the effect of embryonic constituents on transcription and processing can be investigated in vitro. In addition, serious efforts are now being made to develop DNA transformation systems for sea urchin eggs that might permit direct investigation of the role of nucleic acid sequence features in the context of the developing embryo. It should be clear from the foregoing that early embryogenesis is a special, enormously complex process, the actual mechanism of which can scarcely be established by studying anything but embryos themselves. As it has for the past century, the sea urchin egg can be expected to continue to provide an accessible and revealing model for this endeavor.

Summary

Research on the early development of the sea urchin offers new insights into the process of embryogenesis. Maternal messenger RNA stored in the unfertilized egg supports most of the protein synthesis in the early embryo, but the structure of maternal transcripts suggests that additional functions are also possible. The overall developmental patterns of transcription and protein synthesis are known, and current measurements describe the expression of specific genes, including the histone genes, the ribosomal genes, and the actin genes. Possible mechanisms of developmental commitment are explored for regions of the early embryo that give rise to specified cell lineages, such as the micromere-mesenchyme cell lineage.

REFERENCES AND NOTES

1. Leahy PS, Hough-Evans BR, Britten RJ, Davidson EH: J Exp Zool 215:7, 1981; Leahy PS, Tutschulte TC, Britten RJ, Davidson EH, J Exp Zool 204:369, 1978.
2. Hinegardner RT, Rocha Tuzzi MM: "Marine Invertebrates." Washington, DC: National Academy Press, 1981, p 291; Cameron RA, Hinegardner RT: Biol Bull (Woods Hole, Massachusetts) 146:355, 1974.
3. Hinegardner RT: Am Zool 15:679, 1975; Coffaro KA, Hinegardner RT: Science 197:1389, 1977.
4. Davidson E: Unpublished observations.
5. Czihak G: A good description of the developmental origin of the imaginal structures. In Reverberi G (ed): "Experimental Embryology of Marine and Fresh-water Invertebrates." Amsterdam: North-Holland, p 363, 1971.
6. Houk MS, Hinegardner RT: Biol Bull (Woods Hole, Massachusetts) 159:280, 1980.
7. Davidson EH, Thomas TL, Scheller RH, Britten RJ: In Flavell RB (ed): "Genome Evolution." New York: Academic Press, 1982, p 177.
8. Humphreys T: Dev Biol 26:201, 1971.
9. Davidson EH: "Gene Activity in Early Development." New York: Academic Press,

1976, chap 4. This chapter includes a general review of data regarding protein synthesis, maternal mRNA, and its utilization in sea urchin embryos. Species hybrid and actinomycin experiments that foreshadowed the demonstration of maternal mRNA are also reviewed in ibid, chap 2.
10. Goustin AS: Dev Biol 87:163, 1981.
11. Duncan R, Humphreys T: ibid 88:211, 1981.
12. Wilt FH: Cell 11:673, 1977.
13. Costantini FD, Britten RJ, Davidson EH: Nature (London) 287:111, 1980.
14. Duncan R, Humphreys T: Dev Biol 88:201, 1981; Brandhorst BP, Verma DPS, Fromson D: ibid 71:128, 1979.
15. Infante AA, Heilmann LJ: Biochemistry 20:1, 1981.
16. Cognetti G, DiLiegro I, Cavarretta F: Cell Differ 6:159, 1977; Raff RA, Greenhouse G, Gross KW, Gross PR: J Cell Biol 50:516, 1971.
17. Scheller RH et al: Mol Cell Biol 1:609, 1981.
18. Crain WR, Durica DS, Van Doren K: ibid, p 711.
19. Mauron A, Kedes L, Hough-Evans BR, Davidson EH: Dev Biol 94:425, 1982.
20. Ruderman JV, Schmidt DR: ibid 81:220, 1981.
21. Brandhorst BP: ibid 79:139, 1980.
22. Childs G, Maxson RE, Kedes LH: ibid 73:153, 1979; Newrock KM, Cohen LH, Hendricks MB, Donnelly RJ, Weinberg ES: Cell 14:327, 1978; Newrock KM, Alfageme CR, Nardi RV, Cohen LH: Cold Spring Harbor Symp Quant Biol 42:421, 1978; Grunstein M: Proc Natl Acad Sci USA 74:4135, 1978; Poccia D, Salik J, Krystal G: Dev Biol 82:287, 1981. The specific histones for which maternal mRNA's exist are cleavage stage forms of H1, H1A, and H2B; H1α, H2Aα_1, H2Bα_1, and early forms of H3 and H4.
23. Brandhorst BP: Dev Biol 52:310, 1976.
24. Epel D: In Subtelny S, Wessells NK (eds): "The Cell Surface: Mediator of Developmental Processes." New York: Academic Press, 1980, p 169; Curr Top Dev Biol 12:185, 1978; Vacquier VD: Dev Biol 84:1, 1981.
25. Goustin AS, Wilt FH: Dev Biol 82:32, 1981.
26. Regier JC, Kafatos FC: ibid 57:270, 1977.
27. Hille MB, Albers AA: Nature (London) 278:469, 1979; Brandis JW, Raff RA: ibid, p 467; Dev Biol 67:99, 1978.
28. Shen SS, Steinhardt RA: Nature (London) 272:253, 1978; Epel D, Steinhardt RA, Humphreys T, Mazia D: Dev Biol 40:245, 1974.
29. Winkler MM, Steinhardt RA: Dev Biol 84:432, 1981; see also Grainger JL, Winkler MM, Shen SS, Steinhardt RA: ibid 68:396, 1979.
30. Moon RT, Moe KD, Hille MB: Biochemistry 19:2723, 1980; Young EM, Raff RA: Dev Biol 72:24, 1979; Kaumeyer JF, Jenkins NA, Raff RA: ibid 63:266, 1978.
31. Jenkins NA, Kaumeyer JF, Young EM, Raff RA: Dev Biol 63:279, 1978. The "masked message" hypothesis that stimulated these experiments was initially suggested by Spirin AS: Curr Top Dev Biol 1:1, 1966.
32. Wells DE, Showman RM, Klein WH, Raff RA: Nature (London) 292:477, 1981.
33. For *Sp*, Galau GA et al (48), Hough-Evans BR et al (36); for *Ap*, Anderson DM, Galau GA, Britten RJ, Davidson EH: Dev Biol 51:138, 1976; for *Lp*, Wilt FH (12).
34. Reviewed by Thomas TL, Posakony JW, Anderson DM, Britten RJ, Davidson EH: Chromosoma 84:319, 1981.
35. Hough-Evans BR, Ernst SG, Britten RJ, Davidson EH: Dev Biol 69:225, 1979.
36. Hough-Evans BR, Wold BJ, Ernst SG, Britten RJ, Davidson EH: ibid 60:258, 1977.
37. Costantini FD, Scheller RH, Britten RJ, Davidson EH: Cell 15:173, 1978.
38. Posakony JW, Flytzanis CN, Britten RJ, Davidson EH: J Mol Biol (in press); for primary

sequences, Posakony JW, Scheller RH, Anderson DM, Britten RJ, Davidson EH: J Mol Biol 149:41, 1981.

39. Dubroff LM, Nemer M: ibid 95:455, 1975.
40. Cabrera CV, Ellison JW, Moore JG, Britten RJ, Davidson EH: in preparation.
41. Goldberg RB, Galau GA, Britten RJ, Davidson EH: Proc Natl Acad Sci USA 70:3516, 1973; Davidson EH, Hough BR, Klein WH, Britten RJ: Cell 4:217, 1975.
42. Anderson DM, Richter JD, Chamberlin ME, Price DH, Britten RJ, Smith LD, Davidson EH: J Mol Biol 155:281, 1982.
43. An argument adduced in this connection is that all maternal poly(A) RNA is transferred to polysomes after fertilization. However, all that is in fact known is that in embryos most poly(A) is polysomal (11,44), and the matter is obscured by the rapid removal and replacement of poly(A) tracts occurring after fertilization [Dolecki GJ, Duncan RF, Humphreys T: Cell 11:339, 1977; Wilt FH, (12)]. No measurements exclude the possibility that nontranslatable poly(A) RNA from egg could be deadenylated, while other translatable molecules are readenylated in the embryo cytoplasm, since the necessary enzymes are present there [Egrie JC, Wilt FH: Biochemistry 18:269, 1979; Slater I et al (44): Slater DW, Slater I, Bollum FJ: Dev Biol 63:94, 1978].
44. Slater I, Gillespie D, Slater DW: Proc Natl Acad Sci USA 70:406, 1973.
45. Davidson EH, Britten RJ: Science 204:1052, 1979; Lerner MR, Boyle JA, Mount SM, Wolin SL, Steitz JA: Nature (London) 283:220, 1980.
46. An example has been provided by Diaz MO, Barsacchi-Pilone G, Mahon KA, Gall JG [Cell 24:649, 1981], who found that in amphibian oocyte nuclei the normal termination sites of histone genes are ignored, resulting in transcription of RNA's with lengthy 3′ terminal tails including flanking satellite sequences.
47. Bedard P-A, Brandhorst BP: Dev Biol (in press).
48. Galau GA, Klein WH, Davis MM, Wold BJ, Britten RJ, Davidson EH: Cell 7:487, 1976.
49. Galau GA, Britten RJ, Davidson EH: ibid 2:9, 1974.
50. Lev Z, Thomas TL, Lee AS, Angerer RC, Britten RJ, Davidson EH: Dev Biol 76:322, 1980.
51. Lasky LA, Lev Z, Xin J-H, Britten RJ, Davidson EH: Proc Natl Acad Sci USA 77:5317, 1980. In this report it was pointed out that two-dimensional gel analysis of newly synthesized embryo proteins detects the products of synthesis of those embryo transcripts present at $\geqslant 10^4$ transcripts per embryo (ten per late embryo cell on the average), since there are about 10^3 polypeptides resolvable and about this same number of diverse moderate and high prevalence RNA species.
52. Flytzanis CN, Brandhorst BP, Britten RJ, Davidson EH: Dev Biol 91:27, 1982, noted that most prevalent poly(A) RNA species in either egg or embryo are mitochondrial in origin (20). These sequences were not considered in their discussion, which refers only to RNA's coded in the nuclear genome.
53. Bruskin AM, Tyner AL, Wells DE, Showman RM, Klein WH: Dev Biol 87:308, 1981; Bruskin AM, Bedard P-A, Tyner AL, Showman RM, Brandhorst BP, Klein WH: ibid 91:317, 1982.
54. Shepherd GW, Nemer M: Proc Natl Acad Sci USA 77:4653, 1980; in addition, unpublished RNA excess hybridization experiments of B Wold (author's laboratory) with selected single-copy tracer showed a small component of new sequences at the blastula stage.
55. Piperno G, Huang B, Luck DJL: ibid 74:1600, 1977; Regier JC, Mazur GD, Kafatos FC: Dev Biol 76:286, 1980; Wood WB, Revel HR: Bacteriol Rev 40:847, 1976.
56. Bibring T, Baxandall J: Dev Biol 83:122, 1981.
57. Xin J.-H., Brandhorst BP, Britten RJ, Davidson EH, ibid 89:527, 1982.

58. Galau GA, Lipson ED, Britten RJ, Davidson EH: Cell 10:415, 1977.
59. Brandhorst BP, Humphreys T: Biochemistry 10:877, 1971.
60. Busby S, Bakken A: Chromosoma 71:249, 1979; ibid 79:85, 1980.
61. Aronson AI, Chen K: Dev Biol 59:39, 1977 showed that in sea urchin nuclei the polymerase translocaton rate is six to nine nucleotides per second. For comparison, the clone poly(A) RNA sequence studied by Cabrera et al (40) that enters the cytoplasm most rapidly crosses the nuclear boundary of a typical cell about every 6 minutes. Since the cell is diploid, this would require an initiation rate about half the average value calculated for the multiple transcription units (60).
62. Nemer M et al: Dev Genet 1:151, 1979.
63. Hough BR, Smith MJ, Britten RJ, Davidson EH: Cell 5:291, 1975; Davidson EH (9), p 234.
64. Wold BJ, Klein WH, Hough-Evans BR, Britten RJ, Davidson EH: Cell 14:941, 1978.
65. Thomas TL, Britten RJ, Davidson EH: Dev Biol 94:230, 1982.
66. Kleene KC, Humphreys T: Cell 12:143, 1977; Ernst SG, Britten RJ, Davidson EH: Proc Natl Acad Sci USA 76:2209, 1979.
67. Kawabe TT, Armstrong PB, Pollock EG: Dev Biol 85:509, 1981; Chow G, Benson SC: Exp Cell Res 124:451, 1979; Rapraeger AC, Epel D: Dev Biol 88:269, 1981; Glabe CG, Vacquier VD: J Cell Biol 75:410, 1977.
68. Noll H, Matranga V, Cascino D, Vittorelli L: Proc Natl Acad Sci USA 76:288, 1979.
69. Griffith JK, Griffith BB, Humphreys T: Dev Biol 87:220, 1981; Griffith JK, Humphreys T: Biochemistry 18:2178, 1979.
70. For *Sp*, see Emerson CP, Humphreys T: Dev Biol 23:86, 1970, and Galau GA et al (58); for *Lp*, see Humphreys T: in Coward SJ (ed): "Developmental Regulation, Aspects of Cell Differentiation." New York: Academic Press, 1973, p 1, and Surrey S, Ginzburg I, Nemer M: Dev Biol 71:83, 1979.
71. Kane RE: J Cell Biol 66:305, 1975.
72. Kedes LH: Annu Rev Biochem 48:837, 1979; Hentschel CC, Birnstiel ML: Cell 25:301, 1981.
73. Maxson RE, Wilt FH: Dev Biol 83:380, 1981.
74. Maxson RE, Wilt FH: ibid (in press).
75. Maxson RE, Mohun TJ, Gormezano G, Kedes LH: (in preparation).
76. Kuhn O, Wilt FH: Dev Biol 85:416, 1981.
77. Rodgers WH, Gross PR: Cell 14:279, 1978.
78. Ernst SG, Hough-Evans BR, Britten RJ, Davidson EH: Dev Biol 79:119, 1980.
79. Tufaro F, Brandhorst BP: ibid 72:390, 1979.
80. Senger DR, Gross PR: ibid 65:404, 1978.
81. Kitajima T, Okazaki K: Dev Growth Differ 22:265, 1980; Okazaki K: Am Zool 15:567, 1975; Harkey MA, Whiteley AH: Wilhelm Roux Arch Entwicklungsmech Org 189:111, 1980.
82. Hutchins R, Brandhorst BP: Wilhelm Roux Arch Entwicklungsmech Org 186:95, 1979.
83. Giudice G: Dev Biol 5:402, 1962; Giudice G, Millonig G: ibid 15:91, 1967.
84. Schroeder TE: ibid 79:428, 1980; Exp Cell Res 128:490, 1980.
85. Posakony J: unpublished experiment in EH Davidson's laboratory.
86. We are most grateful to colleagues who have given us the benefit of their various perspectives in critically reviewing this manuscript, in particular Fred Wilt of the University of California, Berkeley, Thoru Pederson of the Worcester Foundation, and Lee Hood, James Strauss, Herschel Mitchell, and Barbara Wold of the Division of Biology. Research from this laboratory was supported by NICHD grant HD05753.

Gene Structure and Regulation in Development, pages 171–182

Gene Expression During *Xenopus laevis* Development

Igor B. Dawid, Brian K. Kay, and Thomas D. Sargent
Laboratory of Biochemistry, National Cancer Institute, National Institutes of Health, Bethesda, Maryland 20205

INTRODUCTION

In this article we shall summarize the available information on gene expression and the accumulation of pA^+ RNA in oocytes and embryos of Xenopus laevis, and point out some approaches that are currently being used in this area of research. Xenopus laevis is a suitable organism for the study of developmental and molecular problems and has been much used for this purpose. One advantage of this organism is the accessibility of oogenesis and embryogenesis for study. Oocytes are present in large numbers in mature females and occur in several well-classified stages of development [Dumont, 1972]. Individual stages may be picked manually, but mass isolation procedures have also been described [Higashinakagawa et al, 1977]. Maturation of the oocyte to the mature egg can be induced in vitro and has been studied extensively [eg, Wasserman et al, 1982]. Embryogenesis is amenable for study in Xenopus because fairly large numbers of embryos can be obtained, an established embryology permits easy identification of developmental stages, and the histology of embryos is well known [Nieuwkoop and Faber, 1967]. The large size of Xenopus oocytes and embryos permits microinjection of nuclei, RNA, and DNA. This technology has been a powerful tool in answering some important biological questions (see Gurdon [1974]; also see references below).

Two other advantages of Xenopus should also be mentioned. First, because of its popularity as a laboratory animal, a great deal of information has

I.B. Dawid, B.K. Kay, and T.D. Sargent are presently at Laboratory of Molecular Genetics, National Institute of Child Health and Human Development, National Institutes of Health, Bethesda, MD 20205.

been accumulated about the sequence organization of its genome and some individual genes [Davidson et al, 1973; Reeder, 1980]. Second, as a vertebrate, Xenopus is a useful model system for the study of vertebrate developmental biology.

We should also mention some clear limitations of Xenopus as an experimental animal. While its oocytes go through a lampbrush chromosome stage these structures are rather small, making detailed cytology and cytochemistry very difficult. Such work is much better done in species with larger genomes like the newt [eg, Diaz et al, 1981]. Xenopus embryos are impermeable to radioactive precursors, a serious problem for studies of biosynthesis of macromolecules in development. This problem may be overcome by dissociating the cells of the embryo [Shiokawa and Yamana, 1967; Landesman and Gross, 1968], raising the issue of possible disturbance of normal events, or by microinjection, which is tedious and not adapted to pulse-chase experiments. Most importantly, Xenopus lacks standard genetics, except for the anucleolate frog, which proved most valuable in determining the site of rRNA synthesis and the location of the rRNA genes [Brown and Gurdon, 1964; Wallace and Birnstiel, 1966].

A good deal of information is available on gene expression in Xenopus development, but many problems remain. Three important unresolved questions concern the mechanisms responsible for and the biological significance of the following phenomena: (1) the function of lampbrush chromosomes that synthesize RNA that does not accumulate in oocytes; (2) the presence of repeated sequences on many pA^+ RNA molecules in oocytes; and (3) the differential recruitment of pA^+ RNAs for translation during development. We shall consider these and other problems and also stress some recent experimental approaches that involve the study of the developmental behavior of individual cloned sequences, analysis of spatial distribution of macromolecules, and the fate of DNA introduced into the oocyte or embryo.

GENE EXPRESSION IN OOGENESIS

The growth of the oocyte in Xenopus occupies a period of at least 6 months. During this time the cell grows to a diameter of about 1.5 mm and accumulates many substances for utilization during embryogenesis. Quantitatively, the largest contribution to this accumulation is made by yolk protein, derived from the precursor vitellogenin, which is synthesized in the liver and transported to the ovary [Wahli et al, 1981]. Yolk is stored as "raw material" for use in embryogenesis: lipovitellin and phosvitin are degraded in the embryo and their components are assembled into new macromolecules. The oocyte further produces and stores complex materials that will be used directly during embryogenesis. Prominent among these materials are the components of the protein synthesizing machinery, ribosomes and tRNAs [Brown and Littna, 1964a,b]. We know that the embryo can develop for about 3 days in the total absence of ribosome synthesis [Brown and Gurdon, 1964], even though the normal embryo starts rRNA synthesis earlier. Oocytes also accumulate a large number of mitochondria, which bring to the zygote a large store of mitochondrial DNA and RNA [Dawid, 1965, 1966; Chase and Dawid, 1972; Dworkin et al, 1981]. The mature oocyte contains about 4 μg rRNA, 4 ng mitochondrial DNA, and 20 ng mitochondrial rRNA.

A particularly interesting group of materials present in the egg is a relatively large amount of pA^+ RNA. Different workers report about 40–80 ng pA^+ RNA in the mature oocyte [Darnbrough and Ford, 1976; Ruderman and Pardue, 1977; Dolecki and Smith, 1979; Sagata et al, 1980]. This set of RNAs comprises about 20,000 different sequences, according to measurements by Davidson and Hough [1971] and Perlman and Rosbash [1978]. Oocytes apparently lack a high-abundance class of pA^+ RNA as commonly found in adult or tadpole tissues [Dworkin et al, 1981; Rosbash, 1981].

What is the nature and function of oocyte pA^+ RNA? The simple interpretation is that much of it is mRNA. Since it is clear that only a very small fraction of this RNA is located on polysomes in oocytes or mature eggs [Woodland, 1974], it is often assumed that most is stored maternal mRNA, destined for use by the embryo. This interpretation may well be correct, at least in part, but some difficulties have been pointed out. One problem concerns the observation that about 70% of oocyte pA^+ RNA carries interspersed repeat sequences, whereas somatic (tadpole) pA^+ RNA carries such repeats in only about 15% of its mass [Anderson et al, 1982]. There is no a priori reason why mRNA molecules could not carry repeated elements, as indicated previously by Dina et al [1973, 1974]. If the large fraction of RNA with repeat elements detected by Anderson and co-workers was mRNA or mRNA precursor, it would follow that the mRNA population in tadpoles, where these authors find a much lower fraction of repeat-containing RNAs, would be structurally different than in oocytes; eg, individual RNAs might become smaller by losing their repeat elements. In initial experiments using cloned probes, Golden et al [1980] showed that several individual RNAs have the same size in oocytes and in tadpoles. This finding, although still limited in scope, tends to argue against a precursor role for oocyte pA^+ RNA. Furthermore, oocyte and unfertilized egg RNAs act as efficiently as RNA from later stages in stimulating protein synthesis in vitro [Ballantine et al, 1979; Rosbash, 1981].

The rate of accumulation of RNA in the oocyte has been studied in several aspects. Oocytes accumulate 5S RNA and tRNA very early, whereas the accumulation of 18S and 28S rRNA starts later at the onset of vitellogenesis [Mairy and Denis, 1971; Ford, 1971], and then continues throughout oogenesis [Anderson and Smith, 1978]. Likewise, the accumulation of mitochondrial RNAs proceeds throughout oocyte development [Golden et al, 1980]. However, the situation with respect to pA^+ RNA is rather different in a way that contrasts the intuitive guess one might have had. It has been shown that the total amount of pA$^+$ RNA, and of poly(A) in the oocyte reaches a plateau level quite early in oogenesis, at the beginning of vitellogenesis [Rosbash and Ford, 1974; Cabada et al, 1977; Dolecki and Smith, 1979]. This finding has been extended by Golden et al [1980], who showed that several individual pA$^+$ RNA species, as measured with cloned probes, cease to accumulate after stage II in oogenesis. It is therefore well established that the pA$^+$ RNA population, with most if not all of its individual members, accumulates during early oogenesis up to stage II and thereafter remains essentially constant in amount. These observations are surprising: Even though lampbrush chromosomes loops are actively synthesizing RNA and these structures are most extended in stage III oocytes [MacGregor, 1980], pA$^+$ RNA accumulation ends before their appearance. Furthermore, stage VI oocytes, which no longer have extended lampbrush chromosomes, synthesize RNA at rates similar to those of stage III oocytes [LaMarca et al, 1973; Anderson and Smith, 1978; Dolecki and Smith, 1979]. A further question in this context arises from the observations of Ford et al [1977] that RNA synthesized in early oocytes (stages I or II) is stable during oogenesis for the next 18 months and is found essentially undiminished in mature eggs. This suggests that the pA$^+$ RNA made in stages III to VI turns over differentially; ie, pA$^+$ RNA made early is conserved through to the end of oogenesis and is eventually inherited by the embryo, whereas essentially all the pA$^+$ RNA made during the latter months of oogenesis turns over. The significance of these phenomena is not clear at present.

GENE EXPRESSION IN EMBRYOGENESIS

The Onset of RNA Synthesis and Rates of Accumulation of Classes of RNA

In the previous section we described that the mature Xenopus egg carries a store of 40–80 ng of pA$^+$ RNA, at least some of which is stored mRNA. Protein synthesis, also measurable in terms of the fraction of ribosomes and pA$^+$ RNA that is assembled into polysomes, is very low at fertilization and increases gradually as cleavage proceeds. Woodland [1974] has measured polysome levels in embryos throughout development, thereby providing a basis for all later studies in this area. Ribosome recruitment into polysomes continues during the first 5 hr of development until about 20% are so assembled; a plateau then persists for the next 24 hr, followed by a steady increase that leads to a typical "somatic" polysome loading of about 70% of all ribosomes in feeding tadpoles. Since there is very little accumulation of new mRNA during the first 5 hr (see below), it is clear that a portion of the egg pA$^+$ RNA is recruited into polysomes and thus represents stored mRNA.

The timing of the onset of new RNA synthesis in the embryo has been studied to a considerable extent. Earlier studies suggested that tRNA and heterogenous RNA are first synthesized in the blastula stage whereas rRNA (including 5S RNA) is first produced at gastrula [Brown and Littna, 1964a, 1966a,b]. Recent studies by Shiokawa, Yamana, and their colleagues provide a rather detailed description of RNA synthesis in Xenopus embryos and have somewhat modified our views on the subject. As mentioned above, embryos are not permeable to precursors; therefore, dissociated embryonic cells were used according to methods developed by Shiokawa and Yamana [1967] and Landesman and Gross [1968]. The possibility must be kept in mind that this treatment might alter the synthetic behavior of the embryo cells. Using this method to label embryo cells with ^{3}H-methionine, Shiokawa et al [1981a,b] obtained methylated nucleotides by hydrolysis of high molecular weight RNA and separated them on diethylaminoethyl (DEAE)-Sephadex columns. The data allowed measurement of the amount of label in cap structures and in methylated oligonucleotides characteristic for rRNA. The levels of tRNA and 5S RNA were determined separately by gel electrophoresis, and the specific activity of the S-adenosylmethionine pool was measured. From these data the authors show convincingly that capped RNA, presumably mRNA, is first synthesized in stage 6 morula cells. The first synthesis of tRNA, 5S RNA, and 18S and 28S RNA, is seen in stage 8 blastula cells. It is therefore clear — with the possible reservation mentioned above — that Xenopus embryos have initiated the synthesis of all major classes of RNA by midblastula, when the embryo is composed of about 10,000 cells.

The rate of synthesis of capped RNA in dissociated blastula embryos is very high, especially on a per cell basis. Per embryo, the rate increases only slightly through gastrula and neurula stages, but per cell the rate falls rapidly [Shiokawa et al, 1981a]. rRNA accumulation appears to be constant per cell; consequently, it increases rapidly per embryo and in relation to the accumulation of other RNAs, after blastula. Thus, RNA synthesis in the early embryo starts in an unusual quantitative relation between RNA classes and then "readjusts" to reflect a more typical somatic pattern in the tadpole [Brown and Littna, 1966a; Shiokawa et al, 1981b].

The distribution of newly synthesized RNA in polysomes and in the nonpolysomal compartment has been studied by Shiokawa et al [1981c]. From these data and the rates of accumulation of new RNA [Shiokawa et al, 1981b] the authors calculate the fraction of RNA in the whole embryo and in the polysomal compartment that is new, ie, not maternal. The results suggest that about 40% of the pA^+ RNA in gastrula polysomes and in the whole embryo is newly synthesized, and that essentially the entire pA^+ RNA population has been replaced by neurula. Since a majority of the pA^+ RNA synthesized in embryogenesis has sequence homology to egg pA^+ RNA (see below), these observations suggest that the embryo resynthesizes RNAs that are already present.

Qualitative Changes in pA$^+$ During Development

By measuring the kinetics of reassociation between cDNA and RNA from oocytes and from tadpoles, Perlman and Rosbash [1978] have come to the following conclusions: First, each pA$^+$ RNA population consists of about 20,000 different sequences. Second, the majority of these sequences are shared between ovary and tadpoles (and cultured kidney cells, which were also studied). About 80% of the tadpole cDNA hybridized with ovary RNA, and almost 100% of the ovary cDNA hybridized with tadpole RNA. These results suggest that only a small fraction of tadpole RNAs are newly expressed during development (or increase from very low to substantially higher levels), and very few sequences disappear or greatly decrease. Third, ovary RNA contains no highly abundant pA$^+$ RNAs, whereas tadpole RNA does.

Dworkin and Dawid [1980a,b] and Dworkin et al [1981] have used a different approach to the analysis of RNA populations in embryos. cDNA derived from pA$^+$ RNA of gastrulae and tadpoles was cloned; several hundred cloned sequences were studied by colony hybridization with labeled probes derived from pA$^+$ RNA of different developmental stages and different cell fractions. These studies established the following points. First, gastrula embryos do not contain highly abundant cytoplasmic pA$^+$ RNAs, in agreement with the results of Perlman and Rosbash [1978] and Rosbash [1981]. The most abundant pA$^+$ RNAs in early embryos are mitochondrial RNAs. Second, the pA$^+$ RNA populations studied in oocytes, eggs, and gastrulae are similar, with rather few changes observable. From tailbud stage on, changes become apparent; some RNA species increase in concentration from low or undetectable levels, leading to the presence of a class of highly abundant pA$^+$ RNAs in tadpoles.

The distribution of individual RNA sequences between the polysomal and the nonpolysomal compartment has been studied with cloned cDNA probes [Dworkin and Hershey, 1981]. Abundant RNAs appearing in polysomes during later development were found to be newly synthesized RNAs, rather than being derived from the nonpolysomal compartment of earlier stages. Only a small number of abundant RNAs were studied, so it may be premature to generalize these findings. However, the results imply that the initial wave of polysome assembly during the first few hours of development involves primarily recruitment of maternal or stored mRNAs, and that the next stage of assembly may involve mostly newly synthesized mRNAs, many of which may be qualitatively different from the RNAs in the egg.

Histone Gene Expression: Transcription and Translation

Rapid DNA synthesis and cell division during early development requires the availability of large amounts of histones. Xenopus satisfies its need by storing both histones themselves and histone mRNAs in the egg. This subject has been reviewed by Woodland [1980], and we shall therefore not discuss it as fully as would otherwise be appropriate. Translational control in embryos has been clearly demonstrated for histone mRNAs. The mature oocyte stores histone mRNAs, but synthesizes little histone. The production of nucleosomal histones (H2a, H2b, H3 and H4) starts during oocyte maturation, whereas H1 is synthesized only from the blastula stage onward [Adamson and Woodland, 1974, 1977; Flynn and Woodland, 1980; Woodland, 1980]. Oocyte histone mRNA is largely polyadenylated [Levenson and Marcu, 1976; Ruderman and Pardue, 1977], a fact that may be related to its stability in the oocyte. During embryogenesis most histone mRNAs lack poly(A), as they do in all somatic cells. The stored histone mRNA turns over during early development and by gastrulation appears to be entirely replaced by newly synthesized mRNA [Woodland, 1980].

RNA Turnover in Early Embryos

Turnover data for individual mRNAs or the entire pA^+ RNA population in early embryos are rather sparse. As mentioned above, histone mRNAs turn over quite rapidly, possibly because these RNAs are deadenylated in the embryo. Sea urchin histone mRNAs injected into Xenopus embryos turn over with a half-life of about 3 hr [Woodland and Wilt, 1980]. In contrast, injected mouse globin mRNA is highly stable through embryogenesis [Gurdon et al, 1974]. This observation might imply that the bulk of the endogenous pA^+ mRNA in the embryo should also be rather stable. However, as discussed above, total pA^+ RNA levels in the embryo remain constant through gastrulation and increase only slowly after this time, while accumulation of newly synthesized capped RNA (presumably mRNA) is active from blastula on. These results can only be reconciled by the assumption of RNA turnover. According to Shiokawa et al [1981a], about 40% of the pA^+ RNA would have to turn over by gastrulation to make room for the newly accumulated material. The problem is not resolved at this point, but a combination of the following explanations may obtain. First, all measurements on RNA populations in embryos are difficult and subject to possibly substantial errors. Consequently, there is room for a sizable increase in the pA^+ RNA content, which would reduce the turnover rate needed to accommodate measured accumulation rates. Second, the observed stability of injected pA^+ RNA other than histone mRNAs may not be characteristic for endogenous embryonic RNAs. In fact, different RNA molecules have been shown to have different stabilities in sea urchin embryos [Davidson et al, 1982].

Changing Patterns of Protein Synthesis During Development

Three laboratories have studied protein synthesis during development in a general way by labeling proteins with radioactive amino acids and separating the proteins by two-dimensiuonal gel electrophoresis [Brock and Reeves, 1978; Bravo and Knowland, 1978; Ballantine et al, 1979]. The complex patterns obtained by these investigators are somewhat different from each other, but some conclusions can be drawn from these results. First, protein patterns in oocytes, eggs, or early cleavage embryos are characterized by many rare species without any highly prominent spots. This observation is consistent with the absence of highly abundant mRNAs as revealed by cDNA/RNA reassociation [Perlman and Rosbash, 1978], the analysis of cloned cDNA populations [Dworkin and Dawid, 1980; Dworkin et al, 1981], and the gel electrophoretic analysis of in vitro translation products of isolated mRNA [Rosbash, 1981]. Second, the pattern of proteins synthesized in the embryo changes substantially through development. It is difficult to quantitate such results, but it appears that these patterns change more drastically than might be expected on the basis of the similarity in sequence content of pA^+ RNA populations from different stages, as measured by cDNA/RNA reassociation [Perlman and Rosbash, 1978; see above]. This apparent difference may be due, at least in part, to the fact that gel electrophoresis displays the few hundred most abundant proteins while reassociation looks at the average of all pA^+ RNAs most of which are not highly abundant. Furthermore, only a fraction of the maternal pA^+ RNA is recruited into polysomes during the early stages; thus, differential recruitment (ie, translational control) could account for changing patterns of protein synthesis. At stages past gastrulation the accumulation of new mRNAs becomes significant. Dworkin and Hershey [1981] have shown that several abundant polysomal RNA species from later embryos were rare in egg RNA and thus are qualitatively new. These data are consistent with the observation of new protein spots arising during development as visualized on two-dimensional gels.

Localization of Proteins and RNA in the Oocyte and Embryo

The importance of the spatial organization of the egg for development has been an important subject of study in many animals [eg, Subtelny and Konigsberg, 1979]. It is therefore of interest to examine the distribution of different proteins and mRNAs within the oocyte, egg, and embryo. There have been some recent studies in which this problem has been considered. Ballantine et al [1979] dissected embryos into endodermal and ectodermal halves and analyzed the newly synthesized proteins in these parts by two-dimensional gel electrophoresis. Although most labeled proteins were present in both halves, some differences were observed, especially at the tadpole stage. The clearest difference seen was the muscle-specific α-actin that appeared exclusively in ectodermal halves.

Moen and Namenwirth [1977] have developed a method by which many eggs or embryos were simultaneously oriented and cut into equivalent sections, so that ten fractions were generated. The distribution of unlabeled proteins in these fractions was surveyed and showed substantial fractionation for some proteins. Carpenter and Klein [1982] have applied this method in a study of the distribution of pA^+ RNA in three sections taken normal to the animal-vegetal axis of eggs and gastrulae. The RNA populations in these fractions were very similar, but a few percent of the RNAs were enriched 2- to 20-fold in the vegetal third of the eggs or embryos.

A particularly promising method for the analysis of spatial distribution of RNA is in situ hybridization with a labeled probe [Pukkila, 1975; Lamb and Laird, 1976]. The use of labeled poly(U) for the detection of poly(A) in situ has been exploited by Capco and Jeffery [1982] in a study of Xenopus oocytes. Since free poly(A) has not been detected in this material, it may be concluded with confidence that the assay detects pA^+ RNA. A series of changing localizations of the bulk of the poly(A) was observed in oocytes, but the distribution became homogeneous after maturation.

Information about the localization of individual mRNAs in the egg and embryo will be most interesting, but it will be technically much more difficult to detect any one RNA, as compared to total poly(A). However, the localization of histone mRNA in sea urchin eggs and embryos has been studied by Venezky et al [1981], and similar work in Xenopus may be possible.

OUTLOOK

A considerable amount of information is available on gene expression in oocytes and embryos of Xenopus. Most of this information is at a general level; ie, it refers to entire populations of classes of RNA molecules. It seems clear that the future will bring much more insight into the expression of individual genes or gene families. Such individualized studies will also illuminate the spatial distribution of various RNA molecules, and the developmental patterns of expression of different proteins. In the effort to study individual genes and gene products, recombinant DNA techniques are of critical importance. Genes for known products can be isolated, and their developmental regulation studied, eg, histone, actin, tubulin genes. But, these methods also allow the study of changing gene activity at the individual level without the need for an already known product. cDNA libraries, derived from different developmental stages of Xenopus, have been prepared and characterized to some extent [Dworkin and Dawid, 1980a,b; Golden et al, 1980; Jacob, 1980; Dworkin et al, 1981, Dworkin and Hershey, 1981; Spohr et al, 1981; Knöchel and John, 1982]. The study of developmentally regulated gene expression may be supplemented by analysis of the fate of genes injected into embryos. In Xenopus, the retention of injected genes has been demonstrated up to the stage of the metamorphosed frog, and correct transcription and splicing has been observed [Rusconi and Schaffner, 1981; Bendig, 1981]. Further work along these lines may confidently be expected.

REFERENCES

Adamson ED, Woodland HR (1974): Histone synthesis in early amphibian development: Histone and DNA synthesis are not coordinated. J Mol Biol 88:263–285.

Adamson ED, Woodland HR (1977): Change in the rate of histone synthesis during oocyte maturation and very early development of Xenopus laevis. Dev Biol 57:136–149.

Anderson DM, Chamberlin ME, Britten RJ, Davidson EH (1982) Sequence organization of the poly(A) RNA synthesized and accumulated in lampbrush chromosome stage Xenopus laevis oocytes. J Mol Biol 155:281-309.

Anderson DM, Smith LD (1978): Patterns of synthesis and accumulation of heterogeneous RNA in lampbrush stage oocytes of Xenopus laevis (Daudin). Dev Biol 67:274–285.

Ballantine JEM, Woodland HR, Sturgess EA (1979): Changes in protein synthesis during the development of Xenopus laevis. J Embryol Exp Morphol 51:137–153.

Bendig MM (1981): Persistence and expression of histone genes injected into Xenopus eggs in early development. Nature (London) 292:65–67.

Bravo R, Knowland J (1979): Classes of proteins synthesized in oocytes, eggs, embryos and differentiated tissues of Xenopus laevis. Differentiation 13:101–108.

Brock HW, Reeves R (1978): An investigation of *de novo* protein synthesis in the South African clawed frog, Xenopus laevis. Dev Biol 66:128–141.

Brown DD, Gurdon JB (1964): Absence of ribosomal RNA synthesis in the anucleolate mutant of Xenopus laevis. Proc Natl Acad Sci USA 51:139–146.

Brown DD, Littna E (1964a): RNA synthesis during the development of Xenopus laevis, the South African clawed toad. J Mol Biol 8:669–687.

Brown DD, Littna E (1964b): Variations in the synthesis of stable RNAs during oogenesis and development of Xenopus laevis. J Mol Biol 8:688–695.

Brown DD, Littna E (1966a): Synthesis and accumulation of DNA-like RNA during embryogenesis of Xenopus laevis. J Mol Biol 20:81–94.

Brown DD, Littna E (1966b): Synthesis and accumulation of low molecular weight RNA during embryogenesis of Xenopus laevis. J Mol Biol 20:95–112.

Cabada MO, Darnbrough C, Ford PJ, Turner PC (1977): Differential accumulation of two size classes of poly(A) associated with messenger RNA during oogenesis in Xenopus laevis. Dev Biol 57:427–439.

Capco DG, Jeffery WR (1982): Transient localizations of messenger RNA in Xenopus laevis oocytes. Dev Biol 89:1–12.

Carpenter CD, Klein WH (1982): A gradient of poly(A)$^{+}$RNA sequences in Xenopus laevis eggs and embryos. Dev Biol 91:43–49.

Chase JW, Dawid IB (1972): Biogenesis of mitochondria during Xenopus laevis development. Dev Biol 27:504–518.

Darnbrough C, Ford PJ (1976): Cell-free translation of messenger RNA from oocytes of Xenopus laevis. Dev Biol 50:285–301.

Davidson EH, Hough BR (1971): Genetic information in oocyte RNA. J Mol Biol 56:491–506.

Davidson EH, Hough BR, Amenson CS, Britten RJ (1973): General interspersion of repetitive with non-repetitive sequence elements in the DNA of Xenopus. J Mol Biol 77:1–23.

Davidson EH, Hough-Evans BR, Britten RJ (1982): Molecular biology of the sea urchin embryo. Science 217:17–26.

Dawid IB (1965): Deoxyribonucleic acid in amphibian eggs. J Mol Biol 12:581-599.

Dawid IB (1966): Evidence for the mitochondrial origin of frog egg cytoplasmic DNA. Proc Natl Acad Sci USA 56:269–276.

Diaz MO, Barsacchi-Pilone G, Mahon KA, Gall JG (1981): Transcripts from both strands of satellite DNA occur on lampbrush chromosome loops of the newt Notophthalmus. Cell 24:649–659.

Dina D, Crippa M, Beccari E (1973): Hybridization properties and sequence arrangement in a population of mRNAs. Nature New Biol 242:101–105.

Dina D, Meza I, Crippa M (1974): Relative positions of the "repetitive," "unique" and poly(A) fragments of mRNA. Nature (London) 248:486–490.

Dolecki GJ, Smith LD (1979): Poly(A)$^+$RNA metabolism during oogenesis in Xenopus laevis. Dev Biol 69:217–236.

Dumont J (1972): Oogenesis in Xenopus laevis (Daudin). I. Stages of oocyte development in laboratory maintained animals. J Morphol 136:153–179.

Dworkin MB, Dawid IB (1980a): Construction of a cloned library of expressed embryonic gene sequences from Xenopus laevis. Dev Biol 76:435–448.

Dworkin MB, Dawid IB (1980b): Use of a cloned library for the study of abundant poly(A)$^+$ RNA during Xenopus laevis development. Dev Biol 76:449–464.

Dworkin MB, Hershey JWB (1981): Cellular titers and subcellular distributions of abundant polyadenylate-containing ribonucleic acid species during early development in the frog Xenopus laevis. Mol Cell Biol 1:983–993.

Dworkin MB, Kay BK, Hershey JWB, Dawid IB (1981): Mitochondrial RNAs are abundant in the poly(A)$^+$RNA population of early frog embryos. Dev Biol 86:502–504.

Flynn JM, Woodland HR (1980): The synthesis of histone H1 during early amphibian development. Dev Biol 75:222–230.

Ford PJ (1971): Non-coordinated accumulation and synthesis of 5S ribonucleic acid by ovaries of Xenopus laevis. Nature (London) 233:561–564.

Ford PJ, Mathieson T, Rosbash M (1977): Very long-lived messenger RNA in ovaries of Xenopus laevis. Dev Biol 57:417–426.

Golden L, Schäfer U, Rosbash M (1980): Accumulation of individual pA$^+$RNAs during oogenesis of Xenopus laevis. Cell 22:835–844.

Gurdon JB (1974): "The Control of Gene Expression in Animal Development." Cambridge: Harvard University Press.

Gurdon JB, Woodland HR, Lingrel JB (1974): The translation of mammalian globin mRNA injected into fertilized eggs of Xenopus laevis. Dev Biol 39:125–133.

Higashinakagawa T, Wahn H, Reeder RH (1977): Isolation of ribosomal gene chromatin. Dev Biol 55:375–386.

Jacob E (1980): Characterization of cloned cDNA sequences derived from Xenopus laevis poly(A)$^+$ oocyte RNA. Nucleic Acids Res 8:1319–1337.

Knöchel W, John ME (1982): Cloning of Xenopus laevis nuclear poly(A)-rich RNA sequences: Evidence for post-transcriptional control. Eur J Biochem 122:11–16.

LaMarca MJ, Smith LD, Strobel MC (1973): Quantitative and qualitative analysis of RNA synthesis in stage 6 and stage 4 oocytes of Xenopus laevis. Dev Biol 34:106–118.

Lamb MM, Laird CD (1976): Increase in nuclear poly(A)-containing RNA at syncytial blastoderm in Drosophila melanogaster embryos. Dev Biol 52:31–42.

Landesman R, Gross PR (1968): Patterns of macromolecule synthesis during development of Xenopus laevis. I. Incorporation of radioactive precursors into dissociated embryos. Dev Biol 18:571–589.

Levenson RG, Marcu KB (1976): On the existence of polyadenylated histone mRNA in Xenopus laevis oocytes. Cell 9:311–322.

MacGregor HC (1980): Recent developments in the study of lampbrush chromosomes. Heredity 44:3–35.

Mairy M, Denis H (1971): Recherches biochimiques sur l'oogenese. I. Synthese et accumulation du RNA pendant l'oogenese du crapaud sud-african Xenopus laevis. Dev Biol 24:143–165.

Moen TL, Namenwirth M (1977): The distribution of soluble proteins along the animal-vegetal axis of frog eggs. Dev Biol 58:1–10.

Nieuwkoop P, Faber J (1967): "Normal Tables of Xenopus laevis (Daudin)." 2nd edition. Amsterdam: North-Holland.

Perlman S, Rosbash M (1978): Analysis of Xenopus laevis ovary and somatic cell polyadenylated RNA by molecular hybridization. Dev Biol 63:197–212.

Pukkila PJ (1975): Identification of the lampbrush chromosome loops which transcribe 5S ribosomal RNA in Notophthalmus (Triturus) viridescens. Chromosoma 53:71–89.

Reeder RH (1980): Structure of cloned genes from Xenopus: A review. In Setlow JK, Hollander A (eds): "Genetic Engineering Principles and Methods." Vol I. New York: Plenum Press, pp 93–116.

Rosbash M (1981): A comparison of Xenopus laevis oocyte and embryo mRNA. Dev Biol 87:319–329.

Rosbash M, Ford PJ (1974): Polyadenylic acid-containing RNA in Xenopus laevis oocytes. J Mol Biol 85:87–101.

Ruderman JV, Pardue ML (1977): Cell-free translation analysis of messenger RNA in echinoderm and amphibian early development. Dev Biol 60:48–68.

Rusconi S, Schaffner W (1981): Transformation of frog embryos with a rabbit β-globin gene. Proc Natl Acad Sci USA 78:5051–5055.

Sagata N, Shiokawa K, Yamana K (1980): A study on the steady-state population of poly(A)$^+$RNA during early development of Xenopus laevis. Dev Biol 77:431–448.

Skiokawa K, Yamana K (1967): Pattern of RNA synthesis in isolated cells of Xenopus laevis embryos. Dev Biol 16:368–388.

Shiokawa K, Misumi Y, Yamana K (1981a): Demonstration of rRNA synthesis in pre-gastrular embryos of Xenopus laevis. Dev Growth Differ 23:579–587.

Shiokawa K, Tashiro K, Misumi Y, Yamana K (1981b): Non-coordinated synthesis of RNAs in pre-gastrular embryos of Xenopus laevis. Dev Growth Differ 23:589–597.

Shiokawa K, Misumi Y, Yamana K (1981c): Mobilization of newly synthesized RNAs into polysomes in Xenopus laevis embryos. Wilhelm Roux Arch 190:103–110.

Spohr G, Reith W, Sures I (1981): Organization and sequence analysis of a cluster of repetitive DNA elements from Xenopus laevis. J Mol Biol 151:573–592.

Subtelny S, Konigsberg IR (eds) (1979): "Determinants of Spatial Organization." New York: Academic Press.

Venezky DL, Angerer LM, Angerer RC (1981): Accumulation of histone repeat transcripts in the sea urchin pronucleus. Cell 24:385–391.

Wahli W, Dawid IB, Ryffel GU, Weber R (1981): Vitellogenesis and the vitellogenin gene family. Science 212:298–304.

Wallace HR, Birnstiel ML (1966): Ribosomal cistrons and the nucleolar organizer. Biochim Biophys Acta 114:296–310.

Wasserman WJ, Richter JD, Smith LD (1982): Protein synthesis during maturation promoting factor- and progesterone-induced maturation in Xenopus oocytes. Dev Biol 89:152–158.

Woodland HR (1974): Changes in the polysome content of developing Xenopus laevis embryos. Dev Biol 40:90–101.

Woodland HR (1980): Histone synthesis during development of Xenopus. FEBS Lett 121:1–7.

Woodland HR, Wilt FH (1980): The functional stability of sea urchin histone mRNA injected into oocytes of Xenopus laevis. Dev Biol 75:199–213.

IV. Genomic Instabilities

Gene Structure and Regulation in Development, pages 185–195

Gene Control in Maize by Transposable Elements

Benjamin Burr and Frances A. Burr
Biology Department, Brookhaven National Laboratory, Upton, New York 11973

Transposable Elements That Control Gene Activity

Transposable genetic elements in maize are remarkable for their ability to regulate gene activity when they insert into a structural gene locus. McClintock [1956a] realized that the new combination formed by transposition did not necessarily involve the alteration of the structural gene and, therefore, termed the transposable units "controlling elements." In 1961 McClintock drew attention to parallels in the action of known gene control elements and the maize controlling elements. Examples of *Spm* elements regulating pigment synthesis at the *A1* locus, and *Ac* controlling the level of expression of *Waxy*, are illustrated in McClintock's [1967, 1978] contributions to the 26th and 36th Symposia in this series. It is worth mentioning that McClintock's proposal presupposed the existence of independent regions for coding and gene control prior to the structural description of any gene, and before genetic evidence indicated the presence of distinct operator and promotor sequences for prokaryotic genes.

A persuasive demonstration of McClintock's concept has been provided by Dooner [1981] working with a *Ds*-induced mutation of the *Bronze* locus. *Bronze* is a gene that conditions one of the last steps in the biosynthesis of anthocyanin and encodes the enzyme UDP glucose: flavonoid 3-*O*-glucosyltransferase (UFGT). In the maize endosperm, the anthocyanin pigments develop in the outermost layer, or aleurone. Dooner followed the level of UFGT during endosperm development and found that in normal kernels UFGT activity was low at 18 days postpollination. The enzymic activity rose gradually until day 30, then increased sharply until the end of development at day 45 when the kernels began to dry down. In contrast, the endosperms of *bz-m4* had three to six times the amount of UFGT activity early in development. This activity peaked at 22 days after pollination and then fell to undetectable levels about 40 days postpollination. *Bz* kernels begin to show purple pigmentation 20–25 days after pollination, but paradoxically the *bz-m4* kernels do not, even though the mutant endosperms have more enzymic activity at this time. Dooner examined the mutant to see whether the cellular distribution of the enzyme had been altered and discovered that the enzyme was being produced in the internal cells of the endosperm rather than in the aleurone. In other respects the mutant activity did not differ from that of the wild-type—both had similar apparent molecular weights, substrate affinities and specificities, isoelectric points, and thermal stabilities. The normal activity is repressed in plants homozygous for recessive alleles at the *C* and *R* loci, but UFGT activity was not affected by these genes in *bz-m4*. Therefore, in transposing to the *Bz* locus, *Ds* apparently did not disrupt the *Bz* structural gene. It did, however, alter the expression of the gene with respect to tissue type and the time at which it was expressed in development. The new pattern of expression is orderly and apparently keyed to specific developmental signals, but the structural gene no longer responds to the usual regulatory signals. As a consequence of transposition *Ds* has evidently created a new combination of the structural gene and another control sequence.

Evidence for the regulation of structural genes by elements inserted in their vicinity has been accumulating recently in other eukaryotic systems: 1) ROAM mutations: *Ty* element insertions into a number of yeast loci cause an overproduction of the gene product. Interestingly, this overproduction is under control of the mating type locus and lower levels are observed in a/α diploids [Errede et al, 1981; Young et al, 1981; Roeder and Fink, 1982a]. 2) Retrovirus activation of host genes: Avian leukosis virus itself does not carry a transforming gene, but the virus can cause tumor formation by inserting next to a cellular oncogene and inducing its activation [Hayward et al, 1980; Payne et al, 1981]. In both cases the initial assumption made was that gene activation had resulted from the insertion of a strong promotor 5′ to the gene that had taken over control of that gene. This has, in fact, been shown to be the case [Hayward et al, 1980]. In other instances it has been found that the elements may also be inserted downstream from the gene, or inserted such that the directions of transcription of the element and the gene are in opposite orientations [Young et al, 1981; Varmus, 1982; Roeder and Fink, 1982b]. Other mechanisms for the insertional activation of genes must therefore be considered. In this regard it is interesting that a 72 base pair (bp) repeat, analogous to the enhancer found in simian virus 40 (SV40), is present in the long terminal repeat (LTR) of murine sarcoma virus and can substitute for the SV40 sequence in activating viral gene transcription [Levinson et al, 1982].

The *Ac-Ds* Controlling Element System

The *Ac-Ds* controlling element system is only one of a number of transposable element families that have been described in maize. *Ac*, or *Activator*, is an autonomously regulated element that can transpose and modulate gene action when it inserts into a locus whose activity can be monitored. *Ac* also produces derivatives; one of these is *Ds (Dissociation)* which lacks the autonomous ability to transpose [McClintock, 1962]. In the presence of an active *Ac* anywhere in the genome, *Ds* can transpose, break chromosomes where it resides, cause adjacent deletions, act as a breakpoint for translocations, and undergo a change of state. If the particular *Ds* being studied possesses the property of being able to break chromosomes, then *Ds* can be mapped on appropriately marked chromosomes even if it is not inserted into a known gene locus [McClintock, 1950]. When chromosome breakage occurs in response to *Ds*, the markers distal to *Ds* are lost and this loss may be indicated by the appearance of somatic sectors on the plant. In some tissues the broken chromatids can fuse and initiate the breakage-fusion-bridge cycle. As a consequence of this cyclic fusion and random breakage, there are subsequent secondary losses of more proximal markers. A property that makes *Ds*-induced mutations particularly well-suited to molecular analyses is that, in the absence of *Ac*, these mutations remain perfectly stable.

Ds Mutations of *Shrunken*

Our approach to the molecular study of maize controlling elements was to begin by choosing a gene for which there were controlling element-induced mutations and that made an abundant product. The best candidate appeared to be the *Shrunken (Sh)* gene that encodes the enzyme sucrose synthetase [Chourey and Nelson, 1976]. This enzyme catalyzes the first step in starch biosynthesis in the endosperm. Mutants of the gene are viable but have a reduced starch content, typically resulting in a seed with an extremely concave crown due to the collapsed endosperm. Mutations or their revertants can be easily recognized against a nonmutant background by visual screening. Because of its easily distinguishable phenotype, the gene can serve as a trap for transposable element insertions. Such mutations that might exhibit differences in the structure or arrangement of the transposable elements can thereby readily be collected. The disadvantages of the gene are that somatic sectors on the kernels are difficult to detect and that intermediate levels of gene expression cannot be distinguished from wild-type levels by visual examination. In addition, no insertions that have a positive regulatory effect on the gene have yet been found.

When used as a hybridization probe, our cDNA clone of sucrose synthetase recognized a unique sequence in maize DNA [Burr and Burr, 1981]. Another sucrose synthetase gene product has been reported [Chourey, 1981], but the second gene sequence was not detected in our Southern hybridization experiments under the conditions employed. The cDNA probe was used to isolate a clone of the wild-type gene [Burr and Burr, 1982]. Heteroduplexes of the genomic clone with sucrose synthetase mRNA were examined in the electron microscope. No large intervening sequences were observed but there were indications of several very small ones just at the limit of resolution. A restriction map of the genomic clone was constructed and found to be colinear with that constructed from whole genome Southern blotting data. The transcribed region of the cloned sequence and its direction of transcription were also determined.

McClintock [1952, 1953] isolated three independent *Ds*-induced mutations of *Sh* and showed that revertants could be obtained from them (see Fig. 1). In these revertants she reported that *Ds* had not transposed far from the locus and that it was possible to complete the cycle and generate new *Ds* mutations from them [McClintock, 1956a]. Subclones of the wild-type genomic clone were employed as probes in whole genome Southern blotting experiments to

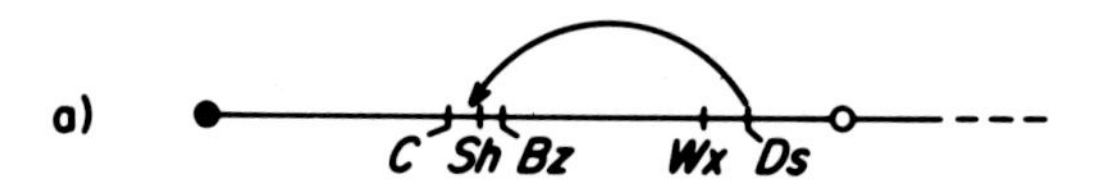

b) *Ds Sh 5245* → *Ds sh 6258* → *Ds Sh** → *Ds sh 6795* → *Ds Sh 6795 Rev.*
→ *Ds sh 5933* → *Ds Sh 5933 Rev.*
Ds Sh 4864 → *Ds sh 6233* → *Ds Sh 6233 Rev.*

c)

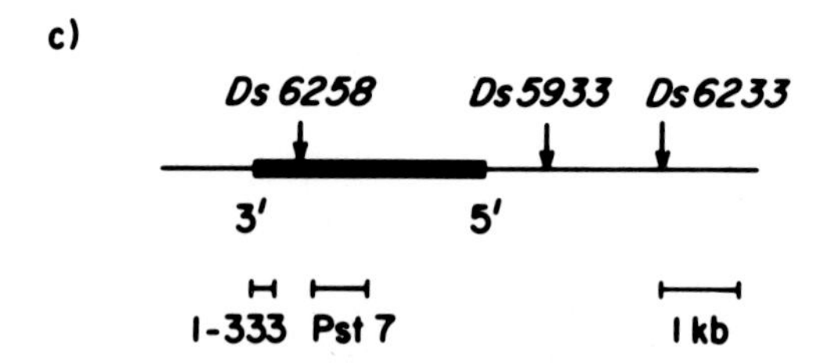

Fig. 1. a) Cytological map of the short arm of chromosome 9 after McClintock [1951]. *Ds* is depicted at its originally observed position. McClintock [1952] isolated two transpositions just distal to *Sh*. These were called *Ds 4864A* and *Ds 5245*; *Ds*-induced *sh* mutants were obtained from these stocks. b) Relationships of the *Ds*-induced *sh* mutations and their revertants. *Ds sh5933* and *Ds sh6258* are independent transpositions from *Ds Sh5245*. The *Ds Sh** revertant of *Ds sh6258* has been lost. c) Schematic diagram of the *Sh* locus showing sites of the original *Ds* insertions. The bar represents the transcribed region; transcription is from left to right. The cDNA clone, 1–333, and the genomic subclone Pst 7 were employed in the preparation of Figure 2.

analyze the nature of four of these mutations [Burr and Burr, 1982]. The *Ds*-induced rearrangements were found to consist of 20-kb insertions into the locus. All appeared to be perfect insertions because no alterations of flanking restriction sites were detected. *Ds sh6258* and *Ds sh5933* were both derived from the same progenitor that carried *Ds* just distal to *Sh*. Despite their common origin, however, they were found to have very different restriction maps and the only similar internal restriction sites seemed to be the *Bam* HI sites located close to both extremities. We have less data on another *Ds*-induced mutation, *Ds sh6233*, but it apparently lacks even these *Bam* HI sites.

Possible clues as to how these elements might have diverged comes from the analysis of *Ds sh6258* and its derivatives. In this mutant the *Ds* element is inserted inside the transcribed region of the gene about two-thirds of the distance from the 5′ end, presumably within a short intervening sequence. The revertants were found to still have *Ds* closely linked with the locus, as judged by the chromosome breakage behavior. A secondary mutation, *Ds sh6795*, was generated from one of these revertants and subsequently another revertant, *Ds Sh6795 Rev.*, was selected from *Ds sh6795*. Both *Ds sh6795* and its revertant have approximately 20 kb of material inserted in the *Sh* gene; moreover, the insertions are located at the same site and are in the same orientation as in the initial mutation, *Ds sh6258*. *Ds sh6795*, however, differs from the progenitor *Ds sh6258* in missing three sites from the 5′ portion of the insert and in having six other sites throughout the insert shifted by 0.5 to 1 kb. These shifts could be the result of small inversions. *DsSh6795 Rev.* differs from its progenitor in having a completely different internal restriction map in the 5′ half of the element. These results suggest that the element never left the site during reversion and backmutation and that changes in the substructure of the element itself occurred which were responsible for either expression or repression of the *Sh* gene.

We have probed poly(A)$^+$ RNA prepared from the *Ds* mutants with fragments from the *Sh* gene that flank the site of the *Ds* insertion (Fig. 2). No poly(A)$^+$ RNA homologous to the *Sh* gene was detected in *Ds sh6258* or *Ds sh6795* when a 3′ probe was used; however, a 5′ probe detected a slightly smaller polyadenylated sequence in *Ds sh6795* but not in *Ds sh6258*. The 3′ probe hybridized to a full-length mRNA of approximately 2.7 kb in two revertants. In *Ds sh6258* the insertion may contain a transcription termination signal but the transcript produced is either not correctly processed or not polyadenylated. Rearrangements of the insert are responsible for the revertants that presumably now allow the synthesis of very large primary transcripts including *Ds* sequences that are then correctly spliced out. We have not yet looked for this very large primary transcript. Another rearrangement of internal *Ds* sequences in *Ds sh6795* was responsible for the synthesis of a shortened transcript that is now correctly processed and polyadenylated.

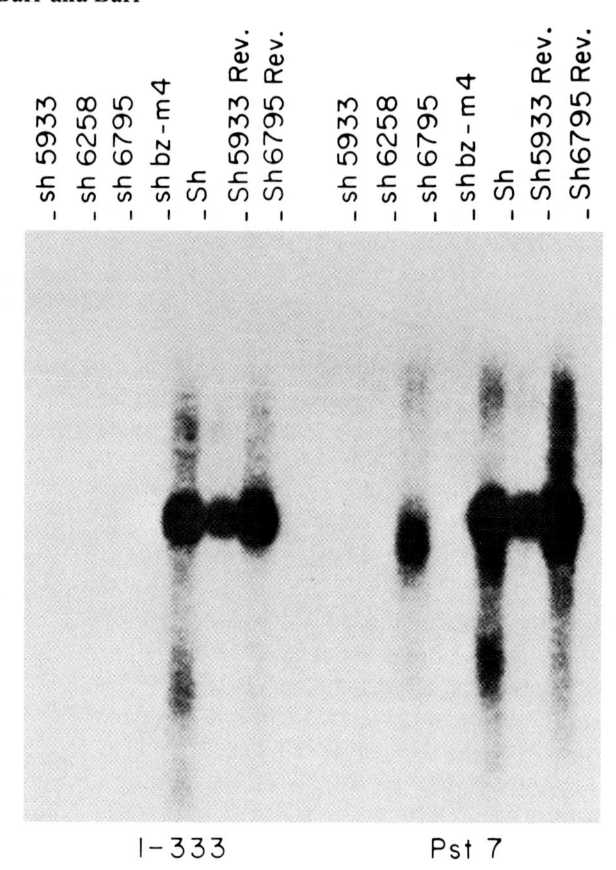

Fig. 2. RNA blots showing hybridization to *Sh* poly(A)$^+$RNA. Poly(A)$^+$RNA was prepared from the seven strains listed above. The mutant *sh bz-m4* is a *Ds*-induced deletion of the *Sh* locus [Dooner, 1981]. RNA was denatured with glyoxal, electrophoresed on a 1% agarose gel, and transferred to nitrocellulose. Hybridization probes were the inserts of clones 1–333 hybridizing downstream from the site of *Ds sh6258* insertion and Pst 7 hybridizing upstream from the site of insertion (see Fig. 1c).

Internal Rearrangements of *Ds* and "Changes of State"

Rearrangements of *Ds* appear to be one way in which the element can regulate the activity of a structural gene. McClintock [1949] had recognized different "states" of *Ds*. In one instance of *Ds* at the *C* locus, she described a *Ds* element that gave predominantly either chromosome breakage or transposition events. These two states were semistable but by selection McClintock found derivatives that had the properties of the alternative state.

McClintock [1965, 1967] described other examples of "changes of state," in particular the mutable alleles of the *A1* gene that are under the control of *Spm*. In this two-element system a receptor at *A1*, a locus that affects anthocyanin expression, responds to the suppressor component of the *Spm* regulatory element by modulating the level of pigmentation and also can react to the mutator component of *Spm* by releasing the gene from control. The latter events will be seen as clonal sectors of darker pigmentation against the modulated background. "Changes of state" are manifested by differences in the degree of pigmentation of the modulated background or the clonal sectors, in the time during kernel development when mutation occurs, and in the frequencies of the mutations. Different states derived from an *a1-m* kernel behave as allelic forms and are able to maintain their separate identities in a heterozygote and in succeeding generations. They will give two types of sectors in response to the same regulatory element.

McClintock [1951] emphasized that changes of state of *Ds* required the presence of an active *Ac* in the same nucleus and that without *Ac* the particular state would remain the same. She furthermore observed that a given state responded in the same manner to different *Ac* elements. The particular state of *Ds* is clearly a property of the element itself but in some way that state can be altered by *Ac*. McClintock [1956b] concluded that the states represented different allelic forms of *Ds* and reflected "structural differences either of the *Ds* element itself or of other components at the locus where it resides."

Transposition of *Ds* most likely occurs by precise excision and reinsertion into a new site. McClintock [1951] was able to observe several cases where the disappearance of *Ds* from its original site was associated with its appearance at a new location. As we have mentioned, dicentric chromatid formation following chromosome breakage allows placement of *Ds* relative to other markers even if *Ds* is not inserted into a known gene. When *Ds* transposes a sufficient distance on a chromosome, it is possible to verify that no genetically active material is left at the original site. Molecular analyses will be able to determine if any portion of the element remains. Preliminary observations of revertants for *Ds sh5933* and *Ds sh6258* indicate that, on the basis of whole-genome Southern-blotting experiments, nothing appears to have been left behind. The limit of resolution of the technique, however, is about 200 nucleotides. From these observations it seems probable that the *Ds* elements derived from the same progenitor stock are related, even though they apparently have undergone structural changes as a result of transposition.

There is ample evidence that dispersed repeated elements in yeast can exchange information by gene conversion [Scherer and Davis, 1980; Klein and Petes, 1981; Jackson and Fink, 1981]. Roeder and Fink [1982a] have shown that the same mechanism permits the replacement of sequences in one *Ty* element with those present in another *Ty*. It is likewise conceivable that changes in the structure of *Ds* at a locus are the result of nonreciprocal exchanges with *Ds* elements located elsewhere in the genome. However, in the absence of an active *Ac* in the same nucleus the *Ds* elements are completely stable and we consider it unlikely that *Ac* could mediate specific gene conversion events. An alternative mechanism envisages extensive internal rearrangements analogous to, but much more complex than, G-loop inversion in bacteriophage *Mu* [Kamp et al, 1978] or phase variation in Salmonella [Zieg et al, 1977]. *Ac* provides a function missing in *Ds* that allows *Ds* to transpose. A similar, or perhaps the identical function, could signal the rearrangements within *Ds*. Roeder and Fink [1982b], recognizing that gene conversion promotes homogeneity of the sequence, proposed that the *Ty* elements themselves are rearranged during transposition. It may be possible to test our idea that the *Ds* elements represent internal rearrangements by cross-hybridizing subclones of related elements and determining whether all the subcomponents are held in common.

Control Sequences Located Some Distance From the Coding Region of the Gene

Changes in the structure of transposable elements that have been inserted into the transcribed region of a gene are only one way in which these entities can regulate gene activity. As mentioned earlier, elements inserted at some distance from the gene can also exert control of gene expression. In the case of the *Sh* gene, *Ds* elements inserted upstream from the gene simply inactivate the gene. If we are correct in our determination of the transcribed region, these insertions are approximately 0.5 and 2 kb from the 5′ end. Clearly these distances are beyond the typical promotor and polymerase entry sites involved in transcription. Either these elements are acting at a distance to exert negative influences over the gene or else they are interrupting as yet undefined sequences adjacent to the wild-type gene that are required for normal developmentally controlled gene expression.

Zachar and Bingham [1982] have defined a domain about 3 kb from the apparent transcribed region of the *white* locus. Four insertional mutations including *white*^{eosin} and *white*^{spotted} are located in this region. The authors propose that these insertions disrupt gene activators that regulate the transcription of the distally located coding sequences. Evidence of a transposable element insertion acting at a distance to produce a negative effect was observed in the *white*^{DZL} mutation. This mutation has a 13–14-kb insertion approximately 9.5 kb from the transcribed region of the gene [Zachar and Bingham, 1982; Levis and Rubin, 1982]. It is located at the edge of, or just outside the proximal breakpoint defining one of, the boundaries of the coding and regulatory sequences of the locus. Revertants of this mutation have suffered partial internal deletions of the elements [Zachar and Bingham, 1982; Levis and Rubin, 1982]. Since portions of the element remain at the locus in the revertants, it is unlikely that they are interrupting normal control sequences.

Control regions of several eukaryotic genes have been probed by deletion and in vitro mutagenesis [Grosschedl and Birnstiel, 1980a,b; Mellon et al, 1981; McKnight and Kinsbury, 1982]. Investigators have found that a TATA box 20–30 nucleotides upstream from the cap site is required for correct and/ or efficient initiation of transcription. One or two modulator sites further upstream, but within 105 nucleotides of the cap site, also appear to be required to maintain the normal levels of transcription in vivo. In the case of the hormonally regulated oviduct genes, a progesterone receptor site has been found 250–300 nucleotides upstream from the 5′ end of the ovalbumin gene [Mulvihill et al, 1982]. Perhaps relevant to the present discussion is the observation that a 72-bp from SV40 or an analogous 244-bp sequence from polyoma can activate the transcription of any gene in its vicinity [Banerji et al, 1981; de Villiers and Schaffner, 1981]. This element works when one copy of the "enhancer" inserted in either orientation is situated up to 3,000 nucleotides 5′ or 3′ to the target gene. Similar elements that activate developmentally regulated expression may be found in the vicinity of other eukaryotic genes. As suggested by Zachar and Bingham [1982], it may be these sequences that are inactivated by insertional mutations some distance from the coding region of a gene. One might speculate that elements inserted at some distance acting positively or negatively on a gene under their control could use a similar method of control. Like the LTRs of RNA tumor viruses [Levinson et al, 1982], they might possess enhancers that could respond to new signals to change the expression of a gene during development. Alternatively, insertional elements might contain sequences that could compete with normal enhancers to negate their effect and thereby abolish or severely reduce the transcription of the gene being controlled.

ACKNOWLEDGMENTS

This work was conducted at Brookhaven National Laboratory, which is operated by Associated Universities, Inc, under the auspices of the US Department of Energy, and was funded in part by National Institutes of Health Grant GM31093.

REFERENCES

Banerji J, Rusconi S, Schaffner W (1981): Expression of a β-globin gene is enhanced by remote SV40 DNA sequences. Cell 27:299–308.

Burr B, Burr FA (1981): Controlling-element events at the *Shrunken* locus in maize. Genetics 98:143–156.

Burr B, Burr FA (1982): *Ds* controlling elements of maize at the *Shrunken* locus are large and dissimilar insertions. Cell 29:977–986.

Chourey PS (1981): Genetic control of sucrose synthetase in maize endosperm. Mol Gen Genet 184:372–376.

Chourey PS, Nelson OE (1976): The enzymatic deficiency conditioned by the *shrunken-1* mutations in maize. Biochem Genet 14:1041–1055.

de Villiers J, Schaffner W (1981): A small segment of polyoma virus DNA enhances the expression of a cloned β-globin gene over a distance of 1400 base pairs. Nucleic Acids Res 9:6251–6264.

Dooner HK (1981): Regulation of the enzyme UFGT by the controlling element *Ds* in *bz-m4*, an unstable mutant in maize. Cold Spring Harbor Symp Quant Biol 45:457–462.

Errede B, Cardillo TS, Wever G, Sherman F (1981): ROAM mutations causing increased expression of yeast genes: Their activation by signals directed toward conjugation functions and their formation by insertion of Ty 1 repetitive elements. Cold Spring Harbor Symp Quant Biol 45:594–602.

Grosschedl R, Birnstiel ML (1980a): Identification of regulatory sequences in the prelude sequences of an H2A histone gene by the study of specific deletion mutants in vivo. Proc Natl Acad Sci USA 77:1432–1436.

Grosschedl R, Birnstiel ML (1980b): Spacer DNA sequences upstream of the TATAAATA sequence are essential for promotion of H2A histone gene transcription in vivo. Proc Natl Acad Sci USA 77:7102–7106.

Hayward WS, Neel BG, Astrin SM (1980): Activation of a cellular *onc* gene by promotor insertion in ALV-induced lymphoid leukosis. Nature (London) 290:475–480.

Jackson JA, Fink GR (1981): Gene conversion between duplicated genetic elements in yeast. Nature (London) 292:306–311.

Kamp D, Kahmann R, Zipser D, Broker TR, Chow LT (1978): Inversion of the G DNA segment of phage Mu controls phage infectivity. Nature (London) 271:577–580.

Klein HL, Petes TD (1981): Intrachromosomal gene conversion in yeast. Nature (London) 289:144–148.

Levinson B, Khoury G, Vande Woude G, Gruss P (1982): Activation of SV40 genome by 72-base pair tandem repeats of Maloney sarcoma virus. Nature (London) 295:568–572.

Levis R, Rubin GM (1982): The unstable w^{DZL} mutation of Drosophila is caused by a 13 kilobase insertion that is imprecisely excised in phenotypic revertants. Cell 30:543–550.

McClintock B (1949): Mutable loci in maize. Carnegie Inst Wash Year Book 48:142–154.

McClintock B (1950): Mutable loci in maize. Carnegie Inst Wash Year Book 49:157–167.

McClintock B (1951): Chromosome organization and gene expression. Cold Spring Harbor Symp Quant Biol 16:13–47.

McClintock B (1952): Mutable loci in maize. Carnegie Inst Wash Year Book 51:212–219.

McClintock B (1953): Mutation in maize. Carnegie Inst Wash Year Book 52:227–237.

McClintock B (1956a): Intranuclear systems controlling gene action and mutation. Brookhaven Symp Biol 8:58–74.

McClintock B (1956b): Controlling elements and the gene. Cold Spring Harbor Symp Quant Biol 21:197–216.

McClintock B (1961): Some parallels between gene control systems in maize and in bacteria. Am Naturalist 95:265–277.

McClintock B (1962): Topographical relations between elements of control systems in maize. Carnegie Inst Wash Year Book 61:448–461.

McClintock B (1965): The control of gene action in maize. Brookhaven Symp Biol 18:162–182.

McClintock B (1967): Genetic systems regulating gene expression during development. In Locke M (ed): "Control Mechanisms in Developmental Processes." Developmental Biology. New York: Academic Press, (Suppl 1):84–112.
McClintock B (1978): Development of the maize endosperm as revealed by clones. In Subtelny S, Sussex I (eds): "The Clonal Basis of Development." New York: Academic Press, pp 217–237.
McKnight SL, Kingsbury R (1982): Transcriptional control signals of a eukaryotic protein-coding gene. Science 217:316–324.
Mellon P, Parker V, Gluzman Y, Maniatis T (1981): Identification of DNA sequences required for transcription of the human α1-globin gene in a new SV40 host-vector system. Cell 27:279–288.
Mulvihill ER, LePennec J-P, Chambon P (1982): Chicken oviduct progesterone receptor: Location of specific regions of high affinity binding in cloned DNA fragments of hormone responsive genes. Cell 28:621–632.
Payne GS, Courtniedge SA, Crittenden LB, Fadly AM, Bishop JM, Varmus HE (1981): Analysis of avian leukosis virus DNA and RNA in bursal tumors: Viral expression is not required for maintenance of the tumor state. Nature (London) 293:323–334.
Roeder GS, Fink GR (1982a): Movement of yeast transposable elements by gene conversion. Proc Natl Acad Sci USA 79:5621–5625.
Roeder GS, Fink GR (1982b): Transposable elements in yeast. In Shapiro J (ed): "Mobile Genetic Elements." (in press).
Scherer S, Davis RW (1980): Recombination of dispersed repeated DNA sequences in yeast. Science 209:1380–1384.
Varmus HE (1982): Form and function of retroviral proviruses. Science 216:812–820.
Young T, Williamson V, Taguchi A, Smith M, Sledziewski A, Russell D, Osterman J, Denis C, Cox D, Beier D (1981): The alcohol dehydrogenase genes of the yeast, Saccharomyces cerevisiae: Isolation, structure, and regulation. In Hollaender A (ed): "Genetic Engineering of Microorganisms for Chemicals." New York: Plenum Press, pp 335–361.
Zachar Z, Bingham PM (1982): Regulation of white locus expression: The structure of mutant alleles at the white locus of D. melanogaster. Cell 30:529–541.
Zieg JM, Silverman M, Hilmen M, Simon M (1977): Recombinational switch for gene expression. Science 196:170–172.

Gene Structure and Regulation in Development, pages 197–212

Approaches to the Study of Mechanisms of Selective Gene Amplification in Cultured Mammalian Cells

Peter C. Brown, Randal N. Johnston, and Robert T. Schimke
Department of Biological Sciences, Stanford University, Stanford, California 94305

Introduction

Amplification of the structural genes coding for important RNA molecules and proteins is but one of the many ways that eukaryotic cells accomplish differential gene expression. These gene amplifications may occur in response to either developmental or environmental signals and ultimately result in an increased number of transcriptionally active genes and in increased production of specific gene products. We wish to distinguish here between generalized gene amplification as seen, for example, in increased ploidy and differential gene amplification in which the representation of specific subsets of DNA is increased relative to total cellular DNA. Examples of differential gene amplification in developmentally regulated systems include chorion protein genes in Drosophila during oogenesis [Thireos et al, 1980; Spradling and Mahowald, 1980]; DNA puffing within polytenized salivary gland chromosomes of Rhynchosciara [Glover et al, 1982]; ribosomal genes in amphibian oocytes [Brown and Dawid, 1968], and in certain mutants of Drosophila [Tartof, 1975]; and in the formations of the macronucleus in ciliated protozoa [Yao et al, 1978].

A second major category of differential gene amplification is in instances of acquired resistance to normally cytotoxic compounds. These compounds are frequently of interest as chemotherapeutic drugs because of the enhanced sensitivity of actively growing cell populations to these agents. Resistant cells emerge after prolonged growth in increasing concentrations of drug and these cells characteristically overproduce the specific intracellular proteins targeted by the compound. Overproduction provides cells with sufficient amounts of unaffected protein, thereby insuring survival in normally toxic concentrations of drug. Although many examples of specific protein overproduction exist, only four examples to date have been shown to result from differential gene amplification. These include CAD amplification in cells resistant to PALA (n-phosphonacetyl-L-aspartate) (CAD is an acronym for the multifunctional enzyme that catalyzes the first three reactions in uridine monophosphate (UMP) biosynthesis) [Wahl et al, 1979]; metallothionein I in cells resistant to cadmium [Beach and Palmiter, 1981]; hypoxanthine phosphoribosyltransferase (HPRT) in HPRT$^-$ revertants selected in hypoxanthine, aminopterin, and thymidine (HAT) [Melton et al, 1982]; and dihydrofolate reductase (DHFR) amplifications in cells resistant to methotrexate (MTX) [Alt et al, 1978]. Overproduction of protein in drug-resistant cells need not result from gene amplification, however, inasmuch as cells resistant to canavanine overproduce the enzyme argininosuccinate synthetase without corresponding amplification of that gene [Su et al, 1981]. Whereas most examples of differential gene amplification in drug resistance are with cultured cells, it has recently been shown that amplification of the DHFR gene had occurred in neoplastic tissue from a patient with chronic myeloid leukemia after relapse during chemotherapy with MTX [Horns et al, 1982].

A third class of differential DNA amplifications is in instances of spontaneously occurring double minute chromosomes, and in homogeneously staining regions within karyotypes of neoplastic tissues [George and Powers, 1982; Hubbell, 1982]. Although we anticipate the assignment of specific genes to the amplified DNA present in these abnormal chromosomal elements, none has been positively identified to date and the function of the amplified DNA, if any, remains unknown.

In this communication we wish to describe experiments that we have recently undertaken and that begin to explore mechanisms of differential gene amplification in cells selected for resistance to the antifolate methotrexate. We have examined variables that affect the frequency of occurrence of DHFR gene amplification in cultured cells. Identification of these factors should provide clues as to the mechanisms of this type of gene amplification (ie, in drug resistance) and may increase the frequency of this relatively rare event to such an extent that it might be studied directly.

Stability of Selective Gene Amplifiction

In certain populations of MTX-resistant cells with amplified DHFR genes, resistance to MTX is stable and is retained even after prolonged growth in MTX-free medium. Analysis of DHFR gene copy number in these cells reveals essentially no loss of DHFR genes during this period. In contrast, other cell lines lose both resistance and DHFR genes with growth in the absence of MTX [Schimke et al, 1981]. Approximately one-half of these unstably amplified DHFR genes are lost within 20 cell doublings in the absence of selection. After 60 or so doublings, MTX-resistance and DHFR copy number approaches that of parental MTX-sensitive cells. It has been well documented that unstably amplified DHFR genes in MTX-resistant cells are localized to extrachromosomal elements called double minute chromosomes [Brown et al, 1981]. Double minutes lack centromeres, and they are partitioned randomly at mitosis [Levan and Levan, 1978]. Thus, instability of resistance can be explained by nondirected segregation of double minutes at mitosis and the inherent instability of acentromeric chromosomal fragments. Stable amplifications, on the other hand, characteristically exist as tandemly duplicated arrays within homogeneously staining regions of oversized marker chromosomes [Nunberg et al, 1978; Dolnick et al, 1979]. Stability results from the alignment of extrachromosomal DNA segments with functional centromeres, as would be predicted from studies with yeast [Stinchcomb, et al, 1982] and other gene transfer experiments [Klobutcher and Ruddle, 1979]. It has been reported, however, that even this stable form of gene amplification may undergo detectable diminutions after long periods of growth in the absence of selection [Biedler et al, 1980].

One might anticipate a correlation between the degree of overall karyotypic stability in a cell line and the tendency of that cell line to display stable or unstable resistance and gene amplification after drug selection. Accordingly, we have compared in Table I these properties for a number of resistant cell lines in which specific gene amplifications have been shown. In general, we find that stable gene amplifications are associated with stable karyotypes, and unstable resistance is most often associated with frequent karyotypic abnormalities. Exceptions to this rule include murine L5178Y cells resistant to MTX, and hamster cells transformed by SV40 and resistant to PALA. Further complicating this issue is the observation that, on occasion, unstably amplified and karyotypically variable cells will, during prolonged drug selection, acquire stable resistance [Kaufman et al, 1981]. At present it remains unclear what mechanisms account for karyotypic instability, but one might invoke as causative agents defective DNA repair systems [Friedberg et al, 1979], viruses [Bender and Brockman, 1981], or transposon-like elements [McClintock, 1978].

Clonal Variation in MTX Resistance and DHFR Gene Amplification

A comparison of inherent MTX sensitivity between cell lines in published reports and in our laboratory revealed widely divergent values. We therefore undertook an examination of MTX-sensitivity among clones of one mouse cell line, NIH-3T6 cells. Ten randomly selected, cloned sublines were tested for MTX sensitivity by determining the relative plating efficiencies of each subline in various concentrations of MTX. These data for four representative

TABLE I. Stability of Resistance in Cell Lines With Selectively Amplified Genes

Cell line[a]	Stable resistance[b]	Stable karyotype[c]	Reference
DHFR			
3T6 (M)	No	No (DM)	Brown et al [1981]
S-180 R_1A (M)	No	No (DM)	Kaufman et al [1979]
L5178Y (M)	Yes	Yes	Dolnick et al [1979]
L5178Y (M)	No	Yes (DM)	Courtenay and Robins [1972]
CHO K_1 (HA)	Yes	Yes	Nunberg et al [1978]
CHL (HA)	Yes	Yes	Biedler et al [1980]
HELA (HU)	No	No (DM)	Masters et al [1982]
EL_4 (M)	No	No (DM)	Tyler-Smith and Bostock [1981]
CAD			
BHK (HA)	Yes	Yes	Kempe et al [1976]
BHK SV-28 (HA)	Yes	No	G. Stark [personal communication]
Metallothionein 1			
S-180 (M)	No	No	Beach and Palmiter [1981]
Friend leukemia (M)	Yes	Yes	R. Palmiter [personal communication]
HEPA (M)	No	No (DM)	R. Palmiter [personal communication]
HPRT			
V-79 (HA)	Yes	Yes	J. Fuscoe [personal communication]
V-79 (HA)	No	No	J. Fuscoe [personal communication]

[a]Letters in parentheses indicate species from which the cell line was derived: M, mouse; HA, hamster; HU, human.
[b]Stability of resistance with prolonged growth in the absence of drug selection.
[c]Stable karyotype is loosely defined as diploid, or nearly so, with only minor differences in chromosome number between cells. DM in parenthesis indicates the presence of double minute chromosomes.

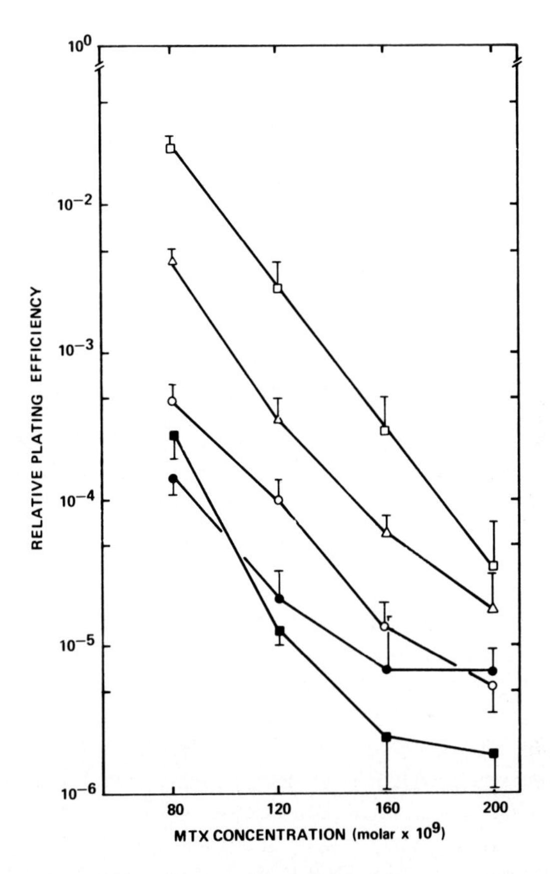

Fig. 1. Clonal variation of resistance to methotrexate (MTX). Predetermined numbers of cells (1×10^3–5×10^5) of clonally derived sublines of NIH 3T6 cells were plated into various concentrations of MTX-supplemented medium in 10-cm petri dishes. Media was changed at 2–3-day intervals and after 14–21 days the resulting colonies (> 50 cells/colony) were fixed, stained, and counted. Colony counts per plate were normalized to the number of viable cells that were initially plated (relative plating efficiency). Standard deviation is indicated by the error bars. Clone 6, □; clone 7, △; clone 5, ○; clone 10, ■; uncloned 3T6 cells, ●.

sublines are shown in Figure 1. Although minor variances in sensitivity were expected, we did not anticipate the magnitude of these differences. The number of surviving colonies varied over two orders of magnitude for sublines 6 and 10 at the concentrations of MTX tested below 200 nM, whereas intermediate values were seen for other sublines which were tested. Similar variation is found among clones of Chinese hamster ovary (CHO) cells as well (not shown).

TABLE II. Variation in the Occurrence of DHFR Gene Amplification in Clonally Derived 3T6 Cells [a]

Concentration of MTX (molar × 10^9)	Proportion of colonies with amplified DHFR genes		
	3T6/5	3T6/6	3T6/7
40	0/10	ND[b]	ND
80	1/9	ND	ND
120	5/9	0/8	1/8
160	8/13	4/9	2/4
200	2/5	2/2	1/1

[a]MTX-resistant colonies from 3T6 sublines 5, 6, and 7 (Fig. 1) were randomly picked, grown for several generations in medium supplemented with the selecting concentration of MTX (40–200 nM) and tested for the presence of amplified DHFR genes by slot hybridization assay [Brown et al, 1982]. In the slot hybridization assay, DHFR gene abundance is determined relative to a nonamplified DNA sequence using appropriate ^{32}P-labeled complementary DNA probes. The proportion of total colonies examined that demonstrated amplified DHFR genes by this assay is shown for each subline and at various concentrations of MTX.
[b]ND, not determined.

The demonstration of DHFR gene amplification in MTX-resistant cells is necessary because MTX-resistant cells have been described that differ from parental cells in respect to decreased transport of MTX [Sirotnak et al, 1981], altered affinity of DHFR for MTX [Haber et al, 1981], as well as by DHFR overproduction (ie, gene amplification). Elsewhere we have described a method for assessing DHFR gene amplification, which is a modification of the dot hybridization technique [Brown et al, 1982]. This technique is capable of accurately detecting DHFR gene amplifications in cells with a minimum of two- to threefold greater DHFR gene copy number than nonamplified parental cells.

We have used this technique (called the slot hybridization assay) to determine whether MTX-resistant subclones of 3T6 clones, 5,6, and 7 showed DHFR gene amplification. MTX-resistant colonies (Fig. 1) were randomly picked, grown for a minimum of 11 cells doublings in MTX, and tested for DHFR gene amplification. These data are shown in Table II. Clone 5 was most extensively characterized and slightly more than half of the colonies selected in 120 nM and 160 nM MTX showed DHFR gene amplification, whereas no amplified colonies were detectable at 40 nM and only one was seen at 80 nM. In striking contrast, clone 6 cells, which displayed the greatest inherent MTX resistance (Fig. 1) showed no amplifications at 120 nM, and clone 7 showed only one amplified colony. At 160 nM, however, amplified colonies in clones 6 and 7 were found in approximately 50% of the subclones analyzed. Thus, we conclude that DHFR gene amplification can occur in each of these three clones; and it seems that the stringency of selection, as measured by relative plating efficiency, determines the occurrence of detectable DHFR gene amplification. In other words, gene amplifications are only apparent when survival values fall below 1×10^{-4}, regardless of the absolute concentration of MTX. We have not at this time determined precisely the reason(s) for clonal variation in inherent sensitivity to MTX in cultured cells. Such variations are clearly of interest to clinicians interested in efficacious treatment of malignancy by chemotherapeutic agents. We assume, however, that the nonamplified resistant colonies evident in Table II represent either transport alterations, enzyme variants with decreased affinities for MTX (or combinations of both), or have a level of DHFR gene amplification below that which we can detect.

Enhancement of DHFR Amplification Frequencies

Methotrexate is a potent antimetabolite which, because of the extremely high affinity of MTX for dihydrofolate reductase, inhibits the reduction of dihydrofolate to tetrahydrofolate. Since tetrahydrofolate is required for de novo synthesis of thymidine, purines, and glycine, depletion of tetrahydrofolate with exposure of cells to MTX results in the rapid cessation of DNA synthesis. When cells are initially exposed to low levels of MTX in vivo, DNA synthesis is transiently inhibited; with time, DNA synthesis and cell growth resumes in at least some cells [Roberts and Wodinsky, 1968]. Similar events may occur in cultured cells and we hypothesize that this transient inhibition of DNA synthesis and subsequent recovery leads to the initiation of extra rounds of DNA synthesis. Reinitiation of DNA synthesis is clearly a violation of conventional rules of DNA synthesis inasmuch as a given segment of DNA is thought to be replicated only once per cell cycle. Reinitiation of DNA synthesis does, however, provide a mechanism for selective gene amplification. Cells that have accumulated extra copies of functional DHFR genes in this manner would predominate with growth in normally toxic concentrations of MTX. This hypothesis is supported by a series of experiments by Woodcock and Cooper [1979] who showed that brief inhibition of DNA synthesis by metabolic inhibitors led to the formation of DNA structures that could only be produced by reinitiation of DNA replication using as a template those strands of DNA that were themselves synthesized just prior to the metabolic block.

In an attempt to mimic the gross effects of MTX, ie, reversible inhibition of DNA synthesis, we exposed cells transiently to hydroxyurea (HU) and subsequently challenged these cells with various concentrations of MTX in order to determine whether pretreatment with HU might increase the frequency of MTX resistance and DHFR gene amplification. HU specifically

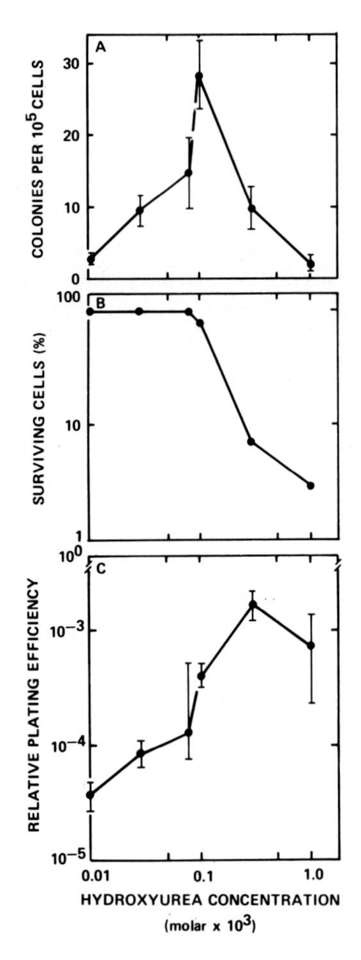

Fig. 2. Pretreatment with hydroxyurea enhances resistance to methotrexate. Exponentially growing cultures of 3T6 clone 5 cells (Fig. 1) were exposed to various concentrations of freshly prepared hydroxyurea (HU) for 14 hours, grown in drug-free medium for 6 hours, trypsinized and plated into medium with 120 nM MTX (panels A,C; 1×10^5 cell/10-cm petri dish), or into normal medium (panel B, $1-5 \times 10^2$ cells/10-cm dish). Colonies were counted after appropriate intervals (see Fig. 1). A) Methotrexate-resistant colonies normalized to the number of cells plated after HU pretreatment with no correction for nonviable cells. B) Surviving cells (colonies) in normal medium without MTX, expressed as a percentage of the total number of cells plated after pretreatment with hydroxyurea. C) Methotrexate-resistant colonies (A) corrected for cell viability (B) after HU pretreatment in the calculation of relative plating efficiency.

inhibits the enzyme ribonucleotide reductase, and thereby eliminates the catalytic reduction of ribonucleoside diphosphate to corresponding deoxyribonucleoside forms, which are necessary for DNA synthesis. In Figure 2 we show the effects of different concentrations of hydroxyurea on the emergence of MTX-resistant colonies. 3T6 clone 5 cells (Fig. 1) were exposed to HU for about one cell cycle, washed free of HU for six hours, and plated into 120 nM MTX, a concentration that yields maximal frequencies of DHFR gene amplification (Table II). The absolute number of MTX-resistant colonies was increased about tenfold when cells were pretreated with 0.1 mM HU (Fig. 2A). Colony counts decline thereafter as the toxicity of the pretreatment increases (Fig. 2B). If colony counts are normalized for this toxicity (ie, expressed as colonies per surviving cell), the frequency of resistant colonies increases to 0.3 mM HU and then declines slightly thereafter (Fig. 2C). Because of the substantial correction factor at 0.3 mM HU (less than 5% survival) in the calculation of relative plating efficiency, we were concerned that HU was merely selecting a subpopulation of cells that was inherently resistant to MTX. That this was not the case was seen in control experiments (not shown), which clearly indicated that previously selected MTX-resistant cells were as sensitive to the toxic effects of HU as were parental MTX sensitive cells.

We next isolated colonies that arose in MTX-supplemented medium after cells were pretreated for 17 hours with 0.3 mM HU and washed free of drug for six hours prior to exposure to MTX. The occurrence of DHFR amplification in various colonies resistant to 120 nM MTX was determined and compared to the occurrence in untreated control cells not exposed to HU though otherwise selected at the same concentration of MTX (Table III). In both control and treated cells about one-half of the colonies isolated showed DHFR gene amplification. Thus, the frequency of DHFR gene amplification detectable by our assay was increased in proportion to the increase in the relative plating efficiency in cells pretreated with hydroxyurea. We conclude, therefore, that the relative amplification frequency of the DHFR gene was increased about 100-fold by pretreatment with hydroxyurea. Complicating this analysis was the simultaneous increase in the frequency of nonamplified, resistant colonies. At present we can only assume that other modes of resistance (ie, decreased MTX transport, or altered affinity of DHFR for MTX) are a result of mutagenic effects of hydroxyurea. In support of this explanation were experiments in which ethyl methane sulfonate (EMS), a known mutagen, was substituted for HU, and in none of the MTX-resistant colonies isolated at 120 nM and 160 nM was the DHFR gene amplified [P. Brown, unpublished results].

These data demonstrate that transient inhibition of DNA synthesis can lead to increased frequencies of cells that are resistant to MTX by amplification

TABLE III. Effects of Hydroxyurea Pretreatment on the Frequency of DHFR Gene Amplification[a]

Treatment	Relative plating efficiency	Amplified colonies / Total examined	Relative amplification frequency
Control	4×10^{-5}	3/5	2×10^{-5}
Hydroxyurea	3×10^{-3}	4/8	2×10^{-3}

[a]Relative plating efficiencies (as in Fig. 2) were determined for 3T6 clone 5 cells that were pretreated with hydroxyurea (0.3 mM for 17 hr). Cells were grown for 6 hr in drug-free medium prior to trypsinization and subsequent selection in medium with 120 nM MTX. Subclones were established from randomly picked, MTX-resistant colonies and tested for DHFR gene amplification by slot hybridization assay (Table II). Relative amplification frequency (4th column) is the occurrence of MTX-resistant colonies for which DHFR gene amplification could be demonstrated. It is the numerical product of relative plating efficiency (ie, overall MTX resistance) and proportion of MTX-resistant subclones that showed DHFR gene amplification by slot hybridization assay.

of the DHFR gene and strongly suggest the key role of DNA replication in the process of gene amplification. These results raise the distinct possibility that MTX, because of its reversible effects on DNA synthesis, may itself enhance the frequency of occurrence of gene amplification in general, and subsequently function as a selecting agent for those cells that have successfully amplified functional DHFR genes. We are currently investigating whether genes other than DHFR are amplified following pretreatment of cells with, in addition to HU, other inhibitors of DNA synthesis.

Spontaneous Amplification of the DHFR Gene

Whereas perturbations in DNA synthesis imposed by HU may enhance the frequency of DHFR gene amplification, we wished to ascertain whether DHFR amplification could occur spontaneously and in the absence of any direct selections by MTX. Parental CHO K1 cells were first incubated overnight with a fluoresceinated derivative of MTX (MTX-F) and processed under conditions in which the amount of bound MTX-F (ie, fluorescence per cell) is proportional to the amount of intracellular DHFR [Kaufman et al, 1978]. The inclusion of thymidine, hypoxanthine, and glycine together is known to reverse the toxic effects of MTX, and these components were added to the incubations with MTX-F in order to eliminate any selective effects of MTX-F. Next, cells labeled with MTX-F were analyzed with the fluorescence activated cell sorter (FACS II, Becton-Dickinson), and the distribution of fluorescence (ie, DHFR) per cell was determined. Third, cells with the greatest fluorescence were sorted and grown an additional 10–15

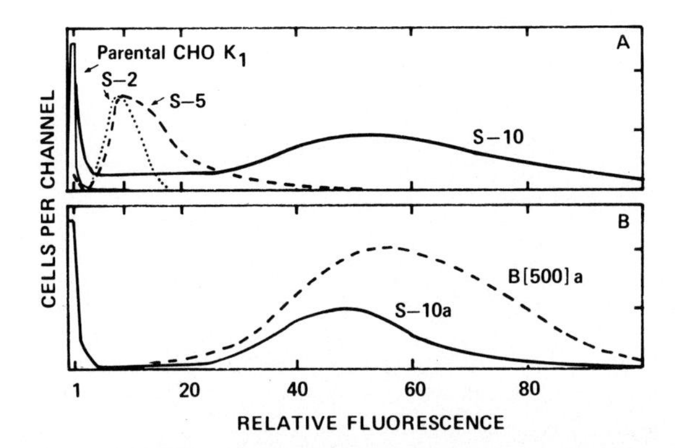

Fig. 3. Spontaneous overproduction of dihydrofolate reductase (DHFR). Intracellular DHFR in Chinese hamster ovary (CHO) K_1 cells was quantitatively labeled with fluoresceinated MTX and the cells were analyzed for DHFR content by fluorescence-activated cell sorting (FACS) [Kaufman and Schimke, 1981]. The most fluorescent cells (1–2% of total) were sorted, grown for 10–15 generations in normal medium without MTX, and re-sorted after labeling with fluoresceinated MTX for a total of 10 cycles. Cellular fluorescence distributions from FACS analysis are shown for the populations of cells after 2 (S-2), 5 (S-5), and 10 (S-10) sequential sorting cycles (A). (B) Cellular fluorescence distributions of a randomly selected clone, S-10 A, grown for 15–20 generations after isolation from S-10 populations (A). B[500]a is the fluorescence distribution of a CHO K_1 population that was selected in increasing concentrations of MTX and is resistant to 0.5 μM MTX.

generations in the absence of MTX or MTX-F. The sorted population was subsequently labeled with MTX-F, re-sorted for greatest fluorescence, and grown for 10–15 generations in the absence of MTX. This re-sorting procedure was repeated for a total of 10 cycles and the results of some of these experiments are shown in Figure 3. After 2,5, and 10 serial cycles, the fluorescence of the resulting populations increased dramatically such that the average fluorescence of the population after ten such sorts was about 50-fold greater than that of the starting population (Fig. 3A).

Figure 3B shows the fluorescence distribution of another CHO cell line that was selected in increasing concentrations of MTX and has about 50 copies of the DHFR gene per cell [Kaufman and Schimke, 1981]. The fluorescence intensity of this known, stably amplified cell line was essentially identical to the fluorescence of the CHO K1 cells after ten sorts. In order to determine the relative stability of the S-10 population, randomly selected clones were grown 15–20 generations and reanalyzed. The fluorescence distribution of a representative clone is shown (S-10a): A significant proportion of these cells had lost fluorescence and had no greater DHFR content than the starting population of CHO K1 cells. Thus, DHFR overproduction

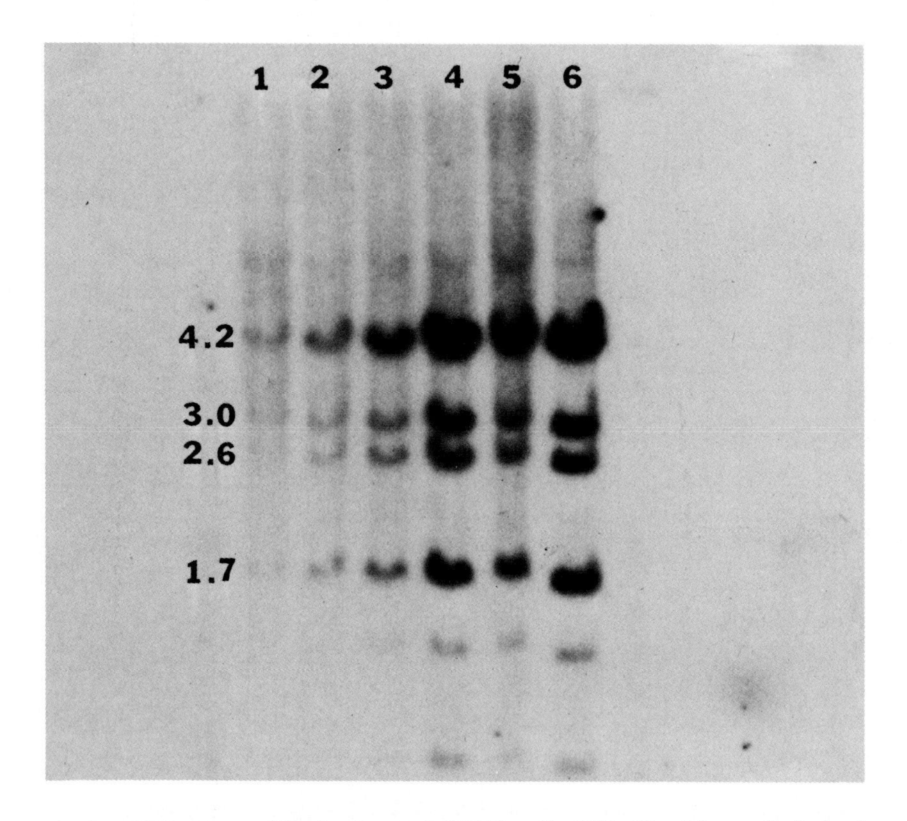

Fig. 4. DHFR gene amplification in sorted CHO K_1 cells. DNA (15 μg) from cells depicted in Figure 4 was digested with EcoRI, electrophoresed, transferred to nitrocellulose, and hybridized with ^{32}P-labeled DHFR cDNA (1×10^7 cpm, specific activity 6×10^8 cpm/μg), and exposed to x-ray film for three days (Kaufman and Schimke, 1981). CHO K_1 parental cells, lane 1; S-2, lane 2; S-5, lane 3; S-10, lane 4. In lanes 5 and 6 are DNA samples from CHO K_1 cells selected in MTX and resistant to 100 nM and 500 nM MTX, respectively [Kaufman and Schimke, 1981].

in these sequentially sorted populations was at least partially unstable. Longer-term experiments are in progress to determine whether the entire population will revert to sensitive cell levels. It is of interest that CHO K1 cells that had been conventionally selected in increasing concentrations of MTX were initially unstable as well [Kaufman and Schimke, 1981].

Most important, the serially sorted populations of cells showed increasing amplifications of the DHFR gene (Fig. 4). DNA samples from cells depicted in Figure 4 were digested to completion with restriction endonuclease EcoRI, electrophoresed in agarose, transferred to nitrocellulose filters, and hybridized with ^{32}P-labeled DNA complementary to mouse DHFR messenger RNA

[Kaufman and Schimke, 1981]. When the intensities of resulting exposures were compared with that of the parental, MTX-sensitive sample (lane 1), it is clear that the levels of DHFR gene amplification in the sorted populations (lanes 2–4) were increased and that these increases were roughly proportional to the increases in DHFR as measured by fluorescence (Fig. 4). In addition to having increased DHFR and amplified DHFR, the sorted populations displayed enhanced resistance to MTX relative to the unsorted parental population (data not shown). The extent and rapidity of the spontaneous increase in DHFR content and gene copy number in this experiment was unexpected. We are forced to the conclusion that even in the absence of direct selection or cell-cycle perturbation by MTX, DHFR gene amplifications occur and are detectable as DHFR enzyme overproduction in individual cells. At present we are attempting to measure the frequency of these spontaneous amplifications of DHFR in order to determine the extent to which this process contributes to DHFR gene amplification observed during conventional selection in MTX.

Discussion and Summary

Our current view of selective gene amplification in cultured cells is as follows. With every DNA replication cycle there is a small, though finite probability that a given gene may be aberrantly replicated more than once. Although all replicons could engage in these aberrant replications, those segments of DNA that were synthesized early in S phase might be more susceptible to reinitiation of synthesis because of the increasing, though incomplete capacity, of the replicative machinery early in S phase. These amplifications would be transient in nature and unstable, and in the absence of selection would eventually be degraded. Thus, the overall constancy of the genome would be maintained. In the presence of a selecting agent such as MTX, cells with additional copies of the DHFR gene (and corresponding increases in enzyme) would be favored and would survive in normally toxic concentrations of the drug. On this background of spontaneous amplifications, one might increase the frequency of reinitiation of synthesis by temporarily inhibiting DNA synthesis with agents such as hydroxyurea. Similarily, the frequency of reinitiation might be increased by mitogenic stimuli such as insulin, growth factors, and tumor-promoting agents [Varshavsky, 1981] in appropriately receptive cells.

Although alternative explanations for gene amplification have been presented [Schimke et al, 1981], we are proceeding on the theory that selective gene amplification results from aberrant processes involved in DNA replication. Whether gene amplification in cultured cells occurs by the same general mechanism as in developmentally regulated systems, we cannot say. We are currently compiling biochemical evidence for enhanced reinitiation of DNA synthesis after hydroxyurea pretreatment and are testing other inhibitors of DNA synthesis for similar effects. In this respect we would ultimately like to know what mechanism exists that ensures that the majority of DNA sequences are replicated only once in the cell cycle and how various compounds and protocols affect this process.

ACKNOWLEDGMENTS

The authors appreciate the technical and clerical assistance of Elise Mosse and Claire Groetsema. This work was supported in part by grants from the American Cancer Society, California Division (PCB); a postdoctoral fellowship from the Natural Sciences and Engineering Research Council of Canada (RNJ); and the National Cancer Institute (CA16318), the National Institute of General Medical Sciences (GM 14931), and the American Cancer Society (NP-148) (RTS).

REFERENCES

Alt FW, Kellems RE, Bertino JR, and Schimke RT (1978): Selective multiplication of dihydrofolate reductase genes in methotrexate-resistant variants of cultured murine cells. J Biol Chem 253:1357–1370.

Beach LR, Palmiter RD (1981): Amplification of the metallothionein-1 gene in cadmium-resistant mouse cells. Proc Natl Acad Sci USA 78:2110–2114.

Bender MA, Brockman WW (1981): Rearrangement of integrated viral DNA sequences in mouse cells transformed by simian virus 40. J Virol 38:872–879.

Biedler JL, Melera PW, Spengler BA (1980): Specifically altered metaphase chromosomes in antifolate-resistant chinese hamster cells that overproduce dihydrofolate reductase. Cancer Genet Cytogenet 2:47–60.

Brown DD, Dawid IB (1968): Specific gene amplification in oocytes. Science 160:272–280.

Brown PC, Beverley SM, Schimke RT (1981): Relationship of amplified dihydrofolate reductase genes to double minute chromosomes in unstably resistant mouse fibroblast cell lines. Mol Cell Biol 1:1077–1083.

Brown P, Tlsty T, Schimke RT (1983): Enhancement of methotrexate resistance and dihydrofolate reductase gene amplification by pretreatment of mouse 3T6 with hydroxyurea. Mol Cell Biol (in press).

Courtenay D, Robins AB (1972): Loss of resistance to methotrexate in L5178Y mouse leukemia grown in vitro. JNCI 49:45–53.

Dolnick BJ, Berenson RJ, Bertino JR, Kaufman RJ, Nunberg JH, Schimke RT (1979): Correlation of dihydrofolate reductase elevation with gene amplification in a homogeneously staining chromosomal region in L5178Y cells. J Cell Biol 83:394–402.

Friedberg LEC, Ehmann UK, Williams JI (1979): Human diseases associated with defective DNA repair. Adv Rad Biol 8:85–174.

George DL, Powers VE (1982): Amplified DNA sequences in Y1 mouse adrenal tumor cells: Association with double minutes and localization to a homogeneously staining region. Proc Natl Acad Sci USA 79:1597–1601.

Glover DM, Zaha A, Stocker AJ, Santelli RV, Pueyo MT, DeToledo SM, Lara FJS (1982): Gene amplification in Rhynchosciara salivary gland chromosomes. Proc Natl Acad Sci USA 79:2947–2951.

Haber DA, Beverley SM, Kiely ML, Schimke RT (1981): Properties of an altered dihydrofolate reductase encoded by amplified genes in cultured mouse fibroblasts. J Biol Chem 256:9501–9510.

Horns RC Jr, Dower WJ, Schimke RT (1983): Gene amplification in a leukemic patient treated with methotrexate (in preparation).

Hubbell HR (1982): Cloning and analysis of double minute DNA from a human colon carcinoid cell line. In Schimke RT (ed): "Gene Amplification." New York: Cold Spring Harbor Laboratory, pp 193–198.

Kaufman RJ, Schimke RT (1981): Amplification and loss of dihydrofolate reductase genes in a chinese hamster ovary cell line. Mol Cell Biol 1:1069–1076.

Kaufman RJ, Bertino JR, Schimke RT (1978): Quantitation of dihydrofolate reductase in individual parental and methotrexate-resistant murine cells. J Biol Chem 253:5852–5860.

Kaufman RJ, Brown PC, Schimke RT (1979): Amplified dihydrofolate reductase genes in unstably methotrexate-resistant cells are associated with double-minute chromosomes. Proc Natl Acad Sci USA 76:5669–5673.

Kaufman RJ, Brown PC, Schimke RT (1981): Loss and stabilization of amplified dihydrofolate reductase genes in mouse sarcoma S-180 cell lines. Mol Cell Biol 1:1084–1093.

Kempe TD, Swyryd EA, Bruist M, Stark GR (1976): Stable mutants of mammalian cells that overproduce the first three enzymes of pyrimidine nucleotide biosynthesis. Cell 9:541–550.

Klobutcher LA, Ruddle FH (1979): Phenotype stabilization and integration of transferred material in chromosome-mediated gene transfer. Nature (London) 280:657–660.

Levan A, Levan G 91978): Have double minutes functioning centromeres? Hereditas 88:81–92.

Masters J, Keeley B, Gay H, Attardi G (1982): Variable content of double minute chromosomes is not correlated with degree of phenotype instability in methotrexate-resistant human cell lines. Mol Cell Biol 2:498–507.

McClintock B (1978): Mechanisms that rapidly reorganize the genome. Stadler Genet Symp 10:25–48.

Melton DW, Brennand J, Ledbetter DH, Konecki DS, Chinault AC, Caskey CT (1982): Phenotypic reversion at the hprt locus as a consequence of gene amplification. In Schimke RT (ed): "Gene Amplification." New York: Cold Spring Harbor Laboratory, pp 59–65.

Nunberg JN, Kaufman RJ, Schimke RT, Urlaub G, Chasin LA (1978): Amplified dihydrofolate reductase genes are localized to a homogeneously staining region of a single chromosome in a methotrexate resistant chinese hamster ovary cell line. Proc Natl Acad Sci USA 75:5553–5556.

Roberts D, Wodinsky I (1968): On the poor correlation between the inhibition by methotrexate of dihydrofolate reductase and of deoxynucleoside incorporation into DNA. Cancer Res 28:1955–1962.

Schimke RT, Brown PC, Kaufman RJ, McGrogan M, Slate DI (1981): Chromosomal and extrachromosomal localization of amplified dihydrofolate reductase genes in cultured mammalian cells. Cold Spring Harbor Symp Quant Biol Vol XLV, pp 785–797.

Sirotnak FM, Moccio DM, Kelleher LE, Goutas LJ (1981): Relative frequency and kinetic properties of transport-defective phenotypes among methotrexate-resistant L1210 clonal cell lines derived in vivo. Cancer Res 41:4447–4452.

Spradling AC, Mahowald AP (1980): Amplification of genes for chorion proteins during oogenesis in Drosophila melanogaster. Proc Natl Acad Sci USA 77:1096–2002.

Stinchcomb DT, Mann C, Davis RW (1982): Centromeric DNA from Saccharomyces cerevisiae. J Mol Biol 158:157–179.

Su T-S, Bock H-GO, O'Brian WE, Beaudet AL (1981): Cloning of cDNA for arginosuccinate synthetase mRNA and study of enzyme over-production in a human cell line. J Biol Chem 256:11826–11831.

Tartof KD (1975): Redundant genes. Annu Rev Genet 9:355–385.

Thireos G, Griffin-Shea R, Kafatos FC (1980): Untranslated mRNA for a chorion protein of Drosophila melanogaster accumulates transiently at the onset of specific gene amplification. Proc Natl Acad Sci USA 77:5789–5793.

Tyler-Smith C, Bostock CJ (1981): Gene amplification in methotrexate-resistant mouse cells III. Interrelationships between chromosome changes and DNA sequence amplification or loss. J Mol Biol 153:237–256.

Varshavsky A (1981): Phorbol ester dramatically increases incidence of methotrexate-resistant mouse cells: possible mechanisms and relevance to tumor promotion. Cell 25:561–572.

Wahl GM, Padgett RA, Stark GR (1979): Gene amplification causes over production of the first three enzymes of UMP synthesis in N-(phosphoacetyl 1-aspartate)-resistant hamster cells. J Biol Chem 254:8679–8689.

Woodcock DM, Cooper IA (1979): Aberrant double replication of segments of chromosomal DNA following DNA synthesis inhibition of cytosine arabinoside. Exp Cell Res 123:157–166.

Yao MC, Blackburn E, Gall JG (1978): Amplification of the rRNA genes in Tetrahymena. Cold Spring Harbor Symp Quant Biol 53:1293–1296.

Gene Structure and Regulation in Development, pages 213–224

Transposable Genetic Elements in *Drosophila*

Gerald M. Rubin, Mary Collins, Roger E. Karess, Robert Levis, Christine Murphy, and Kevin O'Hare
Department of Embryology, Carnegie Institution of Washington, Baltimore, Maryland 21210

Introduction

The *white* locus controls both the level and pattern of pigmentation of the eye. Because *white* is a nonessential gene and has an easily scored phenotype, it provides an attractive experimental system for studying the effects of transposable element insertions on gene expression.

At least three distinct classes of transposable elements have now been found in the Drosophila genome (for review, see Spradling and Rubin [1981]). The major structural features of these elements are depicted in Figure 1. We have shown that elements from all three classes can insert in or near the *white* locus and disrupt its expression. Figure 2 shows a summary of the structures of several mutant *white* alleles that result from transposable element insertions. Our recent studies have revealed a strong correlation between the genetic behavior of each of these insertion mutations and the nature of the responsible transposable element.

The White-Ivory Mutation and Its Revertants

White-ivory (w^i) is a modestly unstable allele of the *white* locus that confers a faint yellow-pink phenotype to the eyes. It is also associated with a reduced frequency of recombination between flanking marker alleles of *white*. w^i reverts to wild-type (w^+), restoring both the original eye color and the normal recombination frequency within the locus at a rate of about 5 per 10^5 chromosomes in homozygous females, and about one-tenth this rate in males and deletion-heterozygote females. (For a review of the genetic properties of w^i, see Green [1976].)

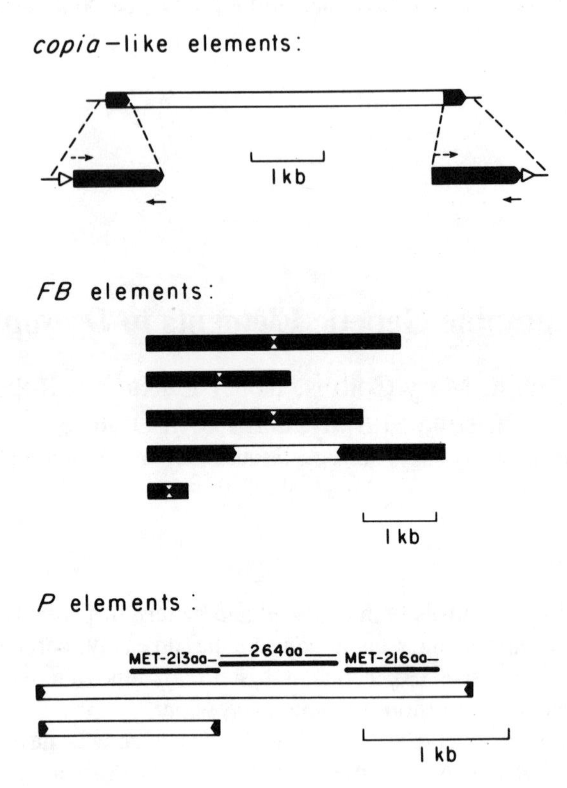

Fig. 1. Summary of the structures of three classes of Drosophila transposable elements. The *copia*-like elements carry long direct terminal repeats. Each repeat makes up about 5% of the length of the element. These repeats are shown at an expanded scale below the element to illustrate the presence of short imperfect inverted repeats at the ends of each long direct repeat and the presence of a few base pairs of duplicated target sequence flanking the element that were present in one copy before insertion. The different genomic copies of the elements of one family are very similar in structure to one another. The FB elements comprise a family of heterogeneous, but cross-homologous sequences ranging in size from a few hundred base pairs to several kb. Each FB element carries long terminal inverted repeats. In some cases the entire element consists of these inverted repeats. In other cases, a central sequence is located between the inverted repeats. The inverted repeated sequences themselves are internally repetitious, having a substructure made up primarily of 31-bp tandem repeats. The number of these 31-bp tandem repeats can differ not only between FB elements, but also between the termini of a single FB element. The P elements have a structure very different from both the *copia*-like and FB elements. P elements carry perfect terminal inverted repeats of 31 bp. A fraction of the P elements (about one-third in the one strain examined) are very similar in sequence to one another and are 2.9 kb in length. The remainder of the P elements are more heterogeneous but all appear to have structures that are consistent with their having been derived from the 2.9-kb element by internal deletions. DNA sequence analysis of the 2.9-kb element revealed three long open translational reading frames, which are indicated.

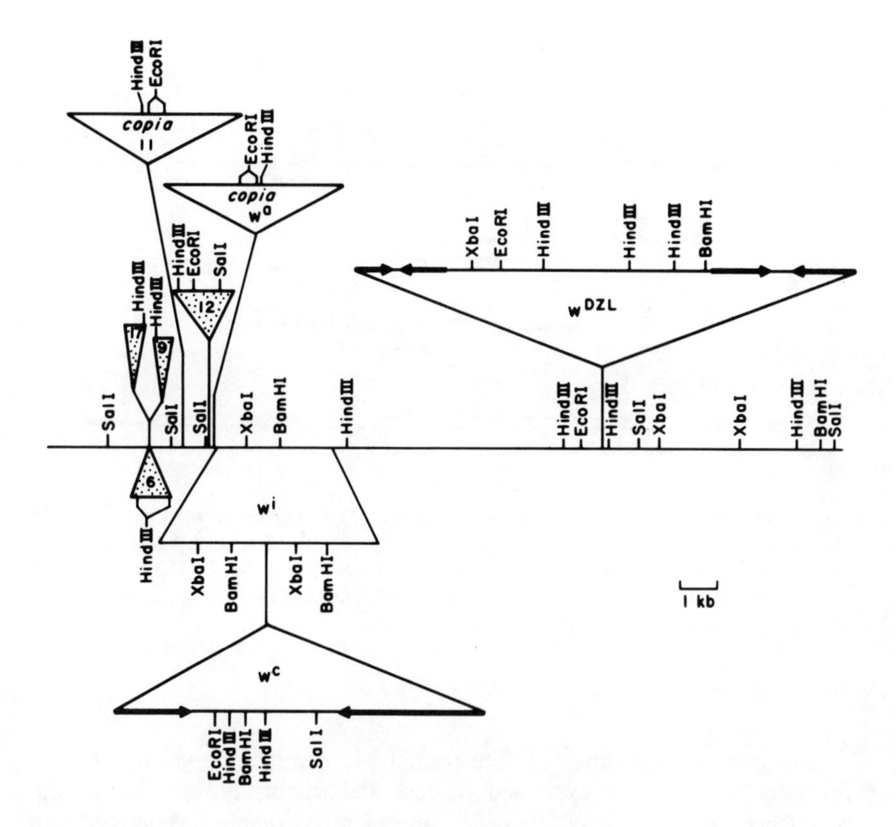

Fig. 2. Summary of the structures of several mutant alleles at the *white* locus. The long horizontal line depicts a restriction site map of the wild-type *white* locus. The *white-ivory* (w^i) allele results from a 3-kb tandem duplication of *white* locus sequences. The *white-crimson* (w^c) allele arose by the insertion of a 10-kb FB element into the w^i duplication. The *white-dominant zeste-like* allele (w^{DZL}) results from a 13-kb insertion consisting of two FB elements flanking a segment of nonrepetitive DNA. The location of FB terminal repeat sequences in the w^c and w^{DZL} insertions is indicated by the heavy arrows. The *white-apricot* (w^a) allele was apparently caused by the insertion of a *copia* element, as was the mutation $w^{hd81b11}$. The insertions causing mutations $w^{\#6}$,$w^{\#12}$, $w^{hd80k17}$ and w^{hd81b9} are members of the P element sequence family.

We have shown by Southern blot analysis and by molecular cloning that the lesion responsible for the *ivory* phenotype is a small tandem duplication of 2.9 kilobases (kb) within the *white* locus (Fig. 3B, Karess and Rubin [1982]). By heteroduplex analysis we delimited the duplicated sequences to between about 610 bases from the lefthand XbaI site, and 540 bases from the right-most XhoI site in Figure 3B.

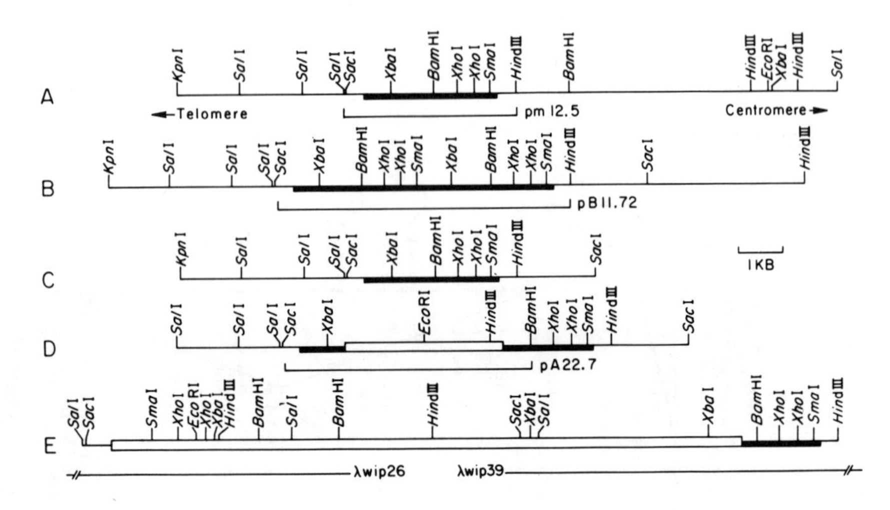

Fig. 3. Physical maps of the *white* locus region from flies of genotype w^+ Canton-S strain (A); w^i (B); w^{i+B}, w^{i+C}, w^{i+D}, w^{i+E} (C); w^{i+A} (D); w^{ip} (E). The black bar indicates the region duplicated in w^i; the white bar indicates non*white* locus DNA. The structures of w^i,w^{i+A}, and most of the w^{ip}, were confirmed by analysis of cloned DNAs, some of which are indicated below their corresponding map positions.

The genomic structures of five phenotypic revertants and one "partial revertant" of w^i were examined [Karess and Rubin, 1982]. Four of the revertants (w^{i+B}, w^{i+C} w^{i+D}, w^{i+E}) proved to be simple deletions of one copy of the duplicated sequences, leaving a single copy as in the wild-type locus (Fig. 3C).

One phenotypically wild-type revertant, w^{i+A}, and the partial revertant w^{ip} had more complex structures and were further analyzed by molecular cloning. The altered phenotypes were accompanied by the introduction of foreign DNA into the site of the *ivory* lesion. w^{i+A} had lost one complete copy of the duplicated sequence, but bore an insertion of 3.4 kb from outside the white locus (Fig. 3D). The partial revertant w^{ip} had an insertion of 14 kb, but had lost one copy of the duplication plus an additional 1.2 kb of wild-type *white* locus sequences (Fig. 3E). Both insertion elements hybridized to a large number of restriction fragments on genomic Southern blots, indicating that they were members of repeated DNA families.

Revertants of w^i to w^+ are found as single individuals and not in clusters, suggesting that w^i reversion is a meiotic event. However, factors known to augment meiotic recombination do not affect the rate of w^i reversion. That reversion is only rarely accompanied by exchange of flanking markers precludes unequal crossing-over as a mechanism. An intrachromosomal "looping out" therefore seems likely.

The 2.9-kb duplication in w^i profoundly suppresses interallelic recombination, but the wild-type w^i revertants, and even the w^{ip} derivative with a 14-kb insertion of nonhomologous DNA, restore the normal recombination frequency. It appears that interallelic recombination rates do not correlate in any simple way with DNA length or extent of homology.

A Structural Analysis of w^c and Its Derivatives

We have been interested in analyzing the structure of the highly unstable *white-crimson* (w^c) allele of the *white* locus. This allele was derived as a partial phenotypic revertant of *white-ivory* (w^i), and confers a light reddish orange eye color as a recessive phenotype. w^c mutates at a frequency of about 1/1,000 X chromosomes to wild-type, w^i, and white-eyed derivatives, as well as to other less common derivatives. Wild-type revertants are phenotypically stable, and w^i derivatives exhibit the same level of instability as the original w^i allele. The white-eyed derivatives can be either phenotypically stable or unstable. Stable derivatives include cytologically detectable deletions with one end-point at the *white* locus. Unstable white-eyed derivatives can mutate to wild-type, w^c, and w^i, as well as to other novel phenotypes. Transpositions of w^c to the autosomes have also been detected. (For a review of the genetic properties of w^c, see Green [1976]).

An analysis of the w^c allele by whole genome Southern blots has indicated that the w^c mutation resulted from the insertion of 10 kb of DNA into the w^i duplication (see Fig. 2, Collins and Rubin [1982]). We suspect that this insertion may contain a promoter, which could account for the partial restoration of *white* locus function. We have isolated sequences from the w^c allele by molecular cloning, and have found by restriction enzyme mapping of these cloned sequences that the only difference between the w^i and w^c alleles is this insertion.

We have also begun to analyze the structure of derivatives of w^c by whole genome Southern blotting [Collins and Rubin, 1982]. In five independently isolated w^i revertants of w^c, phenotypic reversion is accompanied by the apparently precise excision of the w^c insertion. We have also examined six independently isolated phenotypically wild-type derivatives of w^c, and in all six cases reversion is accompanied by loss of the w^c insertion and one copy of the w^i duplication. We have recently begun structural analyses of several stable white-eyed derivatives of w^c. Two of these appear to be simple deletions with one end-point within the w^c insertion, which extend into adjacent white locus sequences. Preliminary analysis of a third white-eyed derivative suggests that it results from an element-internal rearrangement, leaving adjacent *white* locus sequences intact. This derivative is particularly interesting because this rearrangement both decreases the expression of the *white* locus, as judged by phenotype, and because it is mutationally stable, despite the presence of insertion sequences within the locus. Further analysis of this and other derivatives of w^c should help us to determine how the w^c insertion partially restores expression of the *white* locus, and to localize sequences within the w^c insertion that are required for mutational instability.

The Molecular Nature of the w^{DZL} Mutation

We have also investigated the DNA alteration in the *white-dominant zeste-like* (w^{DZL}) mutation. w^{DZL} is an eye color mutation isolated in a wild-type stock by P. M. Bingham, who has gone on to characterize the mutation genetically [Bingham, 1980a]. By recombinational analysis the mutation maps at or near the proximal edge of the *white* locus. The mutation confers a yellow eye color on flies that are homozygous for w^{DZL}, or heterozygous for w^{DZL} and a wild-type chromosome. However, it appears that this partial repression of the *white* locus requires two copies of *white,* either duplicated in tandem or on homologues that are able to synapse. Thus, flies that are hemizygous for w^{DZL} are nearly wild-type in eye color, as are flies that are heterozygous for w^{DZL} and a multiply inverted X chromsome.

We have focused on the w^{DZL} mutation because of its unusual instability; 0.1% to 1% of the progeny of w^{DZL} parents carry new *white* locus alleles [Bingham, 1980b]. Among these exceptional progeny is a class, termed simple revertants, which have a wild-type eye color and are recessive to w^{DZL} in heterozygous females. Other derivatives of w^{DZL} have chromosomal deletions or inversions with one breakpoint at *white.*

We have compared the DNAs cloned from w^{DZL} and wild-type flies and have found that the *white* locus region DNA of w^{DZL} contains a 13-kb insertion not found in this region of the wild-type [Levis and Rubin, 1982]. The position of this insertion relative to other insertions in *white* is shown in Figure 2. We have examined the *white* region DNA of 12 independent simple revertants of w^{DZL}. The insertion is partially excised in each of these revertants, indicating that it is responsible for the w^{DZL} mutation. The insertion that remains in revertants varies in size among different revertants from 1.9 to 6.2 kb, but appears never to include the central 6 kb of the original insertion. We hypothesize that these imprecise excisions occur by deletions between tandem arrays of homologous sequences present at the two termini of the insertion. Many of these revertants remain genetically unstable and the remnant insertions they carry undergo additional rearrangements. This indicates that the sequences of the insertion that remain in these revertants are sufficient to cause the chromosomal rearrangements characteristic of the complete insertion.

The eye color of revertants of w^{DZL} is wild-type, despite the presence of 1.9 to 6.2 kb insertion interrupting the wild-type DNA arrangement. This suggests that this insertion did not occur within the coding sequences of *white*. Consistent with this is the finding that the deletion of sequences proximal to the insertion does not abolish *white* function [Levis et al, 1982a]. Therefore we believe that this insertion is at the proximal edge of, or just outside of the *white* gene.

The Organization of the w^c and w^{DZL} Insertions

The w^c and w^{DZL} mutations are unusually unstable, even in comparison with other insertion mutations. We have therefore undertaken a study of the internal organization of the cloned sequences of the w^c and w^{DZL} insertions and of other sequences within the genome that are homologous to them. This study has revealed a common structural basis for the instability of these two insertions [Levis et al, 1982b].

We have found that although the restriction enzyme cleavage maps of the w^c and w^{DZL} insertions are quite different, they cross-hybridize strongly. The regions of sequence homology were localized by Southern blot hybridizations to be within the terminal portions of both insertions. The terminal regions of both insertions are also homologous to the previously described FB family of transposable elements. FB elements have long, imperfect, inverted, terminally repeated sequences that are themselves made up of a tandem array of simple-sequence DNA [Potter et al, 1980; Truett et al, 1981]. Moreover, the termini of the w^c and w^{DZL} insertions are cleaved at regular intervals by Taq I as are the prototypic FB elements, suggesting that they are also made up of a series of imperfect tandem repeats of simple sequence DNA (Fig. 4).

Although the w^c and w^{DZL} insertions both contain FB-homologous sequences at their termini, the arrangement of these FB sequences differs significantly between the two insertions. The w^c insertion has a single pair of FB-inverted repeats, 2.2 and 3.4 kb in length, flanking a 4.0-kb central segment. The w^{DZL} insertion, on the other hand, has two pairs of FB-inverted repeats, one pair at each end, separated by 6.5 kb of DNA. These two terminal repeat arrangements are indicated by the arrows in Figure 2. This difference in terminal-repeat organization may account for the different modes of excision that we have observed for these two insertions. We hypothesize that the imprecise excision of the w^{DZL} insertion occurs by recombination between the sequences of one terminal repeat and those repeated in direct orientation in the other terminal repeat. The precise excision of the w^c insertion probably occurs by a recombination between the sequences adjacent to the insertion, facilitated by the pairing of the inverted terminal repeats.

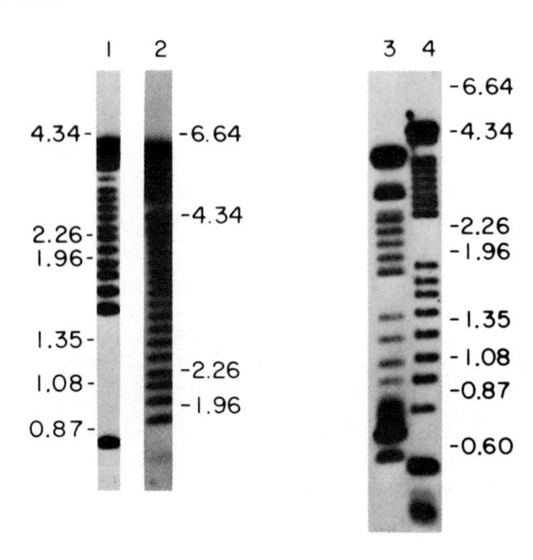

Fig. 4. Autoradiographs of partial Taq I digests of end-labeled fragments derived from the left and right ends of the w^c and ^{DZL} insertions. Lanes 1 and 2 contain end-labeled fragments derived from the left and right ends of the w^c insertion, respectively. Lanes 3 and 4 contain end-labeled fragments derived from the left and right ends of the w^{DZL} insertion, respectively. The positions of Taq I "ladders" in the w^c and w^{DZL} insertions are indicated by the arrows in Figure 2.

The w^c and w^{DZL} insertions have a composite structure unlike most transposable elements; the middle of each insertion has a much different repetition frequency than the terminal repeats. The FB terminal repeat sequences are present at more than 30 sites dispersed throughout the genome. The differences in location of these chromosomal sites in different isolates of D melanogaster indicate that they are transposable. In constrast, the central 6.5 kb of the w^{DZL} insertion is present at only one location besides *white* in the w^{DZL} genome and is a unique sequence in the genome of four wild-type Drosophila strains. It is located at the same spot, near the tip of chromosome arm 2L, in all four wild-type strains tested; we therefore conclude that the sequences of the central region of the w^{DZL} insertion are not normally transposable, but were mobilized by the action of flanking FB elements to transpose to the *white* locus, creating the w^{DZL} mutation. The central sequences of the w^c insertion are present at only five sites in the w^c genome, which differ in location among individuals in the population.

A number of relatively stable mutations of *white* and other loci are now known to be caused by the insertion of repetitive DNA elements. Our discovery that the unstable w^c and w^{DZL} mutations both result from insertions containing FB elements associates this element family with unusual genetic instability. The *copia*-like families of elements have a fundamentally different organization, resembling vertebrate retrovirus proviruses, and seem, in general, not to excise or cause chromosomal rearrangements as often as FB elements.

The Role of P Elements in Hybrid Dysgenesis

Evidence for genetic control of the transposition rate of a family of Drosophila transposable elements comes from studies of hybrid dysgenesis, a "syndrome of correlated genetic traits that is spontaneously induced by hybrids between certain mutually interacting strains, usually in one direction only" [Kidwell et al, 1977]. These traits can include sterility, male recombination, mutation, and chromosomal aberration. Drosophila melanogaster exhibits at least two independent systems of interacting strains: I-R and P-M. Evidence reviewed by Engels [1980] indicates that the genomes of certain strains, called P strains, contain multiple copies of an apparently mobile element known as the P factor. A cross between a male from such a P strain and a female from a strain that lacks P factors, called an M strain, results in hybrid dysgenesis. The P factors do not produce dysgenesis within the P strains, but do so only when placed in a background of an M strain. Many of the mutations produced by hybrid dysgenesis are unstable, are thought to be insertion mutations, and may be due to insertion of the P factor itself.

These and other observations reviewed by Engels (op cit) led to the proposal that the P factor is a transposable element whose rate of transposition is under strict genetic control. During the past two years biochemical evidence that strongly supports this hypothesis has been obtained [Rubin et al, 1982; Bingham et al, 1982]. First, the molecular nature of mutations arising in dysgenic hybrids between P and M Drosophila melanogaster strains has been investigated [Rubin et al, 1982]. Seven independent mutations at the *white* locus were examined and these mutations fell into two classes based on their genetic and structural properties. The five mutations comprising the first class were caused by DNA insertions of 0.5, 0.5, 0.6, 1.2, and 1.4 kb, respectively. The DNA insertions in four of these mutations were examined in detail. Although heterogeneous in size and pattern of restriction enzyme sites, they were homologous in sequence. We refer to members of this sequence family as P elements. Mutations caused by P elements appeared to be stable when maintained in P strains but had reversion rates greater than 10^{-3} when the P strains carrying these mutations were mated to M strains. Phenotypic reversion to wild-type was accompanied by excision of the P element. The two mutations comprising the second class were caused, not by insertion of P elements, but rather by insertion of the 5.0-kb *copia* element. These mutations appeared to be stable under all conditions.

Several other properties of P elements also suggest that they are the causal agents of hybrid dysgenesis [Bingham et al, 1982]. P elements are present in 30–50 copies per haploid genome in all P strains examined and are apparently missing entirely from all M strains examined, with one exception. Members of the P family transpose frequently in P-M dysgenic hybrids; chromosomes descendant from P-M dysgenic hybrids frequently show newly acquired P elements. Finally, the strain-specific breakpoint hotspots for the rearrangement on the X chromosome of the P strain, π_2, are apparently sites of residence of P elements.

Structure of P Elements

Studies of the four P elements inserted into the *white* locus described above led to the proposal that the heterogeneous small P elements have arisen by internal deletion of an intact P element present in P strains, which provides function(s) in *trans* for transposition of the smaller P elements. This hypothesis predicts that P strains harbor a larger, conserved P element and, by analogy with prokaryotic transposable element systems, that whatever signals are required in *cis* for transposition are conserved in the smaller heterogeneous P elements.

A library of λ-phage carrying DNA fragments from a P strain (π_2) was constructed and 12 phage-containing fragments homologous in sequence to the P elements in the *white* locus were characterized [O'Hare and Rubin, in preparation]. Four of the phage carried large, 2.9-kb P elements whose restriction maps were identical (Fig. 5). The complete DNA sequence of one of these elements has been determined [O'Hare and Rubin, in preparation]. It is 2,907 nucleotides long and has 31 base perfect inverted repeats at its ends. In the flanking sequences, a direct repeat of 8 base pairs (bp) is found at both ends of the element. When the same region from a strain lacking P elements was sequenced, only one copy of this 8-bp repeat was found, implying that insertion of the P element led to the duplication. Sequence studies on a second large element revealed a very high degree of conservation, as no base changes were detected in the 2,600 bases determined (shown underlined in Fig. 5). There are three long open reading frames (714, 792, and 654 bases) in the sequence of the large element (see Fig. 5) that could encode several proteins; for example, a P element-specific transposase and other proteins responsible for the regulation of transposase synthesis. All three reading frames are on the same strand and so they could be expressed either by transcription from separate promoters or from one promoter with different splicing patterns. This latter mode of expression seems likely for the second reading frame, as the first ATG codon is 378 bases from its 5′ end whereas there is a sequence similar to acceptor splice sites at its 5′ end.

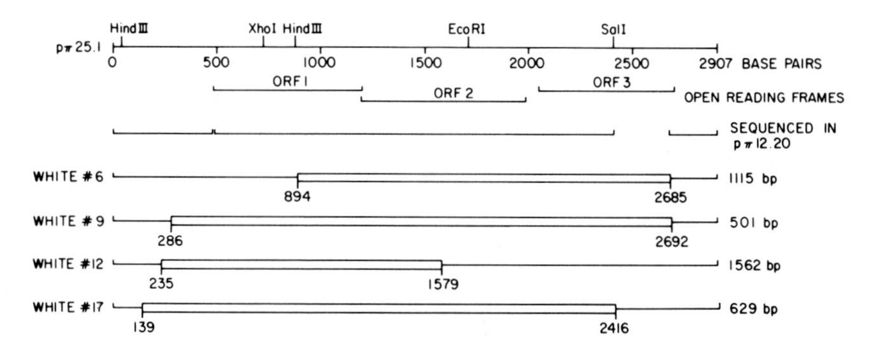

Fig. 5. P element structures. A restriction map of the large P element in plasmid p π25.1, whose DNA sequence has been determined, is shown with the three large open reading frames indicated. The regions of the large P element in plasmid p π12.20 that have been sequenced are underlined. The regions deleted in the four P elements from the *white* locus, determined by comparison with the large P element, are shown as open boxes.

The DNA sequences of the P elements inserted into the *white* locus described above and their flanking sequences have been determined. Three of them (Nos. 6, 9, and 17) are inserted at exactly the same nucleotide with the fourth (#12) defining a second insertion site (see Fig. 2). Both of these sites are within open reading frames, suggesting that the *white* phenotype of these mutations is due to disruption of a protein coding sequence. As for the large elements, an 8-bp sequence is found duplicated at both ends of these elements. The sequence of the parental chromosome into which elements #6 and #12 inserted shows that these sequences are present only once at the site of insertion. Furthermore, the sequence of a wild-type revertant of #6 shows that both the element and one copy of the 8-bp duplication have been eliminated. These results demonstrate that there appears to be considerable sequence specificity for the site of insertion and that the element can excise precisely. The three P elements that are inserted at the same site do not have the same orientation with respect to the *white* sequences. This observation implies that during insertion the target site is brought into conjunction with the ends of the element and not with the internal sequences.

The sequences of each of these four P elements can be derived from that of the 2.9-kb element by a deletion, but the size and location of this putative deletion differs in each case. Elements #12 and #17 show no sequence changes other than the deletion, whereas element #9 has a single T inserted 7 bases after the end of the putative deletion and element #6 has a single A to T substitution at position 33. The deletion endpoints cannot be uniquely defined for elements Nos. 6, 12, and 17 as 1, 2, and 3 bases, respectively, are found in the sequence of the large element at either end of the putative deletions. These repeats seem too small for homologous recombination to be responsible for the generation of the structures observed, although it cannot be known if they were generated in a one-step or multistep process. An attractive alternative mechanism is "slippage repair," which has been proposed for the generation of defective prokaryotic transposable elements. The structure of element #9, in particular, suggests that this type of mechanism may be involved.

Each deletion end-point is unique in the sense that none are shared between different elements (Fig. 5). As each of these elements can excise in the presence of the large P elements, none of them can have lost the sequences necessary for this process. This implies that only the first 138 and the last 16 bases (at most) are required.

REFERENCES

Bingham PM (1980a): The regulation of *white* locus expression: A dominant mutant allele at the *white* locus of Drosophila melanogaster. Genetics 95:341–353.

Bingham PM (1980b): A novel dominant mutant allele at the *white* locus of Drosophila melanogaster is mutable. Cold Spring Harbor Symp Quant Biol 45:519–525.

Bingham PM, Kidwell MG, Rubin GM (1982)The molecular basis of hybrid dysgenesis: The role of the *P* element, a P strain-specific transposon family. Cell 29:995–1004.

Collins M, Rubin GM (1982): Structure of the Drosophila mutable allele, *white-crimson,* and its *white-ivory* and wild-type derivatives. Cell 30:71–79.

Engels W (1982): Hybrid dysgenesis in Drosophila and the stochastic loss hypothesis. Cold Spring Harbor Symp Quant Biol 45:561–565.

Green MM (1976): Mutable and mutator loci. In Ashburner M, Novitski E (eds): "The Genetics and Biology of Drosophila," Vol. 1b. London: Academic Press, pp 929–946.

Karess R, Rubin GM (1982): A small tandem duplication is responsible for the unstable *white-ivory* mutation in Drosophila. Cell 30:63–69.

Kidwell MG, Kidwell JF, Sved JA (1977): Hybrid dysgenesis in Drosophila melanogaster: A syndrome of aberrant traits, including mutation, sterility, and male recombination. Genetics 86:813–833.

Levis R, Rubin GM (1982): The unstable w^{DZL} mutation of Drosophila is caused by a 13 kilobase insertion which is imprecisely excised in phenotypic revertants. Cell 30:543–550.

Levis R, Bingham PM, Rubin GM (1982a): Physical map of the *white* locus of Drosophila melanogaster. Proc Natl Acad Sci USA 79:564–568.

Levis R, Collins M, Rubin GM (1982b): FB elements are the common basis for the instability of the w^{DZL} and w^{c} Drosophila mutations. Cell 30:551–565.

Potter S, Truett M, Phillips M, Maher A (1980): Eucaryotic transposable genetic elements with inverted terminal repeats. Cell 20:639–647.

Rubin GM, Kidwell MG, Bingham PM (1982): The molecular basis of P-M hybrid dysgenesis: The nature of induced mutations. Cell 29:987–994.

Spradling AC, Rubin GM (1981): Drosophila genome organization: Conserved and dynamic aspects. Annu Rev Genet 15:219–264.

Truett MA, Jones RS, Potter SS (1981): Unusual structure of the FB family of transposable elements in Drosophila. Cell 24:753–763.

V. The Control of Gene Activity

Gene Structure and Regulation in Development, pages 227–233

A Procaryotic Model for the Developmental Control of Gene Expression

Janice Pero
Department of Cellular and Developmental Biology, Harvard University, Cambridge, Massachusetts 02138

Biological development requires that specific classes of genes be expressed in defined temporal sequences. To understand how this regulation is achieved at the molecular level, we have investigated a simple model system for development, the temporal control of gene expression in a bacteriophage called SPO1. The relative simplicity of a virus allows one to probe and understand the mechanisms of temporal control at the molecular level. SPO1's lytic cycle is characterized by two temporal switches in gene expression. As described below, these switches are controlled by sequential modifications of the transcriptional machinery of the host cell.

Temporal Control of Phage SPO1 Gene Expression

SPO1 is a virulent phage of Bacillus subtilis whose double-stranded DNA genome (140 kilobases [kb]) encodes about 100 gene products. These proteins appear at regulated intervals after phage infection. *Early* genes are expressed immediately after phage infection and are transcribed by the host RNA polymerase ($E\sigma^{55}$). *Middle* genes are first transcribed about 5 min into the lytic cycle, and finally *late* genes are copied about 13–15 min after infection [Gage and Geiduschek, 1971; Talkington and Pero, 1977]. The transcription of both *middle* and *late* genes is under the control of regulatory genes of the phage. Regulatory gene *28* is an *early* gene whose product is required for *middle* gene expression, whereas regulatory genes *33* and *34* are *middle* genes whose products are both required for *late* gene expression [Fujita et al, 1971]. The products of these regulatory genes function as sigma factors that bind to the bacterial core RNA polymerase and change its promoter-recognition properties [Talkington and Pero, 1978]. Gene *28* encodes a 26,000-dalton sigma factor that directs RNA polymerase to bind *middle* gene promoters [Fox et al, 1976; Duffy and Geiduschek, 1977; Talkington and Pero, 1978], whereas genes *33* and *34* encode 13,000- and 24,000-dalton polypeptides, respectively, which act together as a sigma factor to allow core polymerase to utilize *late* gene promoters [Fox, 1976; Tjian and Pero, 1976]. Thus, in the most simple terms, SPO1's temporal program of gene expression is controlled by a cascade of sigma factors.

Promoters Controlled by Novel Sigma Factors

What is the structure of the different temporal classes of promoters that are controlled by these different sigma factors? To answer that question we have isolated and sequenced 13 phage promoters that are uniquely recognized by either the bacterial RNA polymerase ($E\sigma^{55}$) or one of the phage-modified enzymes ($E\sigma^{gp28}$ or $E\sigma^{gp33\text{-}34}$).

All of the strongest SPO1 *early* gene promoters (about 12) are located within the 12.4-kb terminally redundant ends of the phage genome [Pero et al, 1979; Romeo et al, 1981]. Analysis of the nucleotide sequence of two of these promoters reveals that they closely resemble promoters for Escherichia coli RNA polymerase. The critical contacts between E coli polymerase and its promoters are found in two approximately conserved sequences located 35 bp (−35 region) and 10 bp (−10 region) before the startpoint of mRNA synthesis; the canonical hexamers for these regions are TTGACA and TATAAT, respectively (only the sequence of the nontranscribed strand is shown) [Rosenberg and Court, 1979; Siebenlist et al, 1980]. The phage early promoters have the sequences TTGACT in the −35 region and either CATAAT or AATAAT in the −10 region, each differing by just one base from the prototype E coli sequence [Lee et al, 1980; Lee and Pero, 1981]. (In SPO1 DNA, T is the thymine analog 5-hydroxymethyuracil.) At least 12 bacterial and phage promoters recognized by the principal form of B subtilis RNA polymerase ($E\sigma^{55}$) have now been sequenced. Comparison of these nucleotide sequences reveals the consensus sequences in the −35 and −10 regions of $E\sigma^{55}$ promoters to be identical to those for E coli promoters [Moran et al, 1982].

Middle genes of SPO1 are located in two regions of the phage genome—one large region encompassing the right central half of the genome, and a second smaller region near the left end [Pero et al, 1979]. Based on the regions examined, more than 20 *middle* promoters utilized by σ^{gp28} containing RNA polymerase probably exist [Chelm et al, 1981; Lee and Pero, 1981]. Six of these *middle* promoters have been sequenced [Talkington and Pero, 1979; Lee and Pero, 1981; Costanzo and Pero, 1983]. Comparisons reveal only two regions of homology between these promoters; these regions are centered 35 and 10 bp before the initiation site for RNA synthesis. In the −35 region the sequence T-AGGAGA--A is strikingly conserved with each base being present in at least five of the six promoters; similarly, the sequence TTT-TTT is almost invariant in the −10 region.

Genes that are expressed only *late* after phage infection are located in the left central region of the genome. Within a 10-kb length of this region, a cluster of five *late* promoters recognized by $\sigma^{gp\mathit{33-34}}$ containing RNA polymerase has been located and characterized [N. Hannett, M. Costanzo, J. Jolly, G. Lee and J. Pero, manuscript in preparation]. Like *early* and *middle* promoters, the *late* gene promoters also have conserved sequences in both the −10 and −35 regions. The hexamer GATATT is highly conserved in the −10 region, whereas the heptamer CGTTAGA is much less dramatically conserved in the −35 region. Interestingly, the most actively utilized of the five *late* promoters conforms most closely to the prototype *late* promoter sequence with only a single base mismatch in the sixth position of the −35 region.

Thus, *early, middle,* and *late* gene promoters exhibit distinctive and conserved nucleotide sequences in both their −35 and −10 regions. In other words, each bacterial or phage-modified form of RNA polymerase recognizes a class of promoters characterized by their conformity to particular canonical sequences in the −35 and −10 regions. Since these polymerase forms differ only in the species of sigma with which they are associated, we have argued that in the presence of core polymerase, sigma factors are sequence-specific DNA binding proteins that interact directly with bases in both the −35 and −10 regions [Lee and Pero, 1981; Losick and Pero, 1981]. Furthermore, although different RNA polymerases ($E\sigma^{55}$, $E\sigma^{gp\mathit{28}}$, $E\sigma^{gp\mathit{33-34}}$) "see" different bases, all the enzymes appear to interact with promoters in fundamentally the same manner.

What Is the Structure of These Different Sigma Factors, and How Do They Interact With Core RNA Polymerase?

A first step towards answering these questions has come from the cloning of genes *28, 33,* and *34.* The nucleotide sequence of gene *28* has been determined and from it the amino acid of $\sigma^{gp\mathit{28}}$ has been inferred [Costanzo and Pero, 1983]. $\sigma^{gp\mathit{28}}$ is a 25,707-dalton protein composed of 220 amino acids. Computer analysis showed that it has no extensive homology with E coli σ, the only other sigma factor whose primary structure is known. This was somewhat surprising since both sigmas can interact with the same core RNA polymerases [Shorenstein and Losick, 1973; Achberger and Whiteley, 1980]. When the structure of $\sigma^{gp\mathit{33-34}}$ becomes known, it will be interesting to see if this other phage-coded sigma is evolutionarily related to σ^{gp28}. The clone of gene *28* is now being used to produce large amounts of σ^{gp28}. The availablity of quantities of pure active σ^{gp28} should facilitate studies addressing one of the remaining interesting biological questions about SPO1 gene transcription: How do the different sigma factors replace each other on the core polymerase? Several experiments hint that more is involved than that each sigma is simply displaced by the next appearing "tighter-binding" sigma [Lee, 1981; Chelm et al, 1982]. An attractive model suggested by past experiments with E coli phage T4 [Stevens, 1976] is that SPO1 codes for proteins that bind to either σ^{55} or σ^{gp28} to prevent or weaken their interaction with core polymerase. This would then allow the next newly synthesized sigma factor to interact efficiently with core RNA polymerase. A prediction of this model is that the phage might have two additional regulatory genes—one for the σ^{55}-antagonist and one for the σ^{gp28}-antagonist. Mutants in these genes might fail to turn on *middle* and *late* gene transcription, respectively. Gene *27* might be a candidate for the σ^{gp28}-antagonist. This gene is located adjacent to gene *28,* and mutants in gene *27* block *late* gene transcription [Greene et al, 1982]. It will be interesting to see if future experiments will support such hypotheses.

Other Developmental Systems in Bacteria

Are the molecular mechanisms used for the temporal control of gene expression by our model system, phage SPO1, relevant to other simple developmental systems in bacteria? The emerging answer is yes. When depleted of nutrients, the Gram-positive bacterium B subtilis undergoes a primitive developmental process called sporulation, which results in formation of a dormant cell type known as the endospore. It is now clear that the transcripton of certain genes involved in sporulation is controlled by novel, bacterial-encoded sigma factors. Losick and colleagues [Haldenwang and Losick, 1980; Haldenwang et al, 1981; Johnson et al, 1983] have characterized three new sigma factors (σ^{37}, σ^{32}, σ^{29}) on the basis of their ability to direct core polymerase to recognize and utilize promoters for specific cloned sporulation genes. These promoters differ from those recognized by the major form of RNA polymerase ($E\sigma^{55}$) in vegetative cells. Like the temporally controlled SPO1 promoters, promoters recognized by $E\sigma^{37}$ or $E\sigma^{29}$ have conserved −10 and −35 regions, both of which differ from the conserved −10 and −35 regions found for B subtilis polymerases with other sigma factors [Moran et al 1981; Lang 1982]; (see Table I, which lists all the reported bacterial and phage-coded sigmas in B subtilis and the consensus sequences for two or more promoters recognized by RNA polymerase con-

TABLE I. Consensus Sequences of Promoters Controlled by Bacterial or Phage-Coded Sigma Factors in B subtilis

Sigma[a]	Consensus sequences	
	−35 region	−10 region
σ^{55}	TTGACA	TATAAT
σ^{gp28}	AGGAGA	TTT-TTT
$\sigma^{gp33-34}$	CGTTAGA	GATATT
σ^{37}	AGG-TT	GG-ATTG-TT
σ^{28}	CTAAA	CCGATAT
σ^{29}	TT-AAA	CATATT

[a]References for the sequences are σ^{55} [Moran et al, 1982]; σ^{gp28} [Lee and Pero, 1981]; $\sigma^{gp33-34}$ [N. Hannett, M. Costanzo, J. Jolly, G. Lee and J. Pero, manuscript in preparation]; σ^{37} [Moran et al, 1981]; σ^{28} [Gilman et al, 1981] and σ^{29} [Lang, 1982].

taining the indicated sigma). Although it has not yet been established that the temporal control of sporulation gene expression is controlled by a *cascade* of sigma factors, it is clear that novel bacterial encoded sigma factors play a key role in the transcription of certain developmentally regulated genes. Subsequent research should indicate whether or not sporulation gene expression is controlled by a true cascade of sigmas. Furthermore, it will be interesting to see if procaryotic developmental systems such as bacteroid formation in Rhizobium or spore formation in Myxobacteria also utilize novel sigma factors.

REFERENCES

Achberger EC, Whiteley HR (1980): The interaction of Escherichia coli core RNA polymerase with specificity-determining subunits derived from unmodified and SP82-modified Bacillus subtilis RNAa polymerase. J Biol Chem 255:11957–11964.

Chelm BK, Romeo JJ, Brennan SM, Geiduschek EP (1981): A transcriptional map of the bacteriophage SPO1 genome: III. A region of early and middle promoters (the gene 28 region). Virology 112:572–588.

Chelm BK, Duffy JJ, Geiduschek EP (1982): Interaction of Bacillus subtilis RNA polymerase core with two specificity-determining subunits: Competition between σ and the SPO1 gene 28 protein. J Biol Chem 251:6501-6508.

Costanzo M, Pero J (1983): Structure of a Bacillus subtilis bacteroiophage SPO1 gene encoding an RNA polymerase sigma factor. Proc. Natl Acad Sci USA 80: in press (March).

Duffy JJ, Geiduschek EP (1977): Purification of a positive regulatory subunit from phage SPO1-modified RNA polymerase. Nature (London) 270:28–32.

Fox TD (1976): Identification of phage SPO1 proteins coded by regulatory genes 33 and 34. Nature (London) 262:748–753.

Fox TD, Losick R, Pero J (1976): Regulatory gene 28 of bacteriophage SPO1 codes for a phage-induced subunit of RNA polymerase. J Mol Biol 101:427–433.

Fujita DJ, Ohlsson-Wilhelm BM, Geiduschek EP (1971): Transcription during bacteriophage SPO1 development: Mutation affecting the program of viral transcription. J Mol Biol 57: 301–317.

Gage LP, Geiduschek EP (1971): RNA synthesis during bacteriophage SPO1 development: Six classes of SPO1 RNA. J Mol Biol 57:279–300.

Gilman MZ, Wiggs JL, Chamberlin MJ (1981): Nucleotide sequences of two Bacillus subtilis promoters used by Bacillus subtilis sigma-28 RNA polymerase. Nucleic Acids Res 9:5991–5999.

Greene JR, Chelm BK, Geiduschek EP (1982): SPO1 gene 27 is required for viral late transcription. J Virol 41:715–720.

Haldenwang WG, Losick R (1980): Novel RNA polymerase sigma factor from Bacillus subtilis. Proc Natl Acad Sci USA 77:7000–7004.

Haldenwang WG, Lang N, Losick R (1981): A sporulation-induced sigma-like regulatory protein from B. subtilis. Cell 23:615–624.

Johnson WC, Moran CP Jr, Losick R (1983): Two RNA polymerase sigma factors from Bacillus subtilis discriminate between overlapping promoters for a developmentally regulated gene. Nature (in press).

Lang N (1982): Novel promoters on the Bacillus subtilis chromosome. PhD Thesis, Harvard University, Cambridge, Massachusetts, pp 68–82.

Lee G (1981): Temporally-controlled phage SPO1 promoters. PhD Thesis, Harvard University, Cambridge, Massachusetts, pp 145–152.

Lee G, Pero J (1981): Conserved nucleotide sequences in temporally-controlled phage promoters. J Mol Biol 152:247–265.

Lee G, Talkington C, Pero J (1980): Nucleotide sequences of a promoter recognized by Bacillus subtilis RNA polymerase. Mol Gen Genet 180:57–65.

Losick R, Pero J (1981): Cascades of sigma factors. Cell 25:582–584.

Moran CP Jr, Lang N, Losick R (1981): Nucleotide sequence of a Bacillus subtilis promoter recognized by Bacillus subtilis RNA polymerase containing σ^{37}. Nucleic Acids Res 9:5979–5990.

Moran CP Jr, Lang N, Legrice SFJ, Lee G, Stephens M, Sonenshein AL, Pero J, Losick R (1982): Nucleotide sequences that signal the initiation of transcription and translation in Bacillus subtilis. Mol Gen Genet 186:339-346.

Pero J, Hannett N, Talkington C (1979): Restriction cleavage map of SPO1 DNA: General location of early, middle, and late genes. J Virol 31:156–171.

Romeo JM, Brennan SM, Chelm BK, Geiduschek EP (1981): A transcriptional map of the bacteriophage SPO1 genome: I. The major early promoters. Virology 111:588–603.

Rosenberg M, Court D (1979): Regulatory sequences involved in the promotion and termination of transcription. Annu Rev Genet 13:319–353.

Shorenstein RG, Losick R (1973): Comparative size and properties of the sigma subunits of ribonucleic acid polymerase from Bacillus subtilis and Escherichia coli. J Biol Chem 248:6170–6173.

Siebenlist U, Simpson RB, Gilbert W (1980): E coli RNA polymerase interacts homologously with two different promoters. Cell 20:269–281.

Stevens A (1976): A salt-promoted inhibitor of RNA polymerase isolated from T4 phage-infected E coli. In Losick R, Chamberlin M (eds): "RNA Polymerase," New York: Cold Spring Harbor, pp 617–627.

Talkington C, Pero J (1977): Restriction fragment analysis of the temporal program of bacteriophage SPO1 transcription and its control by phage-modified RNA polymerases. Virology 83:365–379.

Talkington C, Pero J (1978): Promoter recognition by phage SPO1-modified RNA polymerase. Proc Natl Acad Sci USA 75:1185–1189.

Talkington C, Pero J (1979): Distinctive nucleotide sequences of promoters recognized by RNA polymerase containing a phage-coded "σ-like" protein. Proc Natl Acad Sci USA 76:5465–5469.

Tjian R, Pero J (1976): Bacteriophage SPO1 regulatory proteins directing late gene transcription in vitro. Nature (London) 262:753–757.

Gene Structure and Regulation in Development, pages 235–239

Inheritable Expression of Fusion Genes Microinjected Into Mouse Eggs

Richard D. Palmiter and Ralph L. Brinster
Howard Hughes Medical Institute Laboratory, Department of Biochemistry, University of Washington, Seattle, Washington 98195 (R.D.P.), and Laboratory of Reproductive Physiology, School of Veterinary Medicine, University of Pennsylvania, Philadelphia, Pennsylvania 19104 (R.L.B.)

The mouse metallothionein-I (MT-I) gene was isolated, and probes from it were used to show that the endogenous genes are transcriptionally regulated by heavy metals such as cadmium and zinc [Durnam et al, 1980; Durnam and Palmiter, 1981]. By 5′ deletion mapping followed by gene transfer into L cells or mouse eggs, the minimal region required for transcriptional regulation was shown to be within 90 nucleotides of the transcription start site [Mayo et al, 1982; Brinster et al, 1982]. The MT-I regulatory region was fused to the structural genes of either herpes virus thymidine kinase (TK) or rat growth hormone (GH) to generate plasmids pMK and pMGH, respectively, as shown in Figure 1.

Linear DNA fragments that include the fusion gene were isolated from the plasmids and ~500 copies were injected into the male pronucleus of mouse eggs [Brinster et al, 1981]. These eggs were then transferred to the reproductive tracts of pseudopregnant foster mothers. About 15% of the eggs that developed to term gave rise to mice carrying one or more copies of the injected fusion gene as assayed by DNA dot hybridization [Palmiter et al, 1982b].

To ascertain whether the MK fusion gene was expressed, the mice were injected with Cd; 20 hr later a partial hepatectomy was performed and thymidine kinase (TK) activity was determined. Most of the mice (~75%) carrying the MK fusion gene showed elevated TK activity in the liver, sometimes as much as 50 times normal (Table I). The elevated TK activity was inhibited by antisera specific for herpes TK. To ask whether the MK

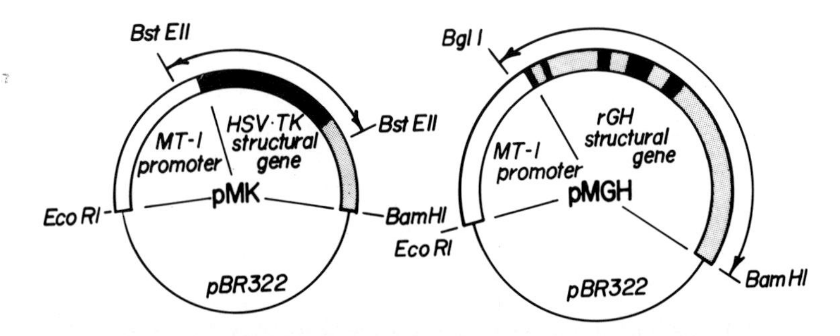

Fig. 1. Plasmids pMK and pMGH. The open regions represent mouse MT-I 5′ sequences; solid regions represent coding portions of herpes thymidine kinase (TK) or rat growth hormone (GH) structural genes; stippled regions represent introns and 3′ flanking regions; the single line represents pBR322. Linear fragments designated by the arrows were isolated from the plasmids and injected into pronuclei of fertilized mouse eggs.

TABLE I. Regulation of MK Gene Expression by Cadmium[a]

Mouse	Number of MK genes	Thymidine kinase activity (cpm/μg/min) −Cd	+Cd
Controls (n = 62)	0	(0.7–1.7)	
Myk-67	2	1.5	68.3
MaK-67	150	3.3	76.5
MyK-84-7	100	3.3	28.0
MyK-103	1	1.2	18.5

[a]$CdSO_4$ (1 mg/kg) was administered 18 hr prior to hepatectomy. Thymidine kinase activity was measured as described by Palmiter et al. [1982b].

genes were regulated, a second partial hepatectomy was performed without prior Cd treatment. In this case the amount of TK activity was low, close to endogenous levels (Table I). Thus, these fusion genes appear to be regulated like the endogenous MT-I genes [Palmiter et al, 1982b].

When mice carrying MK genes were outbred, approximately 50% of the offspring carried the fusion genes and the restriction patterns of the DNA from the offspring generally matched those of the parent. These results suggest that the fusion genes integrated into a single chromosome shortly after injection and that all of the cells of the animal carry the fusion genes. This type of inheritance is observed even when multiple copies of the fusion gene are present, indicating that all of the copies are integrated into the same chromosome, a result that is consistent with the tandem, head-to-tail integration pattern deduced by Southern blotting [Palmiter et al, 1982b].

Offspring of mice carrying MK fusion genes generally express these genes, as shown, for example, in the pedigree of MyK-84 (Fig. 2). Although most of the animals carrying the MK gene express it, the level of expression varies considerably—from substantially more than the parent to substantially less or even no activity. This result has been observed in pedigrees of mice carrying either single or multiple copies of the MK gene. We do not know what causes this variability.

In mice carrying the MGH fusion gene, expression of the foreign gene can be measured by radioimmunoassay of GH in the blood, by GH mRNA levels in the liver, or by growth (Table II). Animals with up to 800 times

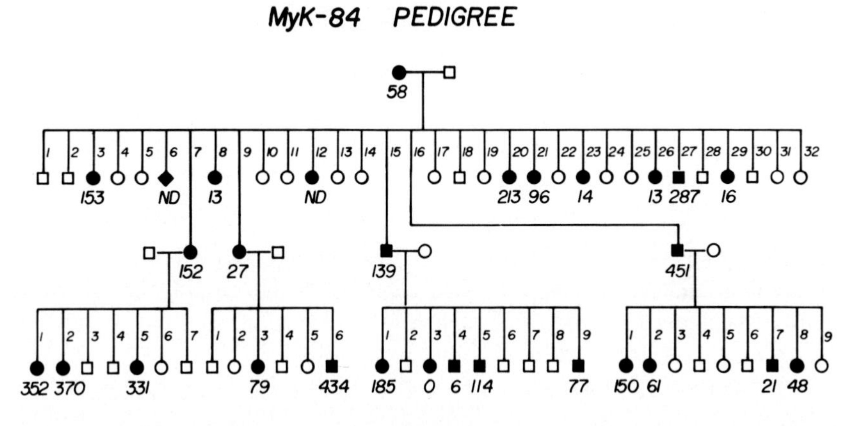

Fig. 2. Pedigree of MyK-84 showing inheritance and expression of MK genes. Animals carrying MK genes (~ 100 copies of a 2.1-kb Bst EII fragment shown in Fig. 1) are indicated by solid symbols. The viral TK activity measured in the liver after challenge with Cd is indicated by the numbers under the symbols.

TABLE II. MGH mRNA, GH Levels, and Growth of Mice Carrying MGH Genes

Mouse	MGH genes (No./cell)	MGH mRNA in liver (molecules/cell)	GH (μg/ml)	Relative growth ratio[a]
Controls	0	0	0.16	1.0
MGH-2	20	800	57	1.87
MGH-19	10	1500	32	1.69
MGH-21	35	3000	112	1.78

[a]See Palmiter et al [1982a], for details.

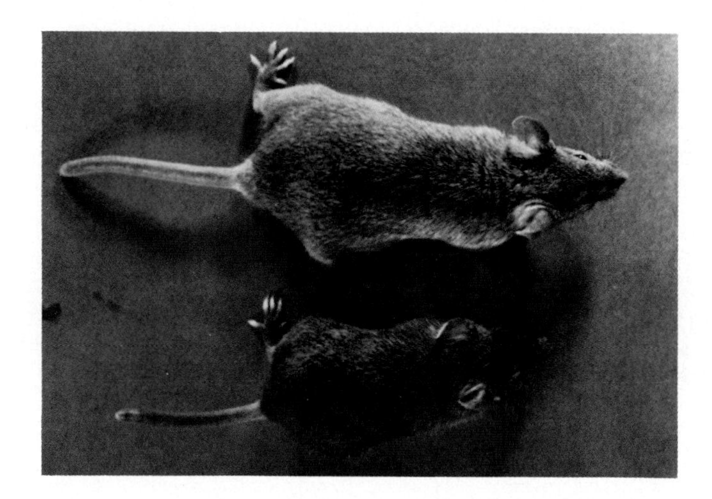

Fig. 3. Growth of mice expressing the MGH gene. A mouse (MGH-2) carrying ~20 copies of the MGH gene (41.2 g) compared to a normal littermate (21.2 g) at age 74 days.

normal serum GH have been obtained. These animals express the MGH gene in the liver and maintain about 3,000 copies of MGH mRNA per cell when on zinc-supplemented diet [Palmiter et al, 1982a]. The chronic stimulation by excess GH nearly doubles the adult weight (Fig. 3).

One animal, carrying about ten copies of the MGH gene, has had 19 offspring, 10 of which carry the MGH gene. Five of these positive offspring were reared on a zinc diet and five were reared on a normal diet. Zinc increased the average GH level in the serum about ten-fold, but all the mice carrying the MGH gene grew larger than their littermates. In this case, even the lower level of GH is probably sufficient to promote growth. We anticipate that other animals developing from this procedure will show zinc-dependent growth if one with a lower level of constitutive expression can be found.

Our experience indicates that when the mouse MT-I promoter is fused to other structural genes, there is a high probability that the gene will be expressed in the liver, and probably in other tissues as well. Expression of these genes is regulated by heavy metals. Moreover, these genes are inherited in a Mendelian manner and are generally expressed in the offspring as well. We have observed heterogeneity of expression in offspring, even extinction of activity. In some cases, the changes in activity correlate with changes in DNA methylation but in other cases there is no obvious change in DNA arrangement of methylation, suggesting that there may be other modifications that affect gene activity.

Gene transfer into the germ line of animals will be an invaluable means of investigating the role of DNA sequences, DNA modifications, and chromosomal position on tissue-specific gene expression. It also lends itself to practical application as a means of producing valuable gene products and as a way of mimicking or correcting genetic defects.

Summary

The promoter region of the mouse metallothionein-I gene was fused to the structural genes of either herpes virus thymidine kinase or rat growth hormone. DNA fragments containing these fusion genes were microinjected into the nuclei of fertilized eggs, which were subsequently transferred to the reproductive tracts of foster mothers. In many of the animals that develop from this procedure, the fusion gene(s) are integrated into one of the chromosomes. Most of the animals express the foreign gene in the liver, where they are regulated by heavy metals that are known to modulate metallothionein-I gene transcription. Thus, animals have been produced that express high levels of either herpes thymidine kinase or rat growth hormone. In the latter case, these animals grow considerably larger than littermates not carrying the fusion gene. Furthermore, offspring of these animals generally inherit the foreign DNA in a Mendelian fashion and the fusion genes are still expressed.

REFERENCES

Brinster RL, Chen HY, Trumbauer M, Senear AW, Warren R, Palmiter RD (1981): Somatic expression of herpes thymidine kinase in mice following injection of a fusion gene into eggs. Cell 27:223–231.

Brinster RL, Chen HY, Warren R, Sarthy A, Palmiter RD (1982): Regulation of metallothionein-thymidine kinase fusion plasmids injected into mouse eggs. Nature 296:39–42.

Durnam DM, Palmiter RD (1981): Transcriptional regulation of the mouse metallothionein-I gene by heavy metals. J Biol Chem 256:5712–5716.

Durnam DM, Perrin F, Gannon F, Palmiter RD (1980): Isolation and characterization of the mouse metallothionein-I gene. Proc Natl Acad Sci USA 77:6511–6516.

Mayo KE, Warren R, Palmiter RD (1982): The mouse metallothionein-I gene is transcriptionally regulated by cadmium following transfection into human or mouse cells. Cell 29:99–108.

Palmiter RD, Brinster RL, Hammer RE, Trumbauer ME, Rosenfeld MG, Birnberg NC, Evans RM (1982a): Dramatic growth of mice that develop from eggs microinjected with metallothionein-growth hormone fusion genes. Nature 300:611–614.

Palmiter RD, Chen HY, Brinster RL (1982b): Differential regulation of metallothionein-thymidine kinase fusion genes in transgenic mice and their offspring. Cell 29:701–710.

Gene Structure and Regulation in Development, pages 241–268
Published 1983 Alan R. Liss, Inc., 150 Fifth Avenue, New York, NY 10011

An Altered DNA Conformation Detected by S1 Nuclease Occurs at Specific Regions in Active Chick Globin Chromatin

Alf Larsen and Harold Weintraub
Hutchinson Cancer Research Center, Seattle, Washington 98104

INTRODUCTION

In the article presented below, we have presented our initial observation that specific sites in chromatin are in a non-B form DNA structure and can therefore be recognized and cleaved by single-strand specific nucleases such as S1 nuclease. The sites cut by S1 nuclease correspond to putative regulatory regions since they are located at the 5′ and 3′ ends of active transcription units; moreover, for tissue specific genes, these sites are also tissue specific.

A major conceptual step occurred when we found that these DNA sequences are also S1 sensitive in supercoiled DNA plasmids harboring those sequences. The altered DNA structure depends on supercoiling since relaxed or linear molecules are resistant to S1 cutting. Most important from the point of view of embryology is the finding (developed more extensively since the publication of this paper) that the actual site of cutting by S1 in supercoiled plasmids is a sensitive function of the physical environment being effected by NaCl, pH, divalent cations, spermidine, and single-strand specific DNA binding proteins. In addition, the degree of supercoiling of the plasmids is crucial: 6 kbp supercoiled molecules with 30 supercoils are cut by S1, but molecules with fewer than 25 are not. The fact that the altered DNA conformation responds in an all-or-none way to "thresholds" (of supercoiling, NaCl, pH, Mg^{+2}, etc.) makes such a response an ideal candidate to "read" an embryonic gradient(s). Thus, DNA in nuclei located in different positions in an embryonic gradient should respond differently in terms of its conformation. Once the DNA responds, I think that the resulting altered DNA structure is stabilized and propagated to daughter cells independent of the initial gradient signal that might have induced it. This property of S1 sites to be stably propagated once induced has recently been demonstrated [M. Groudine and H. Weintraub (1982) Cell 30:131-139]. Thus, the heritability of these sites combined with the ability of the associated DNA sequences to respond specifically to threshold levels of rather non-specific substances offers a novel way of viewing how early determinative events may be elicited and propagated and how differences between daughter cells may be generated.

Differential gene expression is correlated with changes in chromatin structure [Weintraub and Groudine, 1976]. These changes are revealed by the preferential sensitivity to DNAase I of actively expressed, tissue-specific genes [Garel and Axel, 1976; Weintraub and Groudine, 1976]. During avian red cell development, changes in globin chromatin structure appear before these genes become expressed [Weintraub et al., 1982]; hence these changes seem to be a necessary condition for transcription and not a consequence of transcription [see also Weintraub, 1979].

When analyzed carefully, DNAase digestion of nuclei reveals a number of structural features of active chromatin structure. For example, it has been shown that the active chromatin structure detected by a marked sensitivity to DNAase I is induced by two nuclear proteins, HMG (high mobility group) 14 and 17 [Weisbrod and Weintraub, 1979; Gazit et al., 1980; Sandeen et al., 1980]. The active transcription unit defined in part by HMG proteins is embedded in a large chromosome domain (50–100 kb). This domain displays a level of DNAase I sensitivity that is intermediate [Stalder et al., 1980; Weintraub et al., 1981] between the resistant chromatin reflective of inactive genes and the HMG-dependent, DNAase-sensitive chromatin that defines the DNA sequences that are transcribed. The structural state reflecting "intermediate" sensitivity may be related to a "relaxed" higher order chromatin structure [Stalder et al., 1980].

A third feature of chromatin revealed by DNAase I are DNAase I-hypersensitive sites [Scott and Wigmore, 1978; Waldeck et al., 1978; Varshavsky et al., 1979; Wu et al., 1979] where DNAase I cuts at specific localized DNA sequences that often, but not always, map to the 5′ side of genes when they are active, but not when they are inactive [Kuo et al., 1979; Stalder et al., 1980; Wu, 1980; Groudine and Weintraub, 1981; Keene et al., 1981; McGhee et al., 1981; Samal et al., 1981; Weintraub et al., 1981; Wu and Gilbert, 1981]. These DNAase I–hypersensitive regions appear before the activation of the globin genes during chick red cell development [Weintraub et al., 1982], and in certain cases, they map over a rather large (100 bp) region; in the case of the active β^{A} globin gene [McGhee et al., 1981], between -70 and -160 bp 5′ to the CAP site. This region is part of a larger region that is also "hypersensitive" to restriction enzymes [McGhee et al., 1981], as well as to DNAase II and micrococcal nuclease. More recently, it has been shown that DNA sequences such as the SV40 72-bp repeat or the avian leukosis virus long terminal repeat—sequences that seem to activate distantly placed, neighboring genes [Lee et al., 1981; Moreau et al., 1981; Neel et al., 1981]—also contain DNAase-hypersensitive sites [Cremisi, 1981; Groudine et al., 1981].

Here we show that S1 nuclease introduces doublestranded cleavages at specific DNA sequences associated with active α and β globin chromatin in mature avian red cell nuclei. These cleavage sites map only near known DNAase I–hypersensitive sites. The DNA region exhibiting S1 sensitivity in chromatin is also preferentially sensitive to S1 as pure DNA when put under stress in supercoiled, but not relaxed, recombinant DNA plasmids. These results suggest that the sequence of DNA in these hypersensitive regions defines a propensity for DNA to adopt an altered conformation (possibly single-stranded) as revealed by S1 sensitivity. This altered conformation must be induced or stabilized in the appropriate environment, be it a supercoiled plasmid or an active chromosomal transcription unit. The ability of certain DNA sequences to change conformation in response to the environment may be a property useful for generating differences between daughter cells during development.

RESULTS

Globin Chromatin is Cut in Specific DNA Regions by S1 Nuclease

Nuclei from mature red blood cells (obtained from 14 day embryos) containing adult type hemoglobin were digested in S1 buffer (30 mM sodium acetate, 3mM $ZnCl_2$ and 1 mM EDTA, pH 4.5) with and without 5 units of S1 nuclease per μg of chromosomal DNA for 30 min at 37°C. The DNA was purified and cut with the restriction endonuclease Eco RI, separated on 1.2% agarose gels, transferred to nitrocellulose [Southern, 1975] and hybridized to a nick-translated α-globin clone (α-5). The results show (Fig. 1A, lane g) that as a consequence of S1 digestion, a specific subfragment (arrow) is generated in the α-globin chromosomal DNA. The S1 cutting at this (and other) sites does not go to completion; thus even a tenfold excess of S1 (50 units/μg DNA) over our standard condition fails to increase significantly the relative ratio of the subfragment to the parent restriction fragment (Fig. 1A, lane h). Depending on the specific S1-hypersensitive site, the maximal amount of cutting can range between less than 20% and more than 60% (Fig. 2). Also, the subfragment is not observed without subsequent restriction digestion, indicating that the S1 cutting occurs in one unique region. By using a variety of restriction enzymes and probes, we have mapped this particular S1-sensitive site to the 5′ side of α^D (see map, Fig. 1B). This is the approximate position of a red-cell-specific DNAase I–hypersensitive site as well [Weintraub et al., 1981].

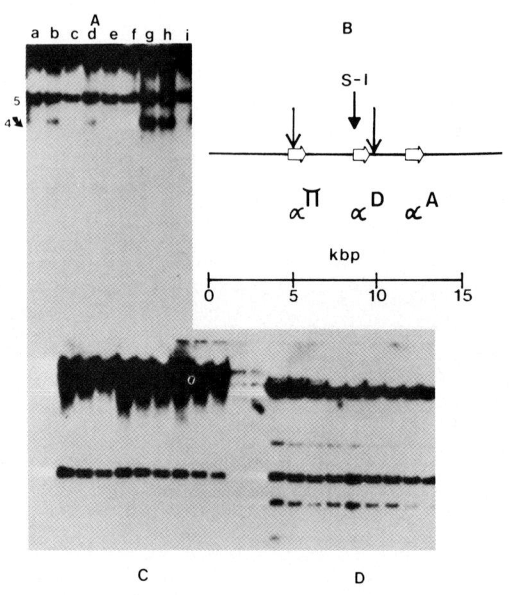

Fig. 1. An S1-Hypersensitive Region at the 5′ Side of α^D. Nuclei from 14 day red cells were digested for 30 min at 37°C with S1 nuclease. (A) Various DNA preparations were digested with Eco RI and blot-hybridized to α-5. Arrow: 4 kb subband generated as a result of S1 digestion. The bar between 0 and 15 kb (B) represents the α-5 probe. (Lane a) S1 nuclease in high salt buffer (3 units S1/μg DNA); (lane b) EDTA-dialyzed S1 nuclease in high salt buffer (3 units/μg); (lane c) control nuclei incubated in high salt buffer; (lane d) (S1-nuclease digestion of embryonic 5 day red cell nuclei in low salt buffer (1 unit/μg; these cells synthesize α^π, α^D and α^A); (lane e) brain nuclei incubated with 10 units/μg S1 in low salt buffer; (lane f) control adult red cell nuclei incubated in low salt buffer without S1; (lane g) nuclei incubated with 5 units/μg S1 in low salt buffer; (lane h) as in (g), but with 50 units/μg S1; (lane i) S1 enzyme preparation dialyzed against EDTA and added to nuclei at 5 units/μg in low salt buffer containing $MgCl_2$ (3 mM) and no $ZnCl_2$. In (g), the same EDTA-treated enzyme is added back to nuclei in 3 mM $ZnCl_2$. (B) Restriction map of α-5 clone showing Eco RI(↓) cleavages. Large RI junction fragments are seen as a dark thick band at top of gel in (A). Location of S1 site being probed is shown. The exact position of this site was determined by using additional restriction enzymes and probes. (C) Same samples as in (A) digested with Bam HI and hybridized to an ovalbumin cDNA clone. (D) As in (C), but digested with Eco RI.

The S1 Cutting is Due to S1 Nuclease and Not to a Contaminating Nuclease Activity

The specific cutting of globin chromatin by S1 nuclease has been extensively characterized. In particular, since it is clear that contaminating nucleases or endogenous nucleases might yield similar results, it is crucial to show that the activity is the single-stranded cutting activity of S1 nuclease in our preparations. The following eleven control experiments have been done, and some of the results from this analysis are shown in Figures 1–3. First, the reaction requires Zn^{+2} but not Mg^{+2}. When the enzyme preparation is dialyzed against EDTA to chelate bound Zn^{+2}, addition of the enzyme preparation to nuclei results in specific cutting when $ZnCl_2$ is provided, but not when $MgCl_2$ (or Ca^{+2}) is provided in the nuclear suspension (Figs. 1A, lane i and 2A, lane e). This suggests that S1 is the activity being observed and that Mg (or CA^{+2})-dependent nucleases (either endogenous or exogenous) are not detected under these conditions. Second, the reaction is inhibited by low levels of phosphate (20 mM; Fig. 2, lane b) and dATP (0.1 mM; Fig. 2, lane c), reagents known to inhibit S1 activity [Wiegand et al., 1975]. Third, the same chromatin preparations that show S1-sensitive sites in the globin domain do not show S1-sensitive sites with the use of an ovalbumin cDNA clone (Figs. 1B and 1C) nor a probe for the inactive *ev*-1 locus coding for the endogenous avian retrovirus (not shown). Fourth, the same specific cutting (Fig. 3, lane k) is observed at pH 7.6 with the purified single-strand-specific endonuclease from T-7 (a gift from R. Wells) or at pH 5.5 with S1 nuclease. This excludes the possibility that the structural feature read by S1 is a chromosome-dependent DNA denaturation phenomenon induced by the low pH required by S1 nuclease. Highly purified S1 preparations obtained from Dr. V. Vogt [Vogt, 1973] also yield the same results with the same level of enzyme (not shown). Fifth, S1 activity is minimal at pH 7.4 [Wiegand et al., 1975]; similarly, the specific cutting in chromatin does not occur in the same buffer at pH 7.4 (Fig. 2, lane d). Sixth, the minimal concentration used to see S1 cutting in chromatin under these conditions is 1 unit of S1/μg chromatin. This is the same concentration that we use to cut single-stranded DNA regions in purified supercoiled plasmids (see below) under identical buffer conditions. Seventh, specific S1 sensitivity is not usually observed in purified chromosomal DNA (not shown); however, in some preparations subbands are observed at about 1/10–1/20 the intensity as observed with nuclei. This cutting in free DNA is most easily explained by endogenous nicks that occur in the chromatin, presumably due to endogenous nucleases. This point is being investigated. Eighth, treatment of nuclei with S1 buffer (pH 4.5) does not irreversibly destroy the ability of DNAase I to introduce DNAase-hypersensitive cuts after subsequent resuspension at pH 7.4 in reticulocyte standard buffer; thus the S1 buffer conditions are not irreversibly altering at least this one aspect of chromatin structure. Ninth, the specific cutting by S1 is not inhibited significantly by prior treatment of nuclei with RNAase A in low salt buffer (Fig. 3, lane h); thus an accessible DNA–RNA hybrid is unlikely to be structurally responsible for the S1 sensitivity of this region. Tenth, low levels of ethidium bromide (Fig. 3, lane f) also do not inhibit S1 cutting; thus superhelical strain may not be important in maintaining these sites in the nuclei as we isolate them (see Discussion). Finally, the

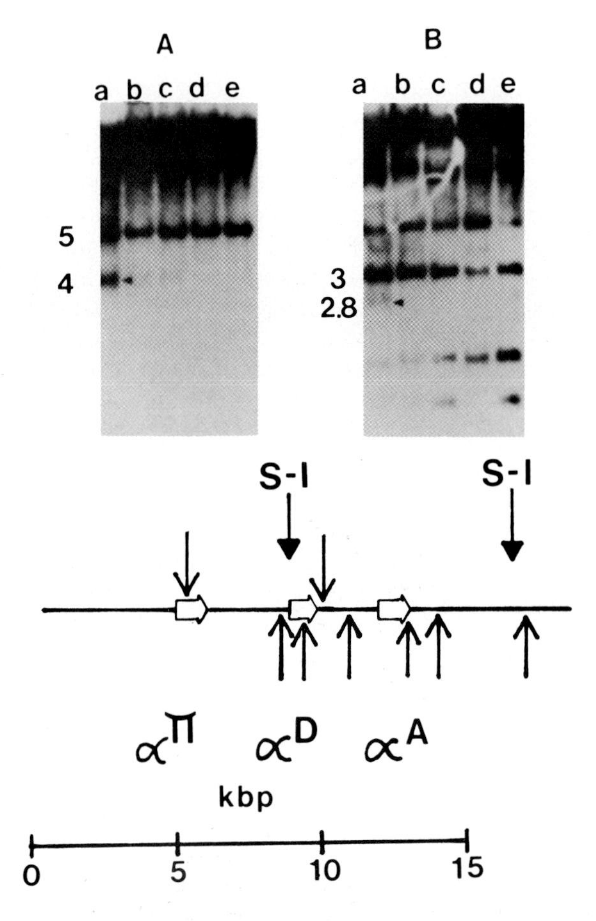

Fig. 2. Known Inhibitors of S1 Nuclease Inhibit the Specific Cleavages in Nuclei. Red cell nuclei were digested under standard conditions (5 units/μg S1 in low salt buffer). (A) and (B) are the same samples. (A) Digested with Eco RI (↓ in map below) and hybridized to α-5. (B) Digested with Bam HI (↑ in map) and hybridized to α-5. As determined from a number of separate restriction digests (see Weintraub et al., 1981), the 4 kb subband in (A) assays the site 5′ to α^{D} (Fig. 1). The less sensitive 2.8 kb subband in (B) assays the site 3′ to α^{A} (see map below). (Lane a) Standard S1 digestion; (lane b) as in (a), but with 20 mM sodium phosphate (ph 4.5) added; (lane c) same, but with 0.1 mM dATP added; (lane d) same, but adjusted to pH 7.4; (lane e) enzyme dialyzed against EDTA and added to nuclei in the presence of 5 mM $MgCl_2$ in the absence of $ZnCl_2$. In (lane a), the same enzyme preparation was added back to nuclei in the presence of 3 mM $ZnCl_2$. (Bottom) Restriction map of α-5 clone. The bar between 0 and 15 kb represents the α-5 probe.

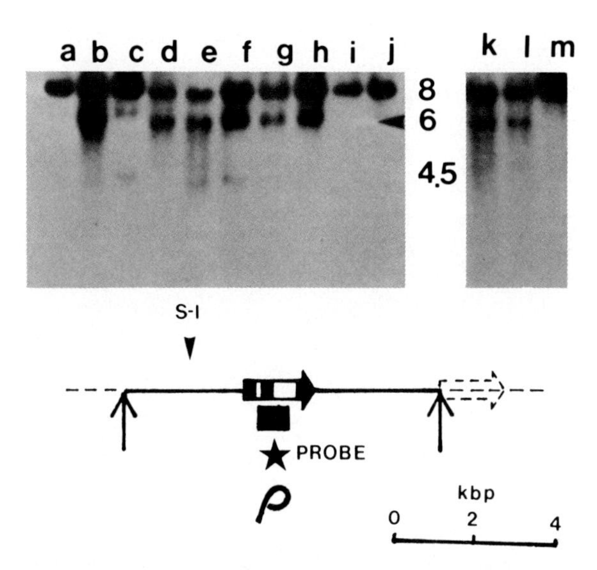

Fig. 3. Digestion of a β-Globin-Gene Hypersensitive Site With S1 Nuclease: Additional Controls. Nuclei were digested with S1 nuclease at 5 units/μg in low salt buffer. The DNA was blot-hybridized to a β^{ρ} probe (dark square in map below) after Bam HI digestion. The site being assayed is indicated by a 6 kb subband and maps to an adult-specific site (black triangle in map below). (Lane a) Control nuclei incubated without S1 (10 μg DNA loaded). (Lane b) Nuclei incubated with 5 units/μg S1 (20 μg DNA loaded). (Lane c) Nuclei pelleted through 2 M NaCl and digested in low salt buffer with 5 units/μg S1 (20 μg DNA loaded). The low level of subband formation in this experiment is similar to what we might expect from endogenous nucleases. The cutting by S1 is significantly reduced; however, further characterization is required to rule out the possibility that some residual S1-sensitive structure is still present in the supercoiled DNA remaining after high salt extraction (Benyajati and Worcel, 1976). (Lane d) Nuclei incubated with 5 units/μg S1 (10 μg DNA loaded). (Lane e) as in (d), but an equal amount of double-stranded 146–160 bp DNA was added (10 μg DNA loaded). (Lane f) As in (d), but with 10 μg/ml of ethidium bromide (20 μg DNA loaded). (Lane g) As in (e), except the added naked DNA was denatured (10 μg DNA loaded). (Lane h) As in (d), but the nuclei were pretreated with RNAase A at low ionic strength (20 μg DNA loaded). (Lanes i and j) Control nuclei with no S1 (5 and 10 μg DNA loaded, respectively). (Lane k) Nuclei treated with the T7 single-strand-specific nuclease (50 mM Tris-HCl, pH 7.6; 0.1 mM DDT; 10 mM $MgCl_2$; and 2.5 mM spermidine; 10 μg of DNA loaded). (Lane l) Corresponding nuclei treated with S1 nuclease in parallel to the experiment shown in (k) (10 μg DNA loaded). (Lane m) Control nuclei incubated in the T7 nuclease digestion buffer, but without enzyme (20 μg DNA loaded). The 4.5 kb band sometimes seen is a cross-hybridizing fragment from β^{A}. The map below shows the probe (★) and the location of the S1 site being probed.

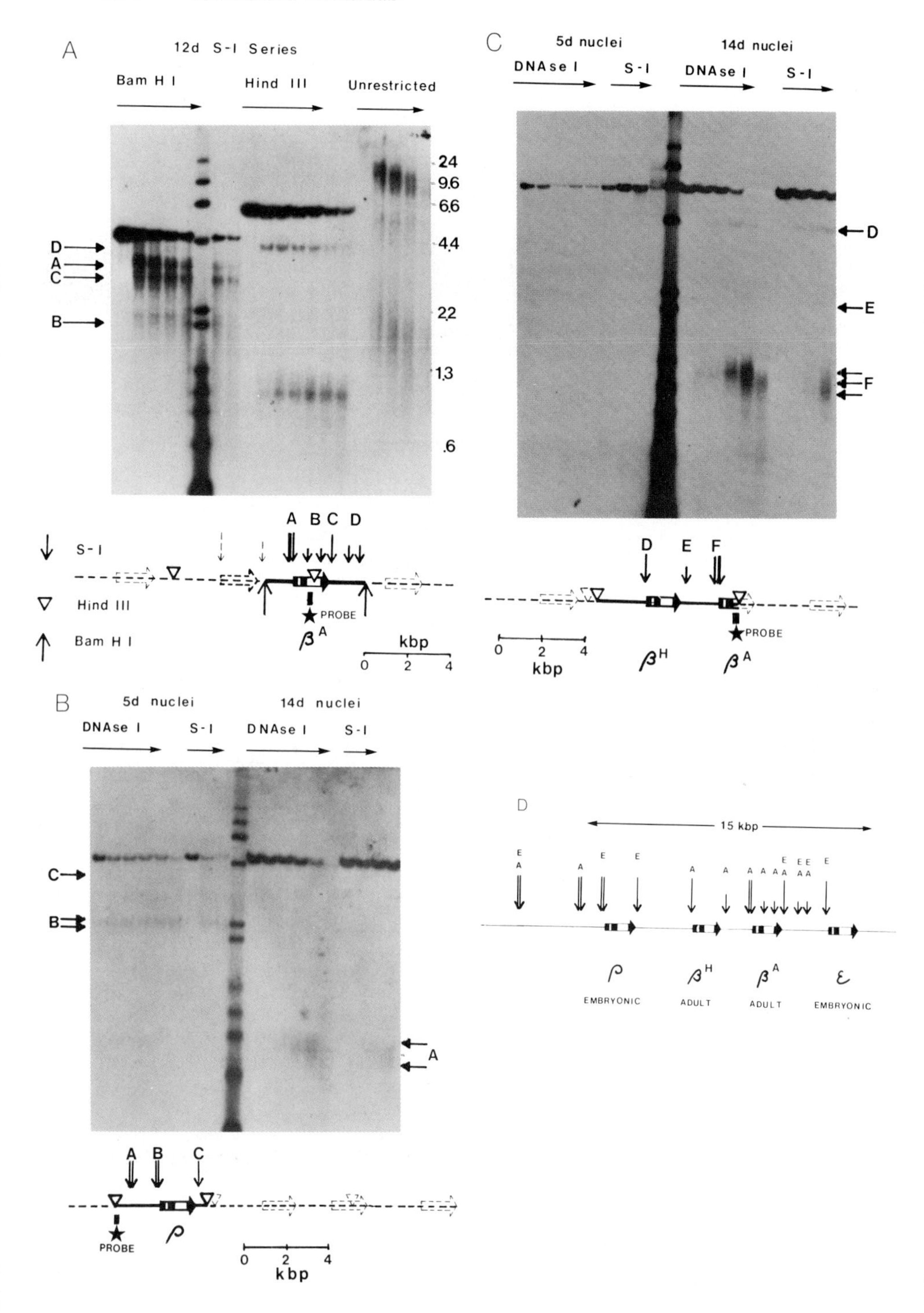
A
12d S-I Series
Bam H I
Hind III
Unrestricted
24
9.6
6.6
4.4
2.2
1.3
.6
D
A
C
B
A B C D
S-I
Hind III
Bam H I
PROBE
kbp
0 2 4
B
5d nuclei
14d nuclei
DNAse I
S-I
DNAse I
S-I
C
B
A
A B C
PROBE
0 2 4
kbp
C
5d nuclei
14d nuclei
DNAse I
S-I
DNAse I
S-I
D
E
F
D E F
PROBE
0 2 4
kbp
D
15 kbp
EMBRYONIC
ADULT
ADULT
EMBRYONIC

low pH used in the S1 buffer is not preferentially depurinating the DNA, since incubation with buffer alone followed by DNA purification, restriction digestion and alkaline hydrolysis fails to detect induced subbands when assayed on denaturing gels (not shown).

While the above results make it very likely that S1 nuclease is the activity observed in these experiments, it is possible that the activity is actually not the one commonly associated with single-stranded DNA cutting. For example, it could be argued that in chromatin, only some sites (for example those hypersensitive to DNAase I) are accessible to S1. These sites would represent only a fraction of the total chromatin (say 0.1%) seen by the enzyme, so that at our usual enzyme concentrations of 5 units/μg total chromatin, the meaningful ratio of S1 to potential sites is actually 10^3 fold greater, and as a result the observed activity is really an associated double-stranded cutting activity that is a result of an effectively high level of enzyme to substrate. To test this possibility, 100 μg double-stranded nucleosome monomer DNA was added to a reaction containing 100 μg nuclei and 500 units S1 in order to increase the level of potential substrate. The results show that the addition of histone-free competing double-stranded DNA does not inhibit the reaction (Fig. 3, lane e), suggesting that the observed S1 activity is, in fact, a single-strand-specific activity. Moreover, under identical, parallel conditions of moderate enzyme excess, addition of an equal amount of single-stranded monomer DNA inhibits the S1-specific cutting by about 50%. (Compare the ratio of parent band to subband in Figure 3, lane g, to that in Figure 3, lane e.)

Cutting by S1 is Cell-Type-Specific

We have assayed all of the prominent DNAase I–hypersensitive sites in the α- and β-globin domain from chick red cells for their S1 sensitivity. All

Fig. 4. Switching of S1-Sensitive Regions in Embryonic (5 day) Versus Adult (14 day) Red Cell Nuclei. (A) Nuclei from adult definitive-line red cells from 12 day embryos were digested with increasing concentrations of S1 nuclease (0, 12, 25, 50 and 100 units/μg for Bam HI; 0, 3, 6, 12, 25, 50 and 100 units/μg for Hind III). Map: (★) Probe. The subbands are correlated with the position of the hypersensitive site.(↓↓) Approximate position of the S1 cuts. Markers are λ Hind III and φX Hae III fragments. (B) Nuclei from 5 day or 14 day red cells were digested with increasing concentrations of DNAase I(0,0.001, 0.003, 0.01, 0.03 and 0.1 μg/ml for 5 day cells; 0, 0.05, 0.1, 0.25, 0.5 and 0.75 μg/ml for 14 day cell) or S1 nuclease (0, 7.5 and 15 units/μg for 5 day cells or 10, 15 and 50 units/μg for 14 day cells). DNA was purified and digested with Hind III and probed with the fragment indicated by the star in the map below. S1 cuts (A, B and C) are also indicated. Markers are λ Hind III and φX Hae III fragments. (C) As in (B), except the probe is a β^A globin intervening sequence fragment. (D) Summary of known S1- or DNAase I-hypersensitive sites. E: embryonic cuts observed in 5 day nuclei. A: adult cuts from definitive cells from 12–14 day nuclei. We have not probed the region 3′ to β^ϵ. The activities of S1 given for the samples shown in this figure are as determined by Bethesda Research Laboratories. For this batch of S1, they are two to five fold lower than our estimates. For other figures, our estimates correspond to those of Bethesda Research Laboratories.

such sites are sensitive to S1. Some of the results for the β locus are shown in Figure 4A and summarized in Figure 4D. We have not observed sites that are sensitive to S1 but not hypersensitive to DNAase I.

Many DNAase I–hypersensitive sites are different in the primitive line of red cells (from 5 day embryos) synthesizing embryonic hemoglobin than in the definitive line of red cells (from 14 day embryos) synthesizing adult hemoglobin [Stalder et al., 1980; Weintraub et al., 1981]. Similarly, we have found that S1-sensitive sites also "switch" when the red cell lineage switches to adult hemoglobin production (Figs. 4B and 4C).

S1-sensitive sites are not associated with inactive genes present in red cell nuclei (for example, ovalbumin and *ev*-1; see above). We digested brain nuclei with S1 and probed with a variety of globin probes, and we found no tissue-specific hypersensitive sites present in globin chromatin derived from brain cells (Fig. 1A, lane e). As a positive control for this experiment, a non-tissue-specific DNAase I–hypersensitive site that maps approximately 7.5 kb 5′ to α^D in all cell types tested [Weintraub et al., 1981] is also sensitive to S1 in brain nuclei (not shown). Thus the S1-sensitive sites that are tissue-specific are also tissue-specific with respect to DNAase I hypersensitivity.

Similar S1-Sensitive Sites Are Present in Supercoiled DNA

The S1 sensitivity of unique sites in chromosomal DNA could in theory be owing to local regions of DNA denaturation or to the formation of "stem-loop" or "hairpin" structures. Recently, Lilley [1980, 1981] and Panayotatos and Wells [1981] have shown that pure DNA, when supercoiled, will relieve some of the supercoil-induced strain either by forming hairpin structures at potential inverted repeat sequences or, alternatively, by locally unwinding specific DNA sequences. Such structures can be assayed by their sensitivity to double-stranded cutting by S1 nuclease in their supercoiled, but not in their relaxed, form. As this might be generally true of any conformation that involves unwinding of the DNA (for example, in chromatin) we therefore decided to test whether the chromosomal regions assayed as S1-sensitive in chromatin are also S1-sensitive in the context of underwound supercoiled recombinant plasmid DNA. Most of the β-globin-hypersensitive sites (2–6 kb inserts) were subcloned into pBR322 or derivatives of pBR322. Supercoiled recombinants were treated with S1 nuclease, producing a mixture of relaxed and linear plasmid molecules. These molecules were digested with the appropriate restriction enzyme (or enzymes), the resulting DNA was hybridized to an end-labeling probe [Nedospasov and Georgiev, 1980; Wu, 1980] and the approximate position of the double-stranded S1 cut was determined by the size of the corresponding subfragments (Fig. 5). For the tested

S1-sensitive sites in the β-globin chromatin domain (α sites have not yet been examined), all have corresponding sites that are sensitive in supercoiled, but not relaxed, plasmids. The results of this analysis and some of the data are shown in Figure 5 and discussed in more detail below. Some sites that are not known to be S1-sensitive in chromatin are nevertheless S1-sensitive in supercoiled plasmids (for example, see Fig. 5a). For technical reasons, the correspondence between S1-sensitive sites in plasmids and S1-sensitive sites in chromatin is probably $\pm$ 50–100 bp at a maximum; this is the approximate resolution of our agarose gel blotting procedure. While it will be reasonably straightforward to eventually obtain the sequence of the S1 cutting sites in recombinant plasmids, our attempts to probe at the nucleotide level the exact site of S1 sensitivity in chromatin have not yet been successful.

Fine Structure of the DNAase I- and S1-Sensitive Region in β^A-Globin Chromatin Compared to the S1-Sensitive Region in β^A-Globin Plasmids

Since McGhee et al. [1981] have analyzed and sequenced the general region of DNAase I sensitivity 250 bp to the 5′ side of β^A, we attempted to map, as precisely as possible, the S1-sensitive region associated with this gene. Some of the results from such an analysis are shown in Figure 6. For most digestions, we tried to use very low levels of enzyme so that we assay only the first cuts introduced into the gene. As a consequence, the observed signal is rather weak. Nevertheless, we think it important that the mapping be based on these very low levels of digestion. There is a major cutting site for DNAase I centered at approximately position −110 (Fig. 6B, lanes 5–7), but the cutting spreads into a region that maps between −70 and −150 bp from the cap site, as first determined by McGhee et al. (1981). Our best estimates, based on ten independent measurements, show that the S1-sensitive site in chromatin is actually a triplet (Fig. 6B, lanes 2 and 3) with apparent cleavage sites occurring at about −10, −110 and −200 bp from the cap site. The S1-sensitive sites in the supercoiled plasmid containing the β^A genomic sequence are also a triplet (Fig. 6C, lanes 1 and 2) with cleavages at approximately −10, −110 and −200; however, the relative intensity of each cleavage is reversed from that observed after S1 treatment of chromatin. Since the relative intensity with which S1 cleaves potential sites in the same supercoiled plasmid is very dependent on buffer conditions (see below), we presume that some aspect of chromosome structure is responsible for altering the relative intensities of the sites in chromatin vis-a-vis the same or similar sites in the plasmid. A nucleotide sequence at the central S1 site (−100) in β^A globin has the potential to form a cloverleaf-type structure. Interestingly, a similar type of structure occurs at the DNAase I–hypersensitive site in polyoma DNA and several independent mutants of polyoma that are capable

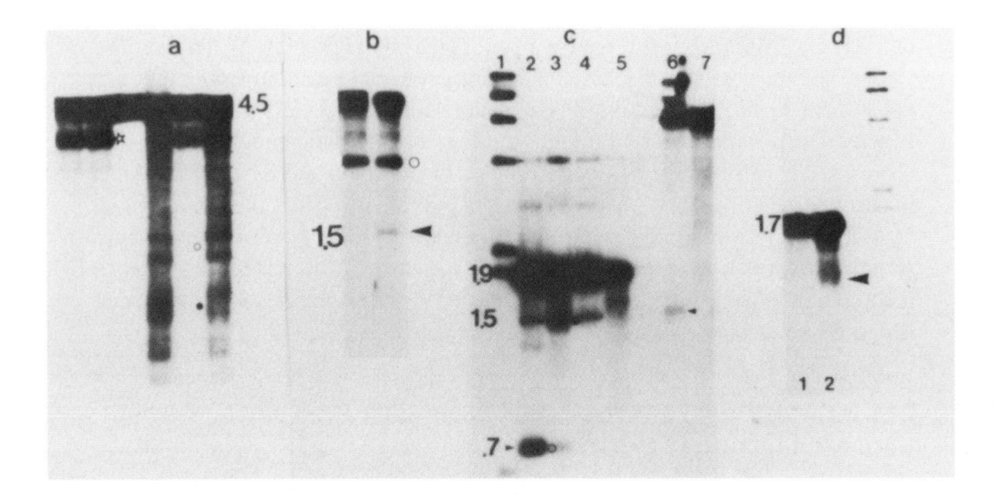

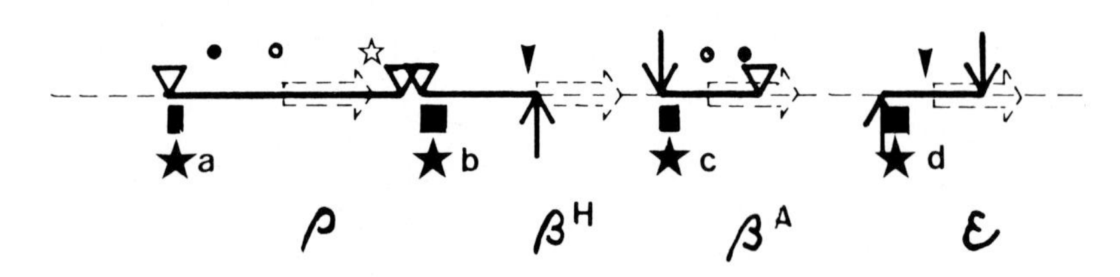

Fig. 5. Sequence-Specific Cutting by S1 Nuclease in Recombinant Plasmids Containing Globin-Related Inserts. Frames (a), (b), (c) and (d) at top correspond to the inserts shown in the β-globin restriction map below. Regions in map denoted by dark lines were subcloned into pBR322 (or Ylp5, a pBR322 plasmid containing a 1.1 kb yeast Ura3 gene [Stinchcomb et al., 1980] in the case of the [b] insert), and the supercoiled recombinant plasmids were purified and digested in various buffer conditions with S1 nuclease so that the supercoils were converted to linears. The DNA was then digested with the appropriate restriction endonuclease to cleave out the insert (a, c and d) or was linearized with Hind III (b) and then blot-hybridized with an end-labeled hybridization probe. (○, ●, ▲, ☆) Particular subband and corresponding hypersensitive site in the map. (a) After digestion of the supercoiled plasmid with S1, the Hind III insert containing β^ρ was cleaved out with Hind III, and hybridization was with the Hind III-Sac I fragment denoted by the black bar and star in the map. (Lanes 1, 2 and 3) S1 digestion in high salt buffer at 50°, 42° and 37°C; (lane 3) control DNA incubated in the absence of S1; (lanes 4 and 6) digestions in low salt buffer at 42°C. One to two units of S1 nuclease per microgram of plasmid was used for all digestions. The site at the 5′ side of β^ρ is clearly a doublet. It also has a doublet character after DNAase 1 digestion of chromatin (Stalder et al., 1980; Figure 4). (b) S1-digested (high salt buffer at 23°C on the left; high salt buffer at 42°C on the right) plasmid DNA was digested with Hind III to linearize the plasmid

of growing in embryonal carcinoma cells have inserted two independent bases that can hydrogen-bond and stabilize this type of structure [Katinka et al., 1981]. In addition, several investigators have recently suggested that cloverleaf DNA structures may be recognized by transcription factors for tRNA synthesis [Koski et al., 1980; Sharp et al., 1981].

Effect of Digestion Conditions Upon the Site at Which S1 Cuts in Plasmids

The relationship between those β-globin sites cut in supercoiled plasmid DNA and the similar sites cut in red cell chromatin is complicated by the fact that in some plasmids carrying particular β-globin regions, additional sites, often visible at higher exposures, are also cleaved by S1. In theory, any one plasmid molecule becomes cut only once by S1, since it relaxes before a second cut will be introduced [Lilley, 1980; Panayotatos and Wells, 1981]. Occasionally, however, we have noticed that two cuts can be introduced into the same molecule, but this is not a general finding. The set of alternative (and mutually exclusive) cutting sites can be very complex and is dependent on conditions of temperature and salt in our experience (pH, divalent cations and so forth have not yet been tested). For example, even in pBR322, where others have observed only one or two major cutting sites, at lower salt or lower temperature and by using an end-labeling hybridization probe, we can observe over 25 specific cuts in pBR322 (Fig. 7). Similar flexibility in cutting is seen in the region of the β domain around β^{ρ} (Fig. 5A). In contrast, the region around β^{A} shows only two predominant clusters of sites that "flip-

---

DNA (in this case the insert was not cut out). A 1.5 kb subband (arrow) appears after S1 digestion (right) at 42°C. The position of this cleavage is indicated by the same arrow in the (b) insert as shown in the restriction map below. A second subband (○) corresponds to a cleavage in the Ura3 gene of the YIp5 vector. (c) The (c) plasmid containing the 5′ side of β^{A} was converted to linears with S1, and after restriction digestion with Hind III and Eco RI and gel separation, it was hybridized with a 5′ probe, shown below by the black square and star. (Lane 1) λ Hind III markers; (lane 2) S1 nuclease digestion in high salt buffer at 42°C; (lane 3) S1 nuclease in high salt buffer at 37°C; (lane 4) S1 nuclease in high salt buffer at 23°C; (lane 5) as in (4), but in low salt buffer; (lane 6) as in (4) but restriction digestion was with Eco RI alone; (lane 7) as in (6), but the plasmid was first linearized with Eco RI *before* S1 digestion. (d) The plasmid containing the 5′ side of β^{ϵ} was treated in the absence (left) and presence (right) of S1 nuclease in high salt buffer at 42°C. The subband at 0.9 kb maps to the 5′ side of β^{ϵ}. (See dark arrow in the summary map below.) (Bottom) Summary map of the β locus showing positions of those S1-sensitive sites in supercoiled plasmids that correspond to S1-sensitive or DNAase I-hypersensitive sites observed in either adult or embryonic red cell nuclei (see Fig. 4). The sites at the 3′ side of β^{A} and β^{H} have not yet been examined in supercoiled plasmids.

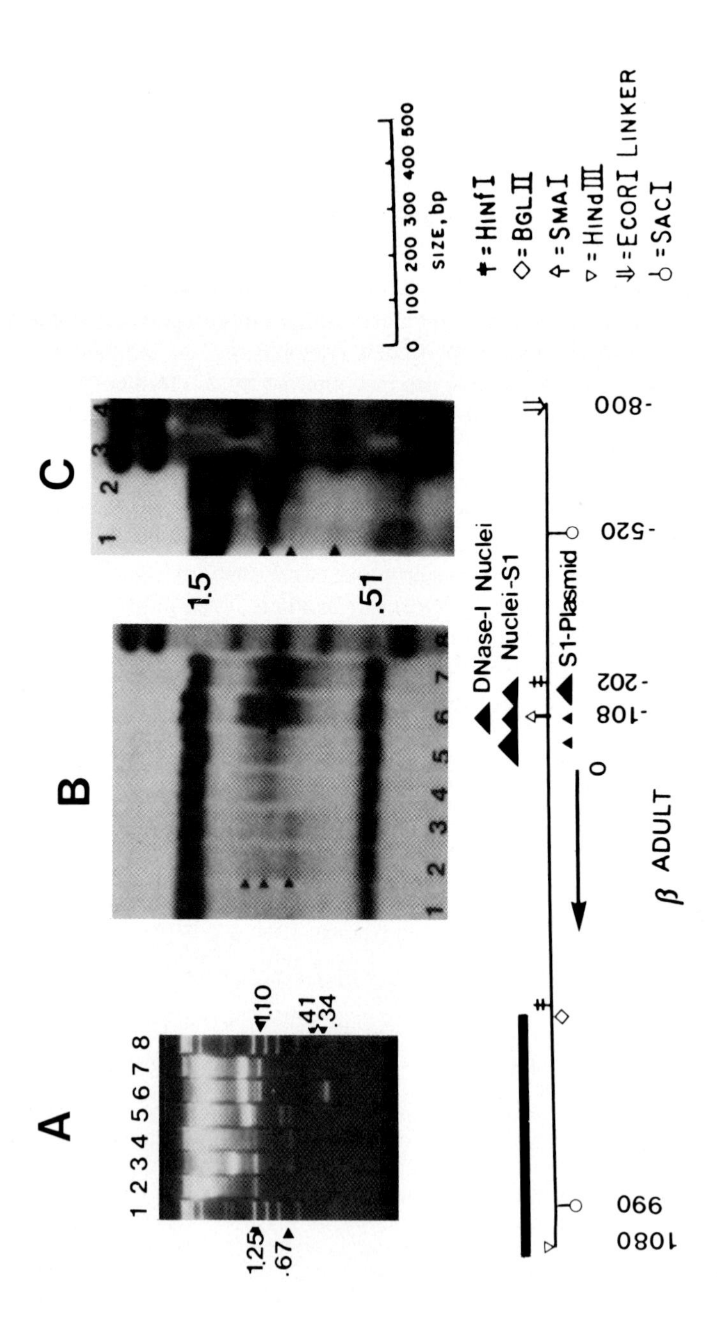

A
1 2 3 4 5 6 7 8
1.25
.67
1.10
.41
.34
B
1 2 3 4 5 6 7 8
1.5
.51
C
1 2 3 4
DNase-I Nuclei
Nuclei-S1
S1-Plasmid
β ADULT
1080
990
0
-108
-202
-520
-800
0 100 200 300 400 500
SIZE, bp
= HINfI
= BGLII
= SMAI
= HINdIII
= ECORI LINKER
= SACI

flop" depending on the digestion conditions (Fig. 5C). The possible biological importance of the "flip-flop" property and its dependence on the environment will be discussed below.

When supercoiled plasmids are digested with S1, an initial single-stranded cut at the sensitive site results in plasmid DNA relaxation. This is a fast process. Cleavage of the second strand across from the initial nick occurs much less rapidly. Thus, at early times in the digestion process, single-stranded breaks are in vast excess to double-stranded breaks [Lilley, 1980; Panayotatos and Wells, 1981]. In contrast, a similar analysis of single- versus double-stranded breaks at low levels of S1 digestion of chromatin has failed to show a marked predominance of single-stranded cuts (not shown). We therefore think that both strands of the presumed single-stranded region recognized by S1 in chromatin are being stabilized (or held open) by chromatin components, possibly by proteins that bind preferentially to single-stranded DNA.

In suggesting a possible relationship between S1-sensitive sites in chromatin and corresponding sites in supercoiled DNA molecules, our data at this point are mainly correlative, and more information is required. While additional S1 sites are present in plasmids, all of the known S1-sensitive

---

Fig. 6. DNAase I– and S1-Hypersensitive Sites at the 5′ Side of β^A Globin Chromatin in Relation to the S1 Cleavage Site at the 5′ Side of β^A in Recombinant Supercoiled β^A Plasmids. (A) The HR16 β^A recombinant plasmid (see map below) was digested with S1 and then with the indicated restriction enzyme. (Lanes 2 and 8) λ Hind III and ϕX 174 Hae III markers; (lane 2) Hind III; (lane 3) Hind III and Eco RI; (lane 4) Eco RI; (lane 5) Eco RI and Bgl II; (lane 6) Eco RI Sac I; (lane 7) Sac I. The indicated sizes represent the length (in kb) of the subfragments produced as a result of S1 digestion. (B) DNA from control nuclei (lane 1) or nuclei cut with S1 (lanes 2 and 3) or DNAase I (lanes 4-7) was purified and digested with Sac I and Hind III and hybridized to the Hind III-Bgl II fragment shown in the map below. (Lane 8) Markers. (C) (Lanes 1 and 2) Two samples of S1-digested plasmid (HR16) cut with Hind III and Sac I and hybridized to the Hind III-Bgl II fragment shown by the thick bar in the restriction map. (Lanes 3 and 4) Markers. (Bottom) The map shows the cleavage sites for DNAase I in nuclei, for S1 in nuclei and for S1 in plasmids. The approximate level of digestion is suggested by the height of the triangles, which also mark the approximate positions (and breadth) of the respective cleavages. The summary represents a "consensus" from ten independent measurements for each cleavage with the use of a variety of probes and restriction enzymes. The range in measured cutting sites in any one comparison was quite broad (~50 bp), even for exact repetitions of the same experiment. By using so many measurements, we presume that the inherent variability in our methods are averaged out and that the indicated sites can probably be considered to be accurate within ~ 10–15 bp. In (B) and (C), most samples are derived from independent digestions, giving an indication of the variability observed from experiment to experiment.

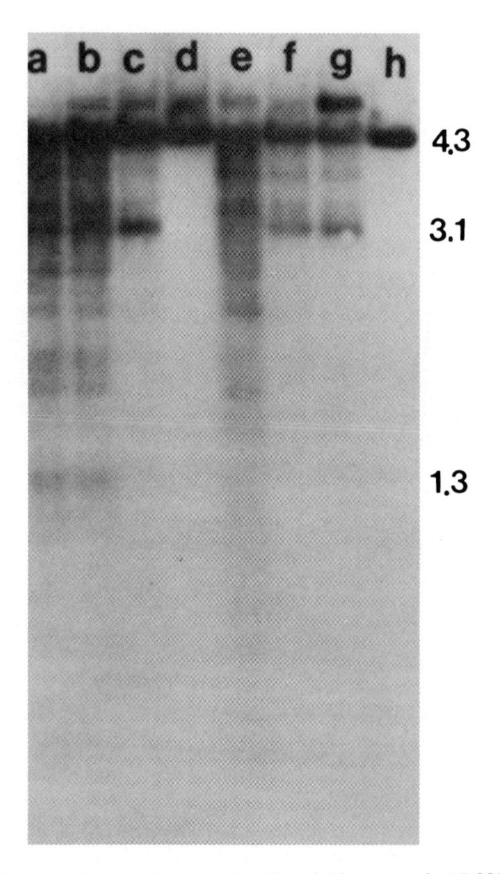

Fig. 7. Effect of Salt and Temperature on the Site of Cleavage of pBR322 by S1 Nuclease. Supercoiled pBR322 was cut with S1 nuclease (1 unit/μg). (Lane a) High salt buffer at 42°C; (lane b) high salt buffer at 37°C (lane c) high salt buffer at 23°C; (lane d) no S1; (lane e) low salt buffer at 37°C; (lane f) low salt buffer at 23°C; (lane g) low salt buffer at 14°C; (lane h) no S1. After S1 cleavage, the DNA was purified and digested with Hind III and probed with a 350 bp Hind III-Bam Hl end-labeling probe. Several of the sites cut by S1 in supercoiled pBR322 are also cut at ~ 1/10–1/100 the frequency in linear pBR322 DNA in the appropriate conditions. The basis for this low frequency cutting is currently under investigation.

regions in chromatin are associated with corresponding S1-sensitive regions in plasmids, given a resolution of about 50–100 bp for most of the sites analyzed thus far. Nevertheless, it is clear that some sites that are not yet known to be S1-sensitive in chromatin are S1-sensitive in plasmids. Whether this structural potential is ever used during development remains to be seen. We should also emphasize that the particular structural feature (or features) of supercoiled DNA recognized by S1 is not at all clear for all S1 cuts. For example, in the case of pBR322, many of the S1 sites observed (Fig. 7) are not AT-rich, nor are they associated with DNA sequences that can form reasonably stable hairpin structures as determined by computer analysis of the known sequences. Whether they can be explained by the DNA slippage mechanism proposed by Hentschel [1982] remains to be determined.

DNAase I- and S1-Hypersensitive Sites in Adenovirus Chromatin

We have analyzed the DNAase I–hypersensitive region and the S1-hypersensitive region in chromatin from two rat cell lines transformed by a single copy of integrated adenovirus (HT14a and T2). We have begun this type of analysis with adenovirus because the major late promoter (L1) is known to contain a very prominent inverted DNA repeat [Ziff and Evans, 1978; Zain et al., 1979]. Also, we have chosen to use transformed cells, since the chromosomal organization of adenovirus chromatin during lytic infection is still not clear, and the presence of multiple viral genomes complicates the analysis. Some of the results from one line (T2) are shown in Figure 8A, which indicates that the region around map position 16.6 in these lines is hypersensitive to both DNAase I and S1 nuclease. This position corresponds to a region containing the promotor for late transcription (L1) as well as the promotor for the IVa_2 gene product (Fig. 8C). The two promotors are separated by approximately 200 bp, and it is impossible at this level of resolution to know whether the hypersensitive region is associated with either or both of these transcription units. At higher levels of S1 or DNAase I, a second subfragment is observed (Fig. 8A, lanes f and e) that maps to position 10 or the promotor for protein IX. Using Sal-digested DNA and hybridizing with the Bal E fragment, we have also mapped S1 amd DNAase I–hypersensitive regions to the E1A and E1B promotor regions (not shown). All of the S1 sites in T2 cell nuclei are also S1-sensitive in HT14a cell nuclei (not shown).

To test whether the region around the late promotor was sensitive to S1 in supercoiled plasmids, we digested plasmids containing either the Bal E fragment or the Sma F fragment (from adenovirus type 2) with S1 nuclease and then cut out the respective inserts with the appropriate restriction enzyme (Fig. 8B). Both of these inserts contain the late promotor, and both yield subfragments that map to exactly position 16.6 (Fig. 8B), where DNA sequence analysis has demonstrated a prominent potential for hairpin formation [Zain et al., 1979]. The IVa_2 promotor is also present in the Bal E and Sma F inserts; however, corresponding S1-sensitive sites are not observed, possibly because the late promotor at 16.6 is so dominant in forming an S1-sensitive site. We have not tested the other adenovirus promotors for their S1 sensitivity in supercoiled plasmids.

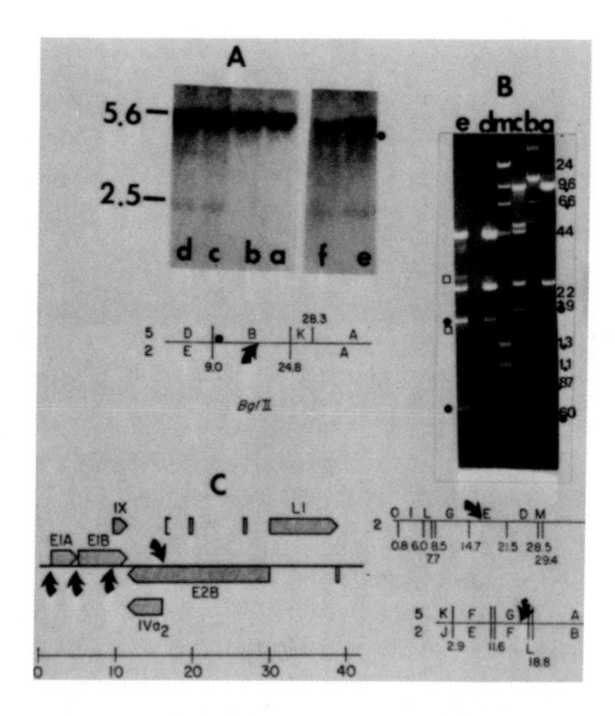

Fig. 8. S1 Sensitivity of the Adenovirus Promotors in Transformed Cells. Data shown comes from T2 cells, a line of rat cells transformed by a single copy of the entire adenovirus genome. Identical results have been obtained in rat cells (HT14a) containing the left-hand 40% of the adenovirus genome. (A) (Lane a) Control nuclei in S1 buffer; (lane b) control nuclei in DNAase digestion buffer; (lanes c and e) nuclei with S1 nuclease; (lanes d and f) nuclei digested with DNAase I. Purified DNA was digested with Bgl II and hybridized with the Bal I E fragment (map position 14.7 to 21.5). The 2.5 kb subband maps to the position of the arrow in the map below, as also was determined by a Sal I digestion and hybridization to Bal I E (not shown). A second subband (●) maps to about map position 10, also confirmed by Sal 1 digestion. (B) Recombinant pBR322 plasmids containing either the Sma I adeno 2F fragment (lanes a and c) or the Bal I E fragment (lanes d and e) were cleaved with S1, and the inserts were cut out with either Sma I or Bal I. m: λ Hind III markers and ϕX Hae III markers. (●) Subbands from the inserts; (□) subbands from pBR. The two digests localize the S1 site to about position 16.6 on the adeno genome (arrow in map below). Note the dominance of the adeno cleavage as compared to the pBR322 cleavage, especially in the Bal E-containing plasmid shown in lane e. (C) Summary of adeno associated DNAase I- and S1-hypersensitive sites in T2 nuclei. The map positions are given below, and the transcripts are given above (Lewis and Mathews, 1981). Arrows: positions of the S1 and DNAase I sites. The E1A and E sites were derived from Sal I-digested DNA hybridized to the Bal E fragment. Sal I introduces its first cut in the left-hand end of adenovirus at position 24.7. L1 refers to the major late promotor (16.6), which produces a transcript that is spliced extensively. The scale is in map units—each is approximately 365 bp.

DISCUSSION

S1 Sensitivity of Specific DNA Sequences in Chromatin and Supercoiled Plasmids

Our results show that specific regions in the active α- and β-globin chromatin domain in avian red cells are sensitive to digestion by S1 nuclease. These regions are located within approximately 50 bp from the known DNAase I–hypersensitive sites; however, they are clearly distinguishable. As with DNAase I–hypersensitive sites, the globin-associated S1 regions are not seen in nonexpressing brain nuclei; moreover, during globin gene switching, both the DNAase I- and S1–hypersensitive regions also switch in a way that reflects the altered pattern of globin gene expression. Potentially expressible genes such as ovalbumin or *ev*-1 do not display hypersensitive sites in erythroid cells where they are not expressed. By using specific inhibitors of S1 nuclease, it is clear that the specific cutting is coming from S1 and not from a contaminating activity; moreover, since under conditions of moderate enzyme excess the cutting is partially inhibited by single-stranded, but not double-stranded DNA, we also assume that it is the single-strand specificity of the enzyme that is being observed. When mapped more precisely (that is, at the 5′ side of the β^A globin [Fig. 6]), the S1-hypersensitive region appears as a triplet (with cuts at -10, -110 and -200 bp) extending over 150–200 bp. For β^A globin, the S1-sensitive region includes the DNAase I–hypersensitive site (centered at approximately -110), though it extends somewhat more 3′ toward the CAP site and somewhat more 5′ as well. Whether this "morphology" applies to other sites remains to be determined.

The functional and causal relationships between DNAase I–hypersensitive sites and S1 sites is still not clear. The two types of sites are clearly related, since they are very near each other, and since they are coordinately present in differentially regulated genes. Nevertheless, the "morphology" of the respective cutting patterns (see for example Fig. 6) is clearly different. Possibly, the two assays monitor different features of the same chromosomal structure. If true, then the structure would be surprisingly large, since the cuts can extend over a distance of some 200–250 bp (Fig. 6; [McGhee et al., 1981]). More likely is the possibility that the DNA is actually folded or coiled in this region and that the actual structure has dimensions not significantly different from a nucleosome. It is possible that the observed S1-sensitive sites are a reflection of the presence of initiated RNA polymerase molecules present on these sequences; however, we do not think this is likely, since these sites are observed in certain red cell precursors transformed by avian erythroblastosis virus (not shown), and, using a run-off nuclear transcription assay, we have shown that no initiated RNA polymerases are active along these genes in these precursors [Weintraub et al., 1982]. For the case of the adenovirus late promotor in transformed cells, there is a clear correlation between cutting by DNAase I and cutting by S1. Moreover, the sequence of this region shows a distinct potential to form a very stable hairpin structure, and when tested, the supercoiled plasmid containing these sequences is also S1-sensitive at or very near these sites.

To try to understand the chromosomal structure that gives rise to the observed S1 specificity, we decided to test whether some of the information might reside in the primary DNA sequence. Pure chromosomal DNA did not show S1-hypersensitive sites, nor did relaxed recombinant plasmids that contained these sequences. However, when purified in their supercoiled form, such plasmids display S1-sensitive sites that correspond to similar sites that are S1-sensitive in chromatin. This correspondence suggests that there may be a general relationship between DNA sequence and its ability to form an S1-hypersensitive site when assembled into the proper structure—either a strained DNA molecule or a particular chromatin environment. At present, however, there are a number of factors that complicate such an interpretation. First, we have not shown in any single case that the S1 site in chromatin is exactly the same at the nucleotide level as the S1 site in supercoiled plasmids. This will eventually require a different technology for mapping these sites in chromatin, and it will be complicated by the fact that the S1 sites in chromatin and in plasmids can be very complex. Second, in several cases, the known DNA sequence does not suggest particularly stable DNA hairpins or AT-rich regions (for example, at the 5′ side of β^A globin or at many of the sites in pBR322 [Fig. 7])—DNA structures known to be S1-sensitive in supercoiled molecules [Lilley, 1980, 1981; Panayotatos and Wells, 1981]. Third, in many recombinant plasmids the insert contains a number of additional S1-sensitive sites that can actually become major cutting sites under the appropriate conditions of salt or temperature (Figs. 5 and 7). Thus, while there is a clear suggestion of a relationship between potential DNA secondary structure in strained plasmid DNA and S1 cutting sites in chromatin, more information must be obtained. Ultimately, the correspondence will require additional examples as well as a better physical understanding of the alternative conformations of DNA and how these may relate to DNA sequence and the physical environment of these molecules. Since most of our experiments have relied heavily if not exclusively on the use of S1 nuclease, we have recently been able to verify, by using a chemical reagent that specifically attacks single-stranded DNA, that many of the sites sensitive to S1 (in chromatin and in plasmids) are also reactive with the chemical reagent [T. Kohwi and H. Weintraub, unpublished results].

Recently, Sims and Benz [1980] have shown that dnaG protein, the primase for G4 replication, binds to an inverted DNA repeat some 110 bp 5′ to the initiation site. This inverted repeat is required for proper initiation;

moreover, the E. coli single-stranded binding protein (DBP) is also required, both for initiation and proper binding to dnaG. These results form an interesting parallel to the emerging picture of transcription initiation described here, since in both cases a presumed alteration of DNA structure is read, at a distance, from the start site, and this structure is dependent on other chromosomal proteins—DBP for the dnaG protein and (presumably) histones and other chromsomal proteins for eucaryotic genes.

S1-Sensitive Regions and Gene Control

While it is not obvious how or if S1- or DNAase I–hypersensitive sites are related to potential gene activation, the correlations are striking, and it would be surprising if the structure that underlies these sensitivities were not one of perhaps several essential requirements for activation. The observation that the DNAase I–hypersensitive sites appear before gene expression [Weintraub et al., 1982] supports this view, although the mechanism by which the assembly of these sites might eventually dictate subsequent gene activation is not at all apparent. It is tempting to think that the altered DNA structure at these S1-hypersensitive sites represents a general feature of DNA recognized by a specific class of regulatory components. This would allow these sites to be easily distinguished from bulk B-form DNA, and for any given protein, it would increase, perhaps by a factor of 10^3 (assuming 0.1% of total chromosomal DNA is in this altered structure), the ratio of a given specific binding site to potential nonspecific binding sites in the nucleus. Thus certain classes of regulatory proteins may be opaque to B-DNA and might only "sense" the structural regions that are S1-hypersensitive. The repertoire of potential S1-sensitive DNA structures could be quite large—denaturable AT-rich regions, simple hairpins, cloverleaf structures or other, as yet unrecognized DNA conformations that involve disruptions of the normal double helix. The theoretical arguments for favoring recognition of altered DNA conformation have been discussed at length by Crick [1971]. As mentioned in the text, the site at the 5′ side of β^A globin and the site near the polyoma early promotor can both theoretically form cloverleaf structures. One wonders whether proteins similar to those that recognize tRNA (for example, synthetases) may also be involved in recognition of these types of hypothetical cloverleaf DNA structures in chromatin. A particular example is the case of the 5S genes and one of its DNA binding regulatory proteins, which is also known to bind 5S RNA [Engelke et al., 1980; Pelham and Brown, 1980].

Formation and Stabilization of S1-Sensitive Regions

How are S1-sensitive regions formed and how are they stabilized? In procaryotes, it is likely that DNA supercoiling is used to control promotor strength and differential expression from a number of genes [see review by

Smith, 1981]. In yeast, K. Nasmythe (personal communication) has presented data to suggest that a similar type of control governs expression from the inactive mating-type locus. The observation that S1-sensitive regions in chromatin correspond to S1-sensitive regions in supercoiled plasmids is, thus far, consistent with this view. The S1 sensitivity of these regions in supercoiled DNA suggests that in some way potential information in the primary DNA sequence is used to induce or stabilize an altered DNA secondary structure. This information might be elicited by DNA supercoiling, by specific proteins or RNAs that bind to these sequences, by general (or specific) proteins that bind to the class of DNA sequences capable of forming S1-sensitive sites or by combinations of the above.

Whether chromatin is under torsional constraint in nuclei is not really known. In most cells, there is a large amount of topoisomerase I, which would relax this strain; however, it is not known if this enzyme is compartmentalized, and a similar excess of this enzyme is found in E. coli, where it is clear that the enzyme activity must be controlled in the cell if it is to be used for regulation. As yet, no DNA gyrase has been described in higher cells, yet the organization of eucaryotic chromatin into constrained loops [Benyajati and Worcel, 1976; Laemmli et al., 1977] suggests that if such an enzyme were present a suitable constrained substrate already exists. An assay for DNA supercoiling based on the inhibition of psoralen binding after in vivo nicking has been presented by Sinden et al. [1981], who have concluded that for the bulk of chromatin, most domains are in fact relaxed. However, as pointed out by the authors, these measurements are of rather low resolution and do not exclude the possibility that a small number of loops (for example, those involved in differential gene expression) are under strain, nor do they exclude the possibility that at some time during development or chromosome replication strain is, in fact, transiently introduced into the system and then stabilized by proteins.

We have made several attempts to determine if the S1 sensitivity of the specific nuclear globin sequences described here results from superhelical strain present in the nuclei as they are isolated. The fact that ethidium bromide does not inhibit S1 cutting (Fig. 3, lane f) suggests that these sites are not under stress. Additional experiments support this interpretation. Thus attempts to relieve the strain (by pretreating nuclei with restriction enzymes or purified nicking-closing enzyme) have not led to a decrease in S1 cutting. Similarly, as discussed in the text, the absence of an excess of single-strand versus double-strand cutting at low levels of S1 also suggests that these sites are not under strain. Finally, we have monitored the cutting frequency of adjacent S1 sites on the same stretch of chromatin 5′ to β^{ρ} (not shown) and find a frequency that does not suggest that the cutting of one site inhibits the cutting of adjacent sites. Thus, while these results are all negative, they do suggest that in isolated nuclei, the S1-sensitive sites are not under the same type of superhelical tension that is typified by supercoiled plasmids. Nevertheless, we should point out the possibility that this type of strain might have been important in establishing these sites during development and that some additional mechanism is responsible for their maintenance. Alternatively, it is possible that the supercoiled domain is very small and resistant to the effects of the relaxation probes we have used. If, for example, the histones were eluted from just one nucleosome in a stable 300 Å chromosome fiber, then two superhelical turns would be released over a stretch of DNA 200 bp in length.

Finally, Behe and Felsenfeld [1981] have shown that methylation of the C residue stabilizes the transition of poly(dCdG) to Z DNA at physiological ionic conditions. In addition, a great deal of evidence has implicated methylated DNA with gene inactivation (Razin and Riggs, 1980). If the S1-sensitive sites that we observe are indeed formed as a result of an induced DNA strain, then the enhanced methylation of inactive chromatin would tend to favor the Z DNA conformation and as a result relieve some of the induced strain, which would then inhibit the formation of S1-sensitive sites. Similarly, the loss of methylation-induced Z structure would enhance local unwinding in a constrained DNA domain and favor the formation of an S1-sensitive site.

The Emergence of S1-Sensitive Sites During Early Development

The position of a given S1 cutting site in a supercoiled plasmid is extremely dependent on the physical and chemical environment (Fig. 7). Sites that are dominant at one salt concentration or temperature are weak or absent at slightly different salt concentrations. Although not tested, we presume that pH, divalent cations, polyamines and specific (and nonspecific) DNA-binding proteins will also be variables in determining which sites are cut. The capacity of supercoiled DNA to respond to the quantitative level of specific ions may be useful for establishing those differences between daughter cells that emerge during early development. Since the S1-sensitive sites are often dramatically dependent on, for example, Na^+ concentration, it is not difficult to imagine that a constrained DNA sequence in a nucleus in one part of an embryonic gradient (for example, a Na^+ gradient) displays a different S1-sensitive site than the same DNA sequence in a nucleus located elsewhere in the gradient. This particular view places much of the burden of generating differences between daughter cells upon the DNA molecule itself and DNA chemistry—DNA secondary and tertiary structure responds differentially to different levels of a very general signal associated with an embryonic gradient. Clearly, one problem with this viewpoint is that while it may adequately explain how differences might be generated, it is unlikely that the gradient of different ionic conditions required to generate this diversity is preserved at late times in development and in different tissues of the adult. Thus a second mechanism would be required for stably propagating the differences initially elicited by such a nonspecific gradient. As discussed separately [M. Groudine and H. Weintraub, manuscript in preparation], S1-sensitive sites seem to have the inherent property of self-propagation once they are assembled.

EXPERIMENTAL PROCEDURES

Cell, Nuclei and Digestions

Erythrocytes were isolated from the circulating blood of 14–16 day old chicken embryos (White Leghorn) by vein puncture. Chick brain was prepared by homogenizing brains from 10–20 embryos in standard citrate saline (0.15 M NaCl; 0.015 M sodium acetate; and 10 mM Tris-HCl, pH 7.4) with a 15 ml loose-fitting Dounce homogenizer. MSB cells (a line of chicken leukemia cells transformed by Marek's disease virus) were grown in RPMI 1640 medium (Gibco) supplemented with 10% fetal calf serum. T2 and HT14a cells were grown in Dulbecco's modified Eagle's medium (Gibco) with 10% fetal calf serum. Nuclei were isolated by suspension in reticulocyte standard buffer (0.01 M Tris-HCl, pH 7.4; 0.01 M NaCl; and 3 mM $MgCl_2$) containing 0.5% Nonidet-P40. The nuclei were washed several times in reticulocyte standard buffer and resuspended in it for DNAase I digestion or in high salt buffer (3 mM $ZnCl_2$; 30 mM sodium acetate, pH 4.5; 300 mM NaCl; and 0.2 mM EDTA) or low salt buffer (3 mM $ZnCl_2$; 30 mM sodium acetate, pH 4.5; 30 mM NaCl; and 0.2 mM EDTA) for S1 digestion. Brain nuclei were homogenized by ten strokes of a loose-fitting Dounce homogenizer before resuspension in reticulocyte standard buffer. S1 nuclease was obtained from Bethesda Research Laboratories, although other S1 preparations from other sources gave identical results. The enzyme was generally used at 1 unit/μg pure DNA or 5 units/μg chromatin. Standard conditions for chromatin digestion are in low salt buffer at 37°C for 30 min. Standard conditions for plasmid digestion are in high salt buffer at 37°C for 30 min, although for most experiments, a variety of reaction conditions were employed.

DNA Isolation and Blotting

DNA was prepared from the various preparations by resuspending nuclei in 10 mM EDTA; 10 mM Tris-HCl, pH 7.4; 0.1% SDS; and 0.1 mg/ml proteinase K and by incubating for 3–5 hr at 37°C. The samples were then extracted several times with an equal volume of phenol (equilibrated to pH 7.0 with standard citrate saline) to chloroform to isoamyl alcohol (1:1:1/24) and several times with chloroform to isoamyl alcohol (1:1/24) and precipitated with 21/2 volumes of 95% ethanol. Purified DNA was digested with restriction enzymes according to the manufacturer's recommendation, and the DNA fragments were separated on agarose gels and blot-hybridized according to Southern (1975) as previously described by Weintraub et al. (1981).

SUMMARY

The single-stranded activity of S1-nuclease cleaves globin chromatin in red cell nuclei in specific regions. The cleavages are observed only in tissues in which the globin genes are active, and they "switch" to reflect the switching pattern of globin-gene expression in embryonic and adult red cells. The positions of the S1 cleavages in the β- and α-globin chromatin correspond to the general region of known DNAase I-hypersensitive sites, but can be distinguished in detail. When DNA segments containing these regions are subcloned into pBR322 and the supercoiled molecules are treated with S1, similar sites are cleaved in the purified supercoiled (but not linear) recombinant plasmid DNA. However, the dominant S1 cutting sites are shifted in the plasmid vis-a-vis the chromatin. We believe that some aspect of DNA sequence is translated into an altered DNA structure in chromatin and that it is this altered structure that is recognized by S1 nuclease and possibly by certain chromosomal proteins. Several physical properties reflected in the S1 digestion of supercoiled plasmids suggest a mechanism for generating differences in daughter cells during development.

Original Summary by Harold Weintraub

ACKNOWLEDGMENTS

This work was supported by grants from the NIH. We are indebted to J.D. Engel, J. Dodgson and J. Lewis for several recombinant plasmids and for restriction map information; to R. Wells for the T7 single-strand-specific nuclease; to S.J. Flint for HT14a cells; to Kim Nasmythe, Mark Groudine and Charles Laird for many helpful suggestions and discussions; and to Helen Devitt for typing the manuscript.

The costs of publication of this article were defrayed in part by the payment of page charges. This article must therefore be hereby marked "*advertisement*" in accordance with 18 U.S.C. Section 1734 solely to indicate this fact.

REFERENCES

Behe M, Fesenfeld G (1981): Effects of methylation on a synthetic polynucleotide: the B-Z transition in poly (dG-m^{5}dC)·poly (dG-m^{5}dC). Proc Nat Acad Sci 78:1619–1623.

Benyajati C, Worcel A (1976): Isolation, characterization and structure of the folded interphase genome of Drosophila melanogaster. Cell 9:393–407.

Cremisi C (1981): The appearance of DNAase I hypersensitive sites at the 5′ end of the late SV40 genes is correlated with transcriptional switch. Nucl Acid Res 9:5949–5964.
Crick FHC (1971): General model for the chromosomes of higher organisms. Nature 234:25–27.
Engelke DR, Ng S-Y, Shastry BS, and Roeder RG (1980): Specific interaction of a purified transcription factor with an internal control region of 5S RNA genes. Cell 19:717–728.
Garel A, Axel R (1976): Selective digestion of transcriptionally active ovalbumin genes from oviduct nuclei. Proc Nat Acad Sci USA 73:3966–3977.
Gazit B, Panet A, Cedar H (1980): Reconstitution of a DNAase I sensitive structure on active genes. Proc Nat Acad Sci USA 77:1787–1790.
Groudine M, Weintraub H (1981): Activation of globin genes during chicken development. Cell 24:393–401.
Groudine M, Eisenman R, Weintraub H (1981): Chromatin structure of endogenous retroviral genes and activation by an inhibitor of DNA methylation. Nature 292:311–317.
Hentschel CC (1982): Homocopolymer sequences in the spacer of a sea urchin histone gene repeat are sensitive to S1 nuclease. Nature 295:714–716.
Katinka M, Vasseur M, Montreau N, Yaniv M, Blangy D (1981): Polyoma DNA sequences involved in control of viral gene expression in murine embryonal carcinoma cells. Nature 290:720–722.
Keene MA, Corces V, Lowenhaupt K, Elgin SCR (1981): DNAase I hypersensitive sites in Drosophila chromatin occur at the 5′ ends of regions of transcription. Proc Nat Acad Sci USA 78:143–146.
Koski RA, Clarkson SG, Kurjan J, Hall BD, Smith M (1980): Mutations of the yeast SUP4 $tRNA^{Tyr}$ locus: transcription of the mutant genes in vitro. Cell 22:415–425.
Kuo MT, Mandel JL, Chambon P (1979): DNA methylation: correlation with DNAase I sensitivity of chicken ovalbumin and conalbumin chromatin. Nucl Acid Res 7:2105–2113.
Laemmli UK (1977): Metaphase chromosome structure: the role of nonhistone proteins. Cold Spring Harbor Symp Quant Biol 42:351–360.
Lee F, Mulligan R, Berg P, Ringold G (1981): Glucocorticoids regulate expression of DHFR in mouse mammary tumor virus chimaeric plasmids. Nature 294:228–232.
Lewis JB, Mathews MB (1981): Viral messenger RNAs in six lines of adenovirus-transformed cells. Virology 115:345–360.
Lilley DMJ (1980): The inverted repeat as a recognizable structural feature in supercoiled DNA molecules. Proc Nat Acad Sci USA 77:6468–6472.
Lilley DMJ (1981): Hairpin-loop formation by inverted repeats in supercoiled DNA is local and transmissible. Nucl Acid Res 9:1271–1288.
McGhee JD, Wood W, Dolan M, Engel JD, Felsenfeld G (1981): A 200 base pair region at the 5′ end of the chicken adult β-globin gene is accessible to nuclease digestion. Cell 27:45–55.
Moreau P, Hen R, Wasylyk B, Everett R, Gaub M, Chambon P (1981): The SV40 72 bp repeat has a striking effect on gene expression. Nucl Acid Res 22:6043–6092.
Nedospasov SA, Georgiev GP (1980): Non-random cleavages of SV40 DNA in the compact minichromosome and free in solution by micrococcal nuclease. Biochem Biophys Res Comm 92:532–539.
Neel BG, Hayward WS, Robinson HL, Fang J, Astrin SM (1981): Avian leukosis virus-induced tumors have common proviral integration sites and synthesize discrete new RNAs: oncogenesis by promotor insertion. Cell 23:323–334.
Panayotatos N, Wells RD (1981): Cruciform structures in supercoiled DNA. Nature 289:466–470.

Pelham HRB, Brown DD (1980): A specific transcription factor that can bind either the 5S RNA gene or 5S RNA. Proc Nat Acad Sci USA 77:4170–4174.
Razin A, Riggs AD (1980): DNA methylation and gene function. Science 210:604–610.
Samal B, Worcel A, Louis C, Schedl P (1981): Chromatin structure of the histone genes of D. melanogaster. Cell 23:401–409.
Sandeen G, Wood W, Felsenfeld G (1980): The interaction of high mobility proteins HMG 14 and 17 with nucleosomes. Nucl Acid Res 8:3757–3778.
Scott WA, Wigmore DJ (1978): Sites in simian virus 40 chromatin which are preferentially cleaved by endonucleases. Cell 15:1511–1518.
Sharp S, DeFranco D, Dingermarn T, Farrell P, Soll D (1981): Internal control regions for transcription of eukaryotic tRNA genes. Proc Nat Acad Sci USA 78:6657–6661.
Sims J, Benz EM (1980): Initiation of DNA replication by E. coli dnaG proteins: evidence that tertiary structure is involved. Proc Nat Acad Sci USA 77:900–904.
Sinden RR, Carlson JO, Pettijohn DE (1980): Torsional tension in the DNA double helix measured with trimethylpsoralen in living E. coli cells: analogous measurements in insect and human cells. Cell 21:773–783.
Smith GR (1981): DNA supercoiling: another level for regulating gene expression. Cell 24:599–600.
Southern EM (1975): Detection of specific sequences among DNA fragments separated by gel electrophoresis. J Mol Biol 98:503–517.
Stalder J, Larsen A, Engel JD, Dolan M, Groudine M, Weintraub H (1980): Tissue-specific DNA cleavages in the globin chromatin domain introduced by DNAase I. Cell 20:451–460.
Stinchcomb DT, Thomas M, Kelly J, Selker E, Davis RW (1980): Eukaryotic DNA segments capable of autonomous replication in yeast. Proc Nat Acad Sci USA 77:4559–4563.
Varshavsky AJ, Sundin O, Bohn M (1979): A stretch of "late" SV40 viral DNA about 400 bp long which includes the origin of replication is specifically exposed in SV40 minichromosomes. Cell 16:453–466.
Vogt VM (1973): Purification and further properties of single-strand specific nuclease from aspergillus oryzae. Eur J Biochem 33:192–200.
Waldeck W, Fohring B, Chowdhury K, Gruss P, Sauer G (1978): Origin of DNA replication in papovavirus chromatin is recognized by endogenous endonuclease. Proc Nat Acad Sci USA 75:5964–5968.
Weintraub H (1979): Assembly of an active chromatin structure during replication. Nucl Acid Res 7:781–792.
Weintraub H, Groudine M (1976): Chromosomal subunits in active genes have an altered conformation. Science 193:848–858.
Weintraub H, Larsen A, Groudine M (1981): α-globin-gene switching during the development of chicken embryos: expression and chromosome structure. Cell 24:333–344.
Weintraub H, Beug H, Groudine M, Graf T (1982): Temperature-sensitive changes in the structure of globin chromatin in lines of red cell precursors transformed by tsAEV virus. Cell 28:931–940.
Weisbrod S, Weintraub H (1979): Isolation of a subclass of nuclear proteins responsible for conferring a DNAase I sensitive structure on globin chromatin. Proc Nat Acad Sci USA 76:630–634.
Wiegand R, Godson N, Radding C (1975): Specificity of the S1 nuclease from aspergillus oryzae. J Biol Chem 250:8848–8855.
Wu C (1980): The 5′ ends of Drosophila heat shock genes in chromatin are sensitive to DNAase I. Nature 286:854–860.

Wu C, Gilbert W (1981): Tissue-specific exposure of chromatin structure at the 5′ terminus of the rat preproinsulin II gene. Proc Nat Acad Sci USA 78:1577–1580.

Wu C, Bingham PM, Livak KJ, Holmgren R, Elgin SCR (1979): The chromatin structure of specific genes: I. Evidence for higher order domains of defined DNA sequence. Cell 16:797–806.

Zain S, Gingeras TR, Bullock P, Wong G, Gelinas RE (1979): Determination and analysis of adenovirus 2 DNA sequences which may include signals for late mRNA processing. J Mol Biol 135:413–433.

Ziff EB, Evans RM (1978): Coincidence of the promotor and capped 5′ terminus of RNA from the adenovirus 2 major late transcription unit. Cell 15:1463–1475.

Index